Lecture Notes in Mathematics

Edited by A. Dold and B. Eckmann

459

Fourier Integral Operators and Partial Differential Equations

Colloque International, Université de Nice, 1974

Edited by J. Chazarain

Springer-Verlag
Berlin · Heidelberg · New York 1975

Editor
Prof. Jacques Chazarain
Institut de Mathématiques
et Sciences Physiques
Université de Nice
Parc Valrose
06034 Nice Cedex/France

Library of Congress Cataloging in Publication Data

Fourier integral operators and partial differential equations.

(Lecture notes in mathematics ; 459)
"Colloque ... réuni à l'Université de Nice, du 20 au 25 Mai 1974."
Bibliography: p.
Includes index.
1. Differential equations, Partial--Congresses. 2. Fourier series--Congresses. 3. Integral operators--Congresses. I. Chazarain, Jacques, 1942- II. Series: Lecture notes in mathematics (Berlin) ; 459.
QA3.L28 no.459 [QA377] 510'.8s [515'.353] 75-19494

AMS Subject Classifications (1970): 35A05, 35A20, 35B99, 35H05, 35J10, 35J70, 35L10, 35L35, 35P20, 35P99, 35S99

ISBN 3-540-07180-6 Springer-Verlag Berlin · Heidelberg · New York
ISBN 0-387-07180-6 Springer-Verlag New York · Heidelberg · Berlin

Printed in Germany
Offsetdruck: Julius Beltz, Hemsbach/Bergstr.

P R E F A C E

Dans le but de faire le point sur la théorie et les applications des "Opérateurs Intégraux de Fourier", ce Colloque a réuni à l'Université de Nice, du 20 au 25 Mai 1974, la plupart des spécialistes de ces questions.

Pour situer le sujet avec l'espoir d'allècher le lecteur non-spécialiste, rappelons brièvement quelques points.

Autour des années 65, s'est développée la théorie des opérateurs pseudo-différentiels, ce qui a permis, entre autres, d'inverser les opérateurs elliptiques et ainsi de substituer aux techniques de majorations, a priori des méthodes plus explicites pour résoudre ces équations.

Mais pour l'étude des opérateurs de type principal qui sont, en un certain sens les plus simples après les elliptiques, les opérateurs pseudo-différentiels s'avèrent insuffisants et de nouvelles méthodes sont élaborées.

Parmi celles-ci, citons principalement :

- l'utilisation des variétés lagrangiennes pour la description globale des solutions asymptotiques.
- l'emploi des transformations canoniques pour transmuer une équation en une autre plus simple.
- la description dans le fibré cotangent des singularités des distributions (ou des hyperfonctions) grâce à la notion de "wave front set" ou spectre singulier ou support essentiel...

Puis, c'est en 70 que paraît l'article de Hörmander où sont synthétisées et généralisées ces diverses techniques pour donner l'outil des Opérateurs Intégraux de Fourier.

Comme en témoigne, par exemple, ce colloque, ce nouvel outil a déjà aidé à soulever un petit coin du voile qui recouvre la théorie des équations aux dérivées partielles ; il semble que l'on est loin d'avoir épuisé le champ de ses possibilités.

R E M E R C I E M E N T S

Ce Colloque a bénéficié de subventions des organismes suivants :

- La Société Mathématique de France
- Le Conseil Général des Alpes Maritimes
- La Municipalité de NICE
- L'Université de NICE.

NICE, Eté 74

Jacques CHAZARAIN.

TABLE DES MATIERES

LISTE DES CONFERENCIERS

. L. BOUTET DE MONVEL	(Université de Paris VII)
. JJ. DUISTERMAAT	(Université d'Utrecht)
. V.W. GUILLEMIN	(Massachusset Institut of technology)
. L. HORMANDER	(Université de Lund)
. J. LERAY	(Collège de France)
. B. MALGRANGE	(Université de Grenoble)
. A. MELIN	(Université de Copenhague)
. L. NIRENBERG	(Courant Institut)
. T. SHIROTA	(Hokkaido University)
. J. SJOSTRAND	(Université de Paris Sud)
. F. TREVES	(Rutgers University)
. A. WEINSTEIN	(Université de Californie)

PROPAGATION DES SINGULARITES DES SOLUTIONS D'EQUATIONS ANALOGUES A L'EQUATION DE SCHRÖDINGER

Louis Boutet de Monvel

Le but de cet exposé est la description d'un résultat de propagation des singularités pour les solutions de certaines équations pseudo-différentielles à caractéristiques doubles . Le résultat est énoncé au §1 . Pour les équations aux dérivées partielles à coefficients constants , le résultat est un cas particulier simple du résultat de L. Hörmander [6] . Une partie de la démonstration (§2) est due à R. Lascar [7] ; je me contenterai ici de donner des indications brèves sur cette partie , en renvoyant pour plus de détails au travail de R. Lascar .

§0. Rappels et Notations

Soit X un ouvert de $\mathbb{R}^n$. On notera $T^{\bullet}X$ le fibré des vecteurs cotangents non nuls sur X , et SX le fibré des vecteurs cotangents unitaires (ou des demi-droites) de $T^{\bullet}X$.

On utilise les notations usuelles pour les espaces de fonctions ou de distributions , et pour les opérateurs différentiels sur X . Les opérateurs pseudo-différentiels qui interviendront ici seront de la forme $a(x,D)$, défini par

$$(0.1) \qquad a(x,D)\, f = (2\pi)^{-n} \int e^{ix.\xi}\, a(x,\xi)\, \hat{f}(\xi)\, d\xi$$

où la fonction symbole $a(x,\xi)$ admet un développement asymptotique (au sens de [4], (2.10)) :

$$(0.2) \qquad a(x,\xi) \sim \sum a_{m-k}(x,\xi)$$

où k parcourt l'ensemble des entiers positifs , et a_{m-k} est C^{∞} pour $\xi \neq 0$, homogène de degré m-k en ξ (resp. quasi-homogène au §2) (m est un nombre réel et on ne perd rien à le supposer entier) .

Le symbole (partie principale) de $A = a(x,D)$ est alors la fonction homogène de degré m :

$$(0.3) \qquad \sigma_A = \sigma_A(x,\xi) = a_m(x,\xi)$$

Rappelons que si A et B sont deux opérateurs pseudo-différentiels de degrés respectifs m et m' (et si A ou B est propre , c'est à dire continu de C_0^∞ dans lui-même et de C^∞ dans lui-même , ce que nous supposerons toujours dans la suite) , le composé $A \circ B$ est un opérateur pseudo-différentiel de degré m+m' et on a les formules :

$$(0.4) \qquad \sigma_{AB} = \sigma_A \ \sigma_B$$

$$(0.5) \qquad \sigma_{[A,B]} = -i \ \{\sigma_A, \sigma_B\}$$

où $[A,B] = AB - BA$, et $\{f,g\}$ est le crochet de Poisson :

$$(0.6) \qquad \{f,g\} = \sum \frac{\partial f}{\partial \xi_j} \frac{\partial g}{\partial x_j} \quad \frac{\partial f}{\partial x_j} \frac{\partial g}{\partial \xi_j}$$

Nous noterons H_f le champ hamiltonien d'une fonction $f(x,\xi)$:

$$(0.7) \qquad H_f = \sum \frac{\partial f}{\partial \xi_j} \frac{\partial}{\partial x_j} - \frac{\partial f}{\partial x_j} \frac{\partial}{\partial \xi_j}$$

de sorte qu'on a $\{f,g\} = H_f \ g$.

Rappelons encore qu'on définit comme suit le spectre singulier (wave front) WF(f) d'une fonction généralisée (distribution) $f \in C^{-\infty}(X)$: on dit que f est C^∞ en un point (x,ξ) (ou dans un voisinage conique de (x,ξ)) s'il existe une fonction $\phi \in C_0^\infty$ non nulle en x telle que la transformée de Fourier $\widehat{\phi f}$ soit à décroissance rapide à l'infini dans un cône ouvert contenant ξ , ou de façon équivalente , s'il existe un opérateur pseudo-différentiel A elliptique en (x,ξ) tel que Af soit C^∞ . Alors WF(f) est l'ensemble (fermé , conique) des points où f n'est pas C^∞ . Le support singulier de f est la projection de WF(f) . Le faisceau des micro-fonctions est le faisceau sur le fibré SX des sphères cotangentes dont la fibre au point (x,ξ) est le quotient de l'espace $C^{-\infty}(X)$ de toutes les fonctions généralisées sur X par le sous-espace de celles qui sont C^∞ en (x,ξ). Un opérateur pseudo-différentiel diminue le spectre singulier , donc définit un endomorphisme du faisceau des microfonctions ; aussi le faisceau des microfonctions est-il bien adapté à l'étude locale sur la sphère cotangente (microlocale) des opérateurs différentiels ou pseudo-différentiels .

Rappelons enfin que la théorie des opérateurs intégraux de Fourier de L. Hörmander [5] , permet d'effectuer des changements très généraux de coordonnées dans T^*X : soient U et V deux ouverts coniques de T^*X , et Φ un isomorphisme canonique (ie. préservant les crochets de Poisson , ou ce qui revient au même, la forme canonique $\sum d\xi_j \wedge dx_j$) homogène de degré 1 de U dans V . On définit alors une classe d'opérateurs $MF(U) \to MF(V)$ (où $MF(U)$ désigne l'ensemble des microfonctions sur la base de U) attachée à Φ (opérateurs intégraux de Fourier associés à Φ sur U) : un tel opérateur transforme une microfonction f de support $(WF(f))$ $\Gamma \subset U$ en une microfonction de support $\Phi(\Gamma)$; d'autre part si un tel opérateur F est elliptique (ce qui implique qu'il est inversible) et si A est un opérateur pseudo-différentiel sur U , le transformé $F A F^{-1}$ est un opérateur pseudo-différentiel sur V , et on a

$$(0.8) \qquad \sigma_{FAF^{-1}} = \sigma_A \circ \Phi$$

Ceci permet de faire l'étude microlocale d'un opérateur pseudo-différentiel en utilisant des coordonnées canoniques bien choisies , et ainsi dans bien des cas de simplifier considérablement les problèmes .

§1. Description du résultat

X désigne toujours un ouvert de $\mathbb{R}^n$. Soit A un opérateur pseudo-différentiel sur X . Nous supposons les caractéristiques de A doubles , et de façon plus précise , que l'ensemble caractéristique car A (ensemble des zéros du symbole σ_A) est un cône C^∞ de codimension d et qu'au voisinage de tout point de ce cône le symbole σ_A peut s'écrire sous la forme

$$(1.1) \qquad \sigma_A = \sum_{1 \leq j,k \leq d} a_{jk}\, u_j\, u_k$$

où les a_{jk} sont des fonctions homogènes C^∞ , la matrice (a_{jk}) est positive non dégénérée , et où les fonctions u_j $(j = 1,\dots,d)$ sont homogènes C^∞ et engendrent l'idéal de définition du cône <u>car A</u> - autrement dit , s'annulent sur <u>car A</u> , et y ont des différentielles linéairement indépendantes .

Nous supposerons de plus que <u>car A</u> est un cône involutif dans T^*X , autrement dit

(1.2) $\{u_i, u_j\} = 0$ sur <u>car A</u> $(i,j = 1,\dots,d)$

Nous supposerons en outre remplie la condition suivante :

(1.3) <u>les différentielles</u> du_j <u>et la 1-forme canonique</u> $\sum \xi_j\, dx_j$ <u>sont linéairement indépendantes sur</u> car A .

Le symbole sous-principal (ou second invariant) $I_2(A)$ est défini comme suit : si (dans notre choix de coordonnées locales) le symbole total de A est

$$a \sim a_m + a_{m-1} + \dots$$

le symbole sous principal est donné par l'expression :

$$(1.4) \qquad I_2(A) = a_{m-1}(x,\xi) - (1/2i) \sum_k \frac{\partial^2 a_m(x,\xi)}{\partial x_k\, \partial \xi_k}$$

sa restriction à <u>car A</u> est invariante par changement de coordonnées , et même par transformation intégrale de Fourier elliptique .

Nous supposons vérifiée la condition suivante :

(1.5) $I_2(A) < 0$ en tout point de <u>car A</u> ([1])

Toutes ces conditions sont satisfaites par l'opérateur de Schrödinger ($A = i\partial/\partial x_n - \sum_{j<n} \partial^2/\partial x_j^{\,2}$) dans la région $\xi_n > 0$: le symbole total de A est

$$\sigma_A = \sum_{j<n} \xi_j^{\,2} - \xi_n$$

L'opérateur de Schrödinger est donc elliptique hors de la droite $\xi_1 = \dots = \xi_{n-1} = 0$,et

---

([1]) On montre (cf. [1]) que si , au contraire , $I_2(A)$ n'est jamais ≤ 0 sur <u>car A</u> , A est hypoelliptique , et que pour toute distribution f , $Af \in H_s$ implique $f \in H_{s+m-1}$ si $m = \deg A$.

hypoelliptique dans la région $\xi_n > 0$, de sorte que le thèoréme 1.8 ci-dessous exprime le fait (connu) que si f est solution de l'équation de Schrödinger (ou simplement si Af est C^∞) WF(f) est réunion de droites parallèles au sous-espace $x_n = 0$, $\xi_n = 0$ dans le cône $\xi_1 = \ldots = \xi_{n-1} = 0$, $\xi_n > 0$.

Quitte à composer notre opérateur A avec un opérateur elliptique convenable B , de symbole positif (ce qui remplace $I_2(A)$ par le produit $\sigma_B\, I_2(A)$ sur car A , et ne change pas l'ensemble des distributions f telles que Af soit C^∞) on peut supposer que $I_2(A)$ induit sur car A une fonction homogène négative arbitrairement donnée , par exemple

(1.6) $I_2(A) = -1$ sur car A

Comme les crochets de Poisson $\{u_i, u_j\}$ s'annulent sur car A , les champs H_{u_j} sont tangents à car A , et leurs restrictions à car A vérifient la condition d'intégrabilité de Fröbenius (on a $[H_{u_i}, H_{u_j}] = H\{u_i, u_j\}$, et comme $\{u_i, u_j\}$ est combinaison linéaire à coefficients C^∞ des u_j , les crochets $[H_{u_i}, H_{u_j}]$ sont , sur car A , combinaisons linéaires à coefficients C^∞ des H_{u_j}). Le cône car A se trouve ainsi muni d'un feuilletage canonique , dont les feuilles sont les variétés intégrales des champs H_{u_j} $(j=1,\ldots,d)$.

Par ailleurs le symbole σ_A -ou plutôt son hessien- définit par restriction une forme quadratique positive non dégénérée sur le fibré tangent normal de car A (à savoir : la forme $\sum a_{jk}\, du_j\, du_k$ sur $T(T^{\cdot}X)/T(\text{car } A)$) . La forme symplectique canonique $\sum d\xi_j \wedge dx_j$ de $T^{\cdot}X$ permet d'identifier le fibré tangent normal de car A au fibré cotangent du feuilletage qui vient d'être défini (canonique) sur car A : par transport de structure , ce feuilletage se trouve ainsi muni d'une métrique Riemannienne (définie au moyen d'une forme quadratique positive non dégénérée sur son fibré cotangent).

(1.7) Nous noterons g_A la métrique ainsi définie sur le feuilletage canonique de car A , lorsque A vérifie (1.6).

Le théorème de propagation annoncé est le suivant :

(1.8) Théorème : Avec les hypothèses et les notations ci-dessus

1. Soit f une distribution . Si Af est C^∞ dans un ouvert conique U de $T^{\cdot}X$, $WF(f) \cap U$ est réunion dans $U \cap$ car A de géodésiques de la métrique g_A

2. Inversement soit (x,ξ) un point caractéristique de A , et C l'arc

de géodésique de la métrique g_A passant par (x,ξ) : si U est un voisinage conique assez petit de (x,ξ), il existe une distribution f telle que Af soit C^∞ dans U, et $WF(f) \cap U = C \cap U$.

(Pour la deuxième assertion, il suffit en fait que $C \cap U$ soit connexe, et que ses bouts soient entièrement contenus dans la frontière de U. En particulier on a un résultat global si la géodésique passant par (x,ξ) "tend vers l'infini à l'infini" -c'est à dire, supposant cette géodésique paramétrée par une fonction $c(t)$: $]T,T'[\to \text{car } A$, pour tout compact $K \subset T^*X$, il existe des nombres $T_1 > T$, $T_1' < T'$ tels que $c(t) \notin K$ si $t < T_1$ ou $t > T_1'$. Ce résultat est analogue à celui qui vaut pour les solutions d'une équation réelle à caractéristiques simples).

Remarquons dès à présent que les assertions du théorème 1.8 sont invariantes par transformation intégrale de Fourier elliptique, car tous les objets qui y figurent sont définis "de façon intrinsèque" en termes du symbole σ_A, de $I_2(A)$ et de la structure symplectique canonique de T^*X. Nous pourrons donc pour les démontrer appliquer à A une transformation intégrale de Fourier elliptique bien choisie.

§2. Démonstration du théorème - première partie.

Observons tout d'abord qu'il résulte des hypothèses (1.2) et (1.3) qu'il existe (localement) une transformation symplectique homogène Φ qui transforme le cône $\text{car } A$ en le cône $\xi_1 = \dots = \xi_d = 0$ de T^*X (cf.[4], ou l'appendice de [1]). Si alors F est un opérateur intégral de Fourier elliptique associé à Φ, et si F^{-1} est une parametrix de F, $F A F^{-1}$ satisfait à nouveau aux hypothèses du §1, et son cône caractéristique est le cône $\xi_1 = \dots = \xi_d = 0$. Quitte à remplacer A par $F A F^{-1}$, et à le composer avec un opérateur elliptique convenable, de symbole positif, on pourra supposer que A est de la forme

$$(2.1) \qquad A = -\sum_{1 \leqslant j,k \leqslant d} A_{jk}\, \partial^2/\partial x_j \partial x_k \quad - \text{Id}$$

ou encore

$$(2.1)\text{bis} \qquad A = -\sum_{1 \leqslant j,k \leqslant d} A_{jk}\, \partial^2/\partial x_j \partial x_k \quad + i\, \partial/\partial x_n$$

où (A_{jk}) est une matrice d'opérateurs pseudo-différentiels de degré -1 (resp. 0) elliptique, à symbole positif non dégénéré (resp. et où, quitte à faire un changement linéaire sur les coordonnées $x_{d+1}, \dots, x_n$ seules, on étudie A au voisinage d'un point où l'on a $\xi_n > 0$, $\xi_1 = \dots = \xi_n = 0$).

car A est donc le cône $\xi_1 = \dots = \xi_d = 0$. Son feuilletage canonique est constitué des sous-espaces (affine) parallèles à l'espace engendré par $\partial/\partial x_1, \dots, \partial/\partial x_d$. Si A est donné par la formule (2.1) et si on pose $a_{jk} = \sigma_{A_{jk}}$, la métrique g_A induit sur le fibré cotangent des feuilles la forme quadratique

$$(2.2) \qquad \sum_{1 \leq j,k \leq d} a_{jk}(x,0,\dots,0,\xi_{d+1},\dots,\xi_n)\, \eta_j \eta_k$$

(où $\eta_1, \dots, \eta_d$ sont les coordonnées d'un vecteur cotangent). Les géodésiques de g_A sont donc les projections sur car A des courbes intégrales du champ de vecteurs (champ hamiltonien de la forme quadratique (2.2)) :

$$(2.3) \qquad 2 \sum_{1 \leq j,k \leq d} a_{jk}\, \eta_j (\partial/\partial x_k) - \sum_{1 \leq i,j,k \leq d} (\partial a_{ij}/\partial x_k)\, \eta_i \eta_j (\partial/\partial \eta_k)$$

Le point essentiel de la démonstration est un résultat de propagation des singularités pour les solutions d'un opérateur à symbole quasi-homogène. Ceci fait l'objet du travail de R. Lascar [7], auquel je renvoie pour plus de détails. Je me conterai ici d'en décrire le résultat, et de montrer comment il s'applique.

Rappelons pour commencer les définitions, et l'allure du calcul symbolique des opérateurs quasi-homogènes. Décomposons l'espace $\mathbb{R}^n$ en produit : $\mathbb{R}^n = \mathbb{R}^p \times \mathbb{R}^q$. On dit qu'une fonction $a(x,\xi,\eta)$ sur $X \times \mathbb{R}^p \times \mathbb{R}^q$ est quasi-homogène de poids m lorsqu'on attribue à ξ le poids r et à η le poids s si l'on a pour tout $\lambda > 0$

$$(2.4) \qquad a(x,\lambda^r \xi, \lambda^s \eta) = \lambda^m\, a(x,\xi,\eta)$$

Nous dirons simplement quasi-homogène s'il n'y a pas risque de confusion. Les nombres r et s sont réels positifs, et pour les applications, ce sont des entiers sans diviseur commun. Nous aurons besoin ici du cas $r = 1$, $s = 2$.

On dit qu'un opérateur pseudo-différentiel $A = a(x,D)$ est quasi-homogène si son symbole total admet un développement asymptotique (au sens de [3] , (2.10))

$$(2.5) \qquad a(x,\xi) \sim \sum_0^\infty a_k(x,\xi)$$

où a_k est quasi-homogène de poids $m_k \to -\infty$ (il suffira ici du cas $m_k = m-k$, k entier positif). A est alors de type $\rho = r/s$, $\delta = 0$ si $r \leqslant s$ (avec la notation de [3], définition 2.1). Dans ce qui suit nous supposerons $r \leqslant s$.

Ces opérateurs donnent lieu à un calcul symbolique analogue à celui des opérateurs pseudo-différentiels "classiques", rappelé au §0 . En particulier ils ont un symbole (partie principale) : si $A = a(x,D)$ où la fonction $a(x,\xi)$ admet le développement asymptotique (2.5), le symbole σ'_A est le premier terme de ce développement :

$$\sigma'_A = a_0(x,\xi)$$

Le symbole est multiplicatif : $\sigma'_{AB} = \sigma'_A \sigma'_B$. L'opérateur $A = a(x,D)$ est quasi-elliptique si son symbole σ'_A est inversible (pour $\xi \neq 0$): A admet alors une parametrix $B = b(x,D)$, qui est un opérateur pseudo-différentiel quasi-homogène , de symbole $\sigma'_B = (\sigma'_A)^{-1}$.

On peut localiser en utilisant ces opérateurs de la même façon qu'au §0 , en remplaçant "ouvert conique" par "ouvert quasi-homogène" (ie. stable par les dilatations $(x,\xi,\eta) \to (x,\lambda^r\xi,\lambda^s\eta)$, $\lambda > 0$) . On a en particulier une nouvelle notion de spectre singulier : une distribution f est C^∞ dans un voisinage quasi-homogène d'un point s'il existe un opérateur quasi-homogène A , elliptique en ce point (ie. σ'_A est inversible en ce point) , tel que Af soit C^∞ ; $WF'(f)$ est alors l'ensemble (fermé , quasi-homogène) complémentaire du plus grand ouvert quasi-homogène dans lequel f est C^∞ . Le lien entre le spectre singulier "quasi-homogène" $WF'(f)$ et le spectre singulier $WF(f)$ dont la définition a été rappelée au §0 est le suivant : si $WF'(f)$ est contenu dans l'ouvert $\eta \neq 0$, $WF(f)$ est la projection de $WF'(f)$ sur le sous espace $\xi = 0$ parallèlement au sous-espace $\eta = 0$. $WF'(f)$ apparaît ainsi comme un raffinement de $WF(f)$ (lorsqu'il est disjoint du cône $\eta = 0$).

Le symbole total du composé $c(x,D) = a(x,D)\ b(x,D)$ est donné par

$$c(x,\xi) \sim \sum (i^{-|\alpha|}/\alpha!)\ (\partial/\partial\zeta)^\alpha a\ \ (\partial/\partial x)^\alpha b$$

(où on a posé $\zeta = (\xi,\eta)$) . Notons que dans cette formule , $(\partial/\partial\xi_j)$ est de poids $-r$ (ie. abaisse le poids de r) , et $(\partial/\partial\eta_j)$ est de poids $-s$. Aussi obtient-on pour le symbole du commutateur de A et B (avec $\sigma'_A = a$, $\sigma'_B = b$)

$$(2.6)\quad \sigma'_{[A,B]} = -i \sum_{1\leqslant j\leqslant p} (\partial a/\partial\xi_j)(\partial b/\partial x_j) - (\partial a/\partial x_j)(\partial b/\partial\xi_j) = -i\{\sigma'_A,\sigma'_B\}'$$

Ainsi s'introduit pour le calcul symbolique un crochet de Poisson partiel, ainsi qu'un hamiltonien partiel :

$$(2.7)\quad H'_f = \sum_{1\leqslant j\leqslant p} (\partial f/\partial\xi_j)(\partial/\partial x_j) - (\partial f/\partial x_j)(\partial/\partial\xi_j)$$

Il en résulte une nouvelle notion de courbe bicaractéristique : les courbes intégrales du champ H'_f situées dans l'hypersurface $f = 0$.

Le résultat de R. Lascar [7] est le suivant :

(2.8) Proposition : Soit A un opérateur pseudo-différentiel quasi-homogène , réel à caractéristiques simples (ie. $H'_{\sigma'_A}$ n'est jamais parallèle au champ $r \sum \xi_j (\partial/\partial\xi_j) + s \sum \eta_j (\partial/\partial\eta_j)$ pour $\sigma'_A = 0$) . Alors si Af est C^∞ dans un ouvert quasi-homogène U , WF'(f) est réunion dans U de courbes intégrales du champ hamiltonien partiel $H'_{\sigma'_A}$ de l'hypersurface $\sigma'_A = 0$.

Pour prouver ce résultat , R. Lascar prouve une inégalité a priori apparentée aux inégalités de Carleman , en adaptant aux opérateurs pseudo-homogènes la méthode de [8] (cf. aussi [2]) (on fait usage en particulier d'une fonction quasi-homogène , strictement convexe par rapport au champ hamiltonien partiel $H'_{\sigma'_A}$) . Je renvoie au travail de R. Lascar pour plus de détails .

Montrons pour terminer comment ce résultat s'applique ici : A est donc donné par la formule (2.1) , et son symbole total est de la forme

$$a = \sum_{1<i,j<d} a_{ij}\, \xi_i\, \xi_j \quad + \quad b$$

où a_{ij} est homogène de degré -1 , b est un symbole de degré 0 , et $b = -1$ pour $\xi_1 = \dots = \xi_d = 0$. Posons alors $\xi = (\xi_1, \dots, \xi_d)$, $\eta = (\eta_{d+1}, \dots, \eta_n)$ et (comme ci-dessus , et bien que celà soit en conflit avec la notation du début) $\zeta = (\xi_1, \dots, \xi_n)$.

(2.9) Lemme : Soit $c(x,\zeta)$ une fonction homogène de degré m , C^∞ pour $\zeta \neq 0$. La série de Taylor

$$c(x,\zeta) \quad \sim \quad \sum \quad (\partial/\partial\xi)^\alpha\, c(x,0,\eta)\ \xi^\alpha/\alpha!$$

fournit un développement asymptotique de c en fonctions quasi-homogènes , dans tout ouvert disjoint du cône $\eta = 0$.

En effet dans ce développement le terme indexé par α est quasi-homogène de poids $m - |\alpha|$. De plus l'erreur dans la formule de Taylor à l'ordre N est de la forme $\sum_{|\alpha| = N} c_\alpha(x,\xi,\eta)\ \xi^\alpha/\alpha!$, où c_α est homogène de degré m-N . Elle est donc majorée par un multiple de $|\eta|^{m-N/2}$ dans tout ensemble de la forme $|\xi| < C|\eta|^{\frac{1}{2}}$,

$x \in K$ (compact) , et il y a une inégalité analogue pour chaque dérivée .

Appliquant le lemme 2.9 successivement aux a_{ij} et à chaque terme du développement asymptotique de b en fonctions homogènes , on voit que notre opérateur A est quasi-homogène (de poids 0) dans tout ouvert quasi-homogène disjoint du cône $\eta = 0$, et on a

$$\sigma'_A = \sum a_{ij}(x,0,\eta)\ \xi_i\ \xi_j - 1$$

Or la matrice (a_{ij}) est positive non dégénérée pour $\xi = 0$, de sorte que A est à caractéristiques simples en tant qu'opérateur quasi-homogène . On a

$$H'_A = 2 \sum a_{ij}(x,0,\eta)\ \xi_i\ (\partial/\partial\xi_j) - \sum (\partial a_{ij}/\partial x_k)\ \xi_i\ \xi_j\ (\partial/\partial\xi_k)$$

La première assertion du théorème 1.8 résulte alors de la proposition 2.8 et de (2.3) , compte tenu du fait que de toute façon A est elliptique dans l'ouvert $\xi \neq 0$, et même hypoelliptique dans l'ouvert quasi-homogène $|\xi|^2 > C|\eta|$ si C est assez grand , et de ce que $WF(f)$ est la projection sur le sous-espace $\xi = 0$ de $WF'(f)$ si $WF'(f)$ est disjoint du cône $\eta = 0$.

§3. Démonstration du théorème - deuxième partie .

Nous noterons dans ce paragraphe $x = (x_1, \dots, x_{n-1})$ les $n-1$ premières variables , et $t = x_n$ la dernière , pour la distinguer des autres ; nous noterons $\xi = (\xi_1, \dots, \xi_{n-1})$ et $\tau (= \xi_n)$ les variables duales . Les fonctions ou opérateurs quasi-homogènes de ce paragraphe le sont lorsqu'on attribue à ξ le poids 1 et à τ le poids 2 . Comme au §2 , formule (2.1)bis , nous supposerons A donné par

$$(3.1) \qquad A = - \sum_{1<i,j<d} A_{ij}\ (\partial^2/\partial x_i \partial x_j) + i\ (\partial/\partial t)$$

où les A_{ij} sont des opérateurs pseudo-différentiels de degré 0 , et où on étudie les propriétés microlocales de l'opérateur A au voisinage d'une demi-droite $\xi = 0$, $\tau = 0$. Quitte à translater , nous pouvons en outre nous placer au dessus du point $x = 0$, $t = 0$.

Nous allons construire une distribution T telle que AT soit C^∞ au voisinage de l'origine , mais que $WF(T)$ soit concentré sur la projection sur le sous-espace $\xi = 0$ $(\tau > 0)$ d'une courbe intégrale du champ de vecteurs

$$(3.2) \qquad 2 \sum a_{ij}(x,t,0,\tau)\ \xi_i\ (\partial/\partial x_j) \quad - \sum (\partial a_{ij}/\partial x_k)\ \xi_i\ \xi_j\ (\partial/\partial \xi_k)$$

au dessus d'un voisinage de l'origine (on a posé $a_{ij} = \sigma_{A_{ij}}$) . (On a ici $I_2(A) = -\tau$ et non $I_2(A) = -1$; mais comme H'_τ est nul , on remplace le champ hamiltonien partiel H'_{σ_A} par un champ proportionnel à (3.2) lorsqu'on remplace A par l'opérateur $(-i\partial/\partial t)^{-1}A$; les courbes intégrales ne sont pas changées , ni leurs projections sur le sous-espace $\xi = 0$).

Nous allons chercher T sous la forme

$$(3.3) \qquad \mathcal{U}T \;=\; f \;=\; (1/2\pi) \int e^{it\tau^2}\ \hat{\hat{f}}(x,\tau)\ d\tau$$

où f est une distribution convenable , et $\hat{\hat{f}}$ désigne sa transformée de Fourier partielle en t . Pour commencer , nous étudions avec plus de détails la transformation $\mathcal{U}$ définie par (3.3) .

Nous noterons $\mathcal{S}'_+$ (resp. $\mathcal{S}'_{++}$) l'espace des distributions tempérées f dont la transformée de Fourier $\hat{f}$ est nulle dans un voisinage conique (resp. parabolique) du demi-espace $\tau \leqslant 0$: on a donc $f \in \mathcal{S}'_+$ (resp. $\mathcal{S}'_{++}$) si supp $\hat{f}$ est contenu dans un ensemble de la forme $\tau > \varepsilon(|\xi|+1)$ (resp. $\tau > \varepsilon(|\xi|^2+1)$) .

La formule (3.3) s'écrit encore , pour $f \in \mathcal{S}'_+$

$$(3.4) \qquad \widehat{\mathcal{U}f}(\xi,\tau) \;=\; \begin{cases} \hat{f}(\xi,\sqrt{\tau})/2\sqrt{\tau} & \text{si } \tau > 0 \\ 0 & \text{si } \tau \leqslant 0 \end{cases}$$

Compte tenu que la fonction de s : $(i/4\pi t)^{1/2} \exp(-is^2/4t)$ est , pour $t \neq 0$, la transformée de Fourier inverse de la fonction (de τ) : $(1/2\pi) \exp(it\tau^2)$ (on choisit la racine carrée d'argument compris entre $-\pi/2$ et $\pi/2$) , il résulte de la formule de Plancherel qu'on a aussi pour $f \in C_0^\infty$

$$(3.5) \qquad \mathcal{U}f(x,t) \;=\; (i/4\pi t)^{1/2} \int e^{-is^2/4t}\ f(x,s)\ ds$$

Il résulte aussitôt de la formule (3.4) que $\mathcal{U}$ définit un isomorphisme de $\mathcal{S}'_+$ sur $\mathcal{S}'_{++}$, et que pour $f \in \mathcal{S}'_+$, on a $f \in \mathcal{S}$ si et seulement si $\mathcal{U}f \in \mathcal{S}$ ($\mathcal{S}$ désignant l'espace de Schwartz des fonctions C^∞ à décroissance rapide). La formule (3.5) montre que si f est à support compact (ou seulement si f est à décroissance rapide à l'infini) , $\mathcal{U}f$ est C^∞ hors de l'hyperplan $t = 0$. En outre si $f \in \mathcal{S}'_+$ coïncide en dehors d'un compact avec une fonction de $\mathcal{S}$, il en est de même de $\mathcal{U}f$ (on le voit en écrivant la formule de Taylor pour l'exponentielle

$\exp(-is^2/4t)$ et en utilisant le fait que si $f \in S'_+$, tous les moments $\int f(x,s)\, s^k\, ds$ sont nuls) .

On a encore les résultats suivants :

(3.6) $\mathcal{U}$ commute aux x_j , $(\partial/\partial x_j)$ et plus généralement à tout opérateur pseudo-différentiel $q(x,D_x)$ de x et D_x seuls .

(3.7) $$i\,(\partial/\partial t)\,\mathcal{U} \;=\; \mathcal{U}(\partial^2/\partial t^2)$$

et , plus généralement si $q'(x,D_x,D_t)$ est un opérateur pseudo-différentiel (quasi-homogène) dont le symbole total $q'(x,\xi,\tau)$ ne dépend pas de t , on a pour toute distribution $f \in S'_+$

$$q'(x,D_x,D_t)\,\mathcal{U}f \;=\; \mathcal{U}\,q(x,D_x,D_t)\,f$$

avec

(3.8) $$q(x,\xi,\tau) \;=\; q'(x,\xi,\tau^2) \quad \text{pour} \quad \tau > 0$$

(3.9) $t\,\mathcal{U}f = \mathcal{U}\,Qf$, si $f \in S'_+$, où Q est un opérateur pseudo-différentiel de degré -1 .

En effet , de la relation $-(\partial/\partial\tau)\,(i\,e^{it\tau^2}/2\tau) = (t + i/2\tau^2)\,e^{it\tau^2}$,on déduit , en intégrant par parties , et compte tenu qu'on a $i\,(\partial\hat{f}/\partial\tau) = \widehat{tf}$

$$t\,\mathcal{U}f = (1/2\pi)\int (-(\partial/\partial\tau)(i\,e^{it\tau^2}/2\tau) - i\,e^{it\tau^2}/2\tau^2)\;\hat{\hat{f}}(x,\tau)\,d\tau =$$

$$= (1/2\pi)\int e^{it\tau^2}\,((i\,\hat{\widehat{tf}}(x,\tau)/2\tau) - (i\,\hat{\hat{f}}(x,\tau)/2\tau^2))\,d\tau$$

d'où (3.9) , avec $\widehat{Qf} = (i/2\tau)\,\widehat{tf} - (i/2\tau^2)\,\hat{f}$.

Utilisant un développement de Taylor des symboles (totaux) au voisinage de $t = 0$, on déduit de (3.8) et (3.9)

(3.10) <u>Si</u> Q' <u>est un opérateur pseudo-différentiel quasi-homogène , et si</u> $f \in S'_+$, <u>on a</u> $Q'\mathcal{U}f = \mathcal{U}\,Qf$ <u>où</u> Q <u>est un opérateur pseudo-différentiel et , pour</u> $\tau > 0$, $\sigma_Q(x,t,\xi,\tau) = \sigma'_{Q'}(x,0,\xi,\tau^2)$

(3.11) <u>Si en outre le symbole total de</u> Q <u>s'annule à l'ordre infini sur</u>

l'hyperplan $t = 0$, $Q'\mathcal{U}f$ est C^∞ .

On a enfin le résultat suivant :

(3.12) Si $f \in \mathcal{S}'_+$ coïncide hors d'un ensemble compact avec une fonction de $\mathcal{S}$, $WF'(f)$ est l'image de $WF(f)$ par l'application $(x,t,\xi,\tau) \to (x,0,\xi,\tau^2)$

preuve : supposons d'abord que la droite (x_0,t,ξ_0,τ_0) $(t \in \mathbb{R})$ ne rencontre pas $WF(f)$. Alors si $\phi \in C^\infty$ est de support assez voisin de x_0 , et si $\chi(\xi,\tau) \in C^\infty$ est homogène pour (ξ,τ) assez grand , de support assez voisin de la demidroite $(\lambda\xi_0,\lambda\tau_0)$, $\lambda \geq 1$, et non nulle sur cette demi-droite , on a $\chi(D_x,D_t)\ \phi(x)\ f \in \mathcal{S}$, donc $\chi(D_x,D_t^2)\ \phi(x)\ f \in \mathcal{S}$, donc $(x_0,0,\xi_0,\tau_0^2) \notin WF'(\mathcal{U}f)$

Inversement , supposons $(x_0,0,\xi_0,\tau_0) \notin WF'(\mathcal{U}f)$. Alors (x_0,t,ξ_0,τ_0^2) n'appartient à $WF'(\mathcal{U}f)$ pour aucun $t \in \mathbb{R}$, et comme f coïncide avec une fonction de $\mathcal{S}$ hors d'un compact , il existe des fonctions χ , ϕ comme ci-dessus telles que $\chi(D_x,D_t^2)\ \phi(x)\ \mathcal{U}f \in \mathcal{S}$, donc $\chi(D_x,D_t)\ \phi(x)\ f \in \mathcal{S}$, et la droite (x_0,t,ξ_0,τ_0) $(t \in \mathbb{R})$ est disjointe de $WF(f)$.

Nous pouvons maintenant achever la preuve de la deuxième partie du théorème 1.8 : A est donné par la formule (3.1) , et d'après ce qui précède (lemme 2.9 et formule (3.10)) on a $A\mathcal{U}f = \mathcal{U}Bf$, où B est (dans un voisinage conique de la demi-droite $x=0$, $t=0$, $\xi=0$, $\tau>0$) un opérateur pseudodifférentiel de symbole

$$\sigma_B = \sum a_{ij}(x,0,0,\tau)\ \xi_i\ \xi_j - \tau^2$$

Le symbole de B est donc réel , à caractéristiques simples , pour $\tau > 0$. D'après [3] il existe une distribution f à support compact , telle que Bf soit C^∞ au voisinage de 0 , et que $WF(f)$ soit concentrée , au dessus d'un voisinage de l'origine , sur une courbe intégrale donnée (dans le demi-espace $\tau > 0$) du champ de vecteurs

$$2 \sum a_{ij}(x,0,0,\tau)\ \xi_i\ (\partial/\partial x_j) - 2\ \tau(\partial/\partial\tau) - \sum (\partial a_{ij}/\partial x_k)\ \xi_i\xi_j\ (\partial/\partial\xi_k) .$$

f est à décroissance rapide dans un voisinage conique du demi-espace $\tau \geq 0$, aussi quitte à modifier f par une fonction de $\mathcal{S}$, on peut aussi bien supposer $f \in \mathcal{S}'_+$ (f coïncidant en dehors d'un compact avec une fonction de $\mathcal{S}$) . On a alors $A(\mathcal{U}f) = \mathcal{U}(Bf) \in C^\infty$, et $WF'(\mathcal{U}f)$ est l'image par l'application $(x,t,\xi,\tau) \to (x,0,\xi,\tau^2)$ de $WF(f)$, donc $WF(\mathcal{U}f)$ est la projection de $WF(f)$ sur le sous-espace $t=0$, $\xi=0$, comme on le désirait .

Ceci achève la démonstration .

[1] Boutet de Monvel L. : Hypoelliptic Operators with Double Characteristics and related Pseudo-differential Operators, Comm. Pure Appl. Math. (1974)

[2] Duistermaat J.J. : On Carleman Estimates for Pseudo-differential Operators , Inventiones Math. 17.1 (1972) 31-43 .

[3] Duistermaat J.J. , Hörmander L. : Fourier Integral Operators II , Acta Math.128 (1972) 183-269 .

[4] Hörmander L. : Pseudo-differential Operators and Hypoelliptic Equations , Amer. Math. Soc. Proc. Symp. Pure Math. 10 (1967) 138-183 .

[5] Hörmander L. : Fourier Integral Operators I , Acta Math. 127 (1971) 79-183 .

[6] Hörmander L. : On the Existence and Regularity of Solutions of Linear Partial Differential Equations , l'Enseignement Mathématique , 2ème série 17(1971) 99-163

[7] Lascar R. : Propagation des Singularités des Solutions d'Equations Pseudo-différentielles Quasi-homogènes (thèse de troisième cycle) , à paraître .

[8] Unterberger A. : Résolution des Equations aux Dérivées Partielles dans des Espaces de Distributions d'Ordre de Régularité Variable , Ann. Inst. Fourier Grenoble 21,2 (1971) 85-128 .

Louis Boutet de Monvel
Université de Paris VII
2, place Jussieu , 75005 Paris

ON THE SPECTRUM OF POSITIVE ELLIPTIC OPERATORS AND PERIODIC BICHARACTERISTICS*

J.J. DUISTERMAAT

INTRODUCTION

Let X be a compact boundaryless C manifold and let P be a positive self-adjoint pseudodifferential operator of order $m > 0$ on X. For technical reasons we will assume that P operates on half-densities rather than functions. (We will denote the half-density bundle by X by $\Omega_{1/2}$.) We will also assume that P is a classical pseudodifferential operator in the sense that on every coordinate patch its total symbol $\sigma_P(x,\xi)$ admits an asymptotic expansion

$$\sigma_P(x, \xi) \sim \sum_{j=0}^{\infty} p_{m-j}(x, \xi)$$

with $p_{m-j}(x,\xi)$ homogeneous of degree m-j. We recall that the principal symbol p of P is equal to p_m on local coordinates, and the subprincipal symbol is equal to $p_{m-1} - \frac{1}{2i} \Sigma \frac{\partial p}{\partial x_j \partial \xi_j}$.

Let $\lambda_1, \lambda_2, \ldots$ be the eigenvalues of P. It was remarked by Chazarain in [6] and by ourselves in [11] that the sum $\Sigma e^{-i \sqrt[m]{\lambda_k} t}$ is well-defined as a generalized function of t and that if T is in its singular support then the Hamiltonian vector field

---

* Introduction d'un article en collaboration avec V.W. GUILLEMIN et à paraître dans Invent. Math.

$$H_q = \frac{\partial q}{\partial \xi_j}\frac{\partial}{\partial x_j} - \frac{\partial q}{\partial x_j}\frac{\partial}{\partial \xi_j} \quad , \quad q = \sqrt[m]{p}$$

has a periodic integral curve of period T. The purpose of this article is to analyze the nature of the singularities at these T. The analysis of Hörmander [16] of the "big" singularity at T = 0 leads to an asymptotic expansion of the form

$$\Sigma\rho(\mu - \mu_j) \sim (2\pi)^{-n} \, \Sigma \, c_k \mu^{n-1-k}, \quad \mu_j = \sqrt[m]{\lambda_j} \tag{0.1}$$

as $\mu \to +\infty$, for an appropriate class of Schwartz functions, ρ. The c_k's are the integrals over the cosphere bundle of polynomial expressions in the symbol of P and its derivatives, and are independent of ρ. (See 2.16). In §2 we show that they are related to the residues at the poles of the zeta function of P and to the coefficients occuring in the asymptotic expansion of the trace of the heat kernel at t = 0. From this we obtain rather easily results of Seeley [22] on the zeta function and Minakshisundaram-Pleyel [18] on the trace of the heat kernel, (just for scalar operators, however). We note in passing that the asymptotic expansion of the trace of the heat kernel involves logarithmic terms unless P is a differential operator. The existence of these terms seem to have been neglected in the literature. Section 2 concludes with a priori estimates for the spectrum which follow from (0.1) and which are used in Section 3.

There we analyze how the right hand side of (0.1) is affected when $\rho(\mu)$ is replaced by $\rho_M(\mu) = \rho(\frac{\mu}{M})$ and M allowed to tend to infinity. This leads to some results concerning operators, P, for which all H_q solution curves are periodic with the same period. Specifically we show that if the H_q flow is periodic with period T there exists a constant β such that most of the spectrum of $\sqrt[m]{P}$ is concentrated near the lattice points $\frac{2\pi}{T}k + \beta$,

$k = 1, 2, \ldots$. We show that conversely if this "clustering" occurs then the H_q flow is periodic. In fact we show that if a few pathological examples are excluded then for non-periodic H_q flow the spectrum is rather equally distributed. Also the spectral estimate of Hormander [16] can be slightly improved in this case, and an error term of order $O(\lambda^{\frac{n-1}{m}})$ replaced by an error term of order $o(\lambda^{\frac{n-1}{m}})$.

In Section 4 we begin our analysis of the singularities of $\Sigma\, e^{-i \sqrt[m]{\lambda_k} t}$ at periods $T \neq 0$. Our main result, Theorem 4.5, is that whenever the map, $\exp T H_q : T^*X\backslash 0 \to T^*X\backslash 0$, has a clean fixed point set (in the sense of Bott), then an asymptotic expansion of the form (0.1) is valid in a neighborhood of T; moreover, the leading term in this asymptotic expansion can be computed from such data as the length of the period and the eigenvalues of the Poincaré' map. Chazarain obtains results similar to ours in [6] but without the explicit formula for the leading term. As a corollary of theorem 4.5 we obtain the following residue formula in case all the periodic H_q solution curves of period T are isolated and non-degenerate:

$$(0.2) \quad \lim_{t\to T} (t-T)\Sigma e^{-i \sqrt[m]{\lambda_k} t} = \Sigma \frac{T\gamma}{2\pi} i^{\sigma_\gamma} |I - P_\gamma|^{-\frac{1}{2}}$$

the sum taken over all integral curves, γ, of period T. Here $T\gamma$ is the length of the primitive integral curve determined by γ σ_γ is a Maslov factor (explained in Section 6) and P_γ the Poincare' map around γ.

For the proof of theorem 4.5 we need some results concerning composition of Fourier integral operators under "clean intersection" assumptions. These results generalize results of Hörmander [17, Ch. 4]. They are discussed in Section 5 and proved in Section 7. Similar results have been announced by Weinstein at the Conference on Fourier Integral Operators in Nice, May 1974.

If all the periodic H_q solution curves are isolated and non-degenerate and only one such curve, γ, or two such curves, γ and $-\gamma$, occur for each period*, then from (0.2) one can determine $|I-P_\gamma^k|$ for all k. (Just apply (0.2) to the k-fold iterate of γ.) It turns out that these data almost suffice to determine P_γ itself. In fact it determines all the eigenvalues of P_γ of modulus $\neq 1$ and, up to multiplication by roots of unity, all the eigenvalues of modulus 1. This result is due to Harold Stark; and he has generously allowed us to publish it here in an appendix.

Many of the results of this paper extend to operators operating on vector bundles providing the eigenvalues of the symbol are of constant multiplicity. We hope to discuss these results in a future article. We will content ourselves here with mentioning a typical result concerning the Laplace operator on k-forms. For this operator the residue formula (0.2) is still valid except that the residue associated with γ is

$$\frac{T_\gamma}{2\pi}\; i^{\sigma_\gamma}\, |I - P_\gamma|^{-\frac{1}{2}} \;\text{trace}\; H_\gamma : \Lambda^k \to \Lambda^k$$

H_γ being holonomy along γ.

In conclusion we would like to thank Iz Singer and Michael Atiyah for helping us to clarify the relations among heat equation, wave equation and zeta function asymptotics; and we would like to thank Harold Stark for proving for us the result described above concerning the Poincaré' map. Our main inspiration for writing this paper was the beautiful article of

---

* This is the generic case if P is a differential operator; for pseudo-differential operators generically only one periodic solution curve occurs for each period.

Hörmander [16] on the spectral function of an elliptic operator. We would also like to thank Alan Weinstein for helpful conversations concerning the material in Section 3. Formal resemblances with the methods used by Colin de Verdière [7] and Cotsaftis [8] were an incentive to the computation some of the coefficients in the asymptotic expansions in Theorem 4.5.

R E F E R E N C E S

[1] G.K.Andersson, Analytic wave front sets for solutions of linear differential equations of principal type, Trans. Am. Math. Soc. 177 (1973), 1-27.

[2] V.I.Arnol'd, On a characteristic class entering in quantization conditions, Funct. Anal. Appl. 1 (1967), 1-13.

[3] M.F.Atiyah and R.Bott, A Lefschetz fixed point formula for elliptic complexes I, Ann. of Math. 86 (1967), 374-407.

[4] M.F.Atiyah, R. Bott and V.K. Patodi, On the heat equation and the index theorem, Inv. Math. 19 (1973), 279-330.

[5] R.Bott, On the iteration of closed geodesics and the Sturm intersection theory, Comm. Pure Appl. Math., 9 (1956), 176-206.

[6] J. Chazarain, Formule de Poisson pour les variétés riemanniennes, Inv. Math. 24 (1974), 65-82.

[7] Y. Colin de Verdière, Spectre du laplacien et longueurs des géodésiques périodiques II, Comp. Math. 27 (1973), 159-184.

[8] M. Cotsaftis, Une propriété des orbites périodiques des systèmes hamiltoniens non-linéaires, C. R. Acad. Sc. Paris 275, Série A (1973), 911-914.

[9] J.J. Duistermaat and L. Hörmander; Fourier integral operators II, Acta Math. 128 (1972), 184-269.

[10] J.J. Duistermaat; Fourier Integral Operators, Courant Institute Lecture Notes, New York 1973.

[11] J.J. Duistermaat and V.W. Guillemin, The spectrum of positive elliptic operators and periodic geodesics, Proc. A.M.S. Summer Institute on Differential Geometry, Stanford 1973 (to appear).

[12] J.J.Duistermaat, On the Morse index in variational calculus, to appear in Advances in Math..

[13] I.M. Gelfand and G.E.Shilov, Generalized Functions,I, Academic Press, New York 1964.

[14] V. Guillemin and S. Sternberg, Geometric Asymptotics, A.M.S. Publications (in press).

[15] G.E. Hardy and E.M. Wright, An Introduction to the Theory of Numbers, 4th ed., Oxford, Clarendon Press 1960.

[16] L. Hörmander, The spectral function of an elliptic operator, Acta Math. 121 (1968), 193-218.

[17] L. Hörmander, Fourier integral operators I, Acta Math. 127 (1971), 79-183.

[18] S. Minakshisundaram and Å. Pleijel, Some properties of the eigenfunctionsof the Laplace operator on Riemannian manifolds, Canadian J. Math. 1 (1949), 242-256.

[19] L.Nirenberg, Lectures on Linear Partial Differential Equations, Regional Conference Series in Mathematics, No 17, Conf. Board of the Math. Sc. of the A. M. S., 1972.

[20] M. Sato, Regularity of hyperfunction solutions of partial differential equations, Proc. Nice Congress, Vol. 2, Gauthiers-Villars, Paris 1970, pp.785-794.

[21] M. Sato, T. Kawai and M. Kashiwara, Microfunctions and Pseudo-Differential Equations, Lecture Notes in Math. No 287, Springer-Verlag 1973, pp. 265-529.

[22] R.T. Seeley, Complex powers of an elliptic operator, A.M.S. Proc. Symp. Pure Math. 10 (1967),288-307. Corrections in: The resolvent of an elliptic boundary problem, Am. J. Math. 91 (1969), 917-919.

[23] J.-P. Serre, A Course in Arithmetic, Springer-Verlag, New York, Heidelberg, Berlin, 1973.

[24] A. Weinstein, Fourier integral operators, quantization and the spectra of Riemannian manifolds, to appear in the Proc. of the C.N.R.S. Colloque de Geometrie Symplectique et Physique Mathematique, Aix-en-Provence, June 1974.

[25] W. Klingenberg and F. Takens, Generic properties of geodesic flows, Math. Ann. 197 (1972), 323-334.

[26] L. Hörmander, Linear differential operators, Proc. Nice Congress, Vol. 1, Gauthiers-Villars, Paris 1970, pp. 121-133.

[27] L.Hörmander, Linear Partial Differential Operators, Springer-Verlag, Berlin, Göttingen, Heidelberg 1963.

CLEAN INTERSECTION THEORY AND FOURIER INTEGRALS

Victor GUILLEMIN

1. INTRODUCTION

The purpose of this talk is to report on some recent work of myself and J.J. Duistermaat (*). Let X be a compact, boundaryless n-dimensional manifold, and suppose given on X a positive self-adjoint elliptic differential operator P. For simplicity we'll assume that P operates on half-densities, and that its sub-principal symbol is zero. (See [5], page 200.) Replacing P by $\sqrt[m]{P}$, m = degP, we can assume that P is of order 1. We will henceforth make all these assumptions without explicity saying so.

Let

$$e(t) = \sum_{\lambda \in \operatorname{spec} P} e^{i\lambda t}$$

We will call this the <u>Hörmander spectral function</u> of P. It turns out that its not really a function at all. It is, however, a well-defined distribution. It was recently observed by ourselves [4] and by J.Chazarain [2] that if $T \in$ singsupp e(t) then there exists a closed bicharacteristic on $T^*X - O(X)$ of length $|T|$. In case the closed bicharacteristics of X are non-degenerate, Duistermaat and I obtained rather explicit information about the singularities of e(t). In particular we showed that

$$\lim (t - T)\, e(t)$$

exists, and we gave a simple formula for it. (See Theorem 2 of [4].)

In [2] Chazarain proves a similar result under somewhat more general hypotheses. He assumes that the closed bicharacteristics of length $|T|$ form a "nice" i-dimensional manifold and then shows that

$$\lim_{t \to T+} (t - T)^{\frac{i+1}{2}} e(t)$$

exists and gives criteria for when it is non-zero.

(*) We learned from Alan Weinstein that he has results on clean intersections and Fourier integrals which are rather similar to those described here. For details consult his article in this volume.

In this talk we will give a geometric formulation of the "niceness" assumption that Chazarain seems to be making in [2] , and also compute this limit explicitly.

Before stating this result I'll have to review some elementary symplectic geometry. Let M be a manifold, and let $\Phi : M \to M$ be a diffeomorphism

Definition A submanifold $Z \subset M$ of fixed points of Φ will be called clean if for eacl $z \in Z$ the set of fixed points of $d\Phi_z : T_zX \to T_zX$ equals the tangent space to Z.

Remark This definition is apparently due to Bott. (See [1] .)

Suppose that X is a symplectic manifold and Φ a symplectic diffeomorphism. We will show that a clean submanifold of fixed points possesses a canonical nowhere vanishing smooth density. To see this we will need a little symplectic linear algebra

Lemma 1 Let V be a symplectic vector space with two-form Ω. Let $P : V \to V$ be a symplectic linear mapping. Then Ker(I - P) and coker(I - P) are canonically paired by Ω.

Proof If $v \in \ker(I - P)$, then $v = Pv$, so $v = P^{-1}v$, so $v \in \ker(I - P^{-1})$. But to say that $v \in \ker(I - P^{-1})$ is equivalent to saying that $\Omega(v, I - Pw)$ equals zero for all w, or that $v \in \mathrm{Im}(I - P)^{\perp}$. Hence ker(I - P) and coker(I - P) are canonically paired. Q.E.D.

Now consider the exact sequence

$$(1.1) \qquad o \to \ker \to V \xrightarrow{I - P} V \to \mathrm{coker} \to o$$

Letting $| \;|^{\alpha}$ be the functor that assigns to each vector space V its one-dimensional space of α densities

$$|(\ker)|^{\frac{1}{2}} \otimes |V|^{-\frac{1}{2}} \otimes |V|^{\frac{1}{2}} \otimes |(\mathrm{coker})|^{-\frac{1}{2}} \cong 1$$

Since $|V|^{\frac{1}{2}} \otimes |V|^{-\frac{1}{2}} \cong 1$ and $|\mathrm{coker}|^{\frac{1}{2}} \cong |\ker|^{\frac{1}{2}}$, we get $|\ker| \cong 1$, so we conclude

Lemma 2 If P is a symplectic mapping of V into V, then ker(I - P) possesses a canonical density.

Applying this to $Z \subset M$, a clean submanifold of fixed points of a symplectic diffeomorphism, we obtain

Corollary Z possesses a canonical nowhere vanishing density.

Remark Suppose that P in lemma 2 satisfies

$$(I - P)^{\#} : V/\ker \xrightarrow{\cong} V/\ker \quad , \quad \ker = \ker(I - P) \ .$$

Then by Lemma 1 , Ω restricted to $\ker(I - P)$ is non-singular, and hence $\ker(I - P)$ is a symplectic space. Using (1.1) it is not hard to see that the density on $\ker(I - P)$ described by Lemma 2 is just the symplectic density times the factor $|\det(I-P^{\#})|^{-\frac{1}{2}}$.

Let $\rho : M \to R$ be a smooth function with derivative everywhere unequal to zero. Let H_ρ be the associated Hamiltonian vector field, and $\Phi_t : M \to M$ the flow it generates. Let Z be a submanifold consisting of periodic orbits of Φ of period T. Then Z is a fixed point set for the sympletic mapping $\Phi_T : M \to M$. Assume that Z is a clean fixed point set. Then the submanifold

$$Z_1 = Z \cap (\text{energy surface } \rho = 1)$$

is a clean fixed point set for the map Φ_T restricted to the energy surface. Letting μ be the canonical density on Z , we get a canonical density μ_1 on Z_1 by requiring that μ_1 on Z_1 times $|d\rho|$ in the normal direction equals μ. In particular if Z_1 is compact, we can integrate μ_1 over Z_1 and get a number which we will call $\mathrm{vol}(Z_1)$. We can now state

Theorem I Let $\rho(x, \xi)$ be the symbol of the operator $P(x,D)$, and let H_ρ be the associated bicharacteristic vector field on $M = T^*X$. Suppose that the set of all closed bicharacteristics of length $|T|$ lying on the energy surface $\rho=1$ form a union of connected submanifolds

$$Z_1 \cup \cdots \cup Z_k$$

of dimension i, and that each of the Z_r's is a clean fixed point set for Φ_T. Then

$$\lim_{t\to T} [t-T+o\sqrt{-1}]^{\frac{i+1}{2}} e(t) = \sum_r e^{\sqrt{-1}\,\frac{\pi}{2}\sigma_r}\left(\frac{\sqrt{-1}}{2}\right)^{\frac{i+1}{2}} \Gamma\left(\frac{i+1}{2}\right)\mathrm{vol}(Z_r)$$

σ_r being the Maslov index of any bicharacteristic γ on Z_r. (This is an integer independent of γ. For further details see [3].)

Remark The theorem announced in [4] (concerning the case $i = 1$) can be easily deduced from Theorem 1 plus the remark following Lemma 2.

2. CLEAN INTERSECTION THEORY

The proof of Theorem 1 will require some facts concerning composition of Fourier integrals under hypotheses less restrictive than those considered in [8]. First, however, we'll need some elementary facts about symplectic vector spaces.

Let V and W be symplectic vector spaces, let Γ be a Lagrangian subspace of $V \times W$, and let Λ be a Lagrangian subspace of W. Let $\Lambda^{\#} = \Gamma \circ \Lambda$. ($\Lambda^{\#}$ is the set of vectors $v \in V$, such that there exists $(v, w) \in \Gamma$ with $w \in \Lambda$.)

Lemma 1 $\Lambda^{\#}$ is a Lagrangian subspace of W.

Proof Let ρ and π be the projections of Γ on v and W respectively. Consider the diagram

$$\begin{array}{ccc} \Gamma & \leftarrow & F \\ \pi\downarrow & & \downarrow \\ W & \overset{i}{\leftarrow} & \Lambda \end{array}$$

F being the fiber product. Associated with this diagram is an exact sequence

$$(2.1) \qquad o \to F \to \Gamma \oplus \Lambda \overset{\tau}{\to} W \to \operatorname{coker} \tau \to o$$

where $\tau(a, b) = \pi(a) - i(b)$. $\Lambda^{\#}$ can be defined as the image of the composite map

$$F \to \Gamma \overset{\rho}{\to} V .$$

Denoting this composite map by α, we get an exact sequence

$$(2.2) \qquad o \to \ker \alpha \to F \overset{\alpha}{\to} \Lambda^{\#} \to o$$

We will now show that $\ker \alpha$ and $\operatorname{coker} \tau$ are dually paired by the symplectic structure on W. Note first of all that $\ker \alpha$ consists of all pairs (a, w) in the fiber product for which $\rho(a) = o$. We can write $a \in V \times W$ as a pair (v', w'). To say that (a, w) is in the fiber product says that $w' = w$, and to say $\rho(a) = o$ says that $v' = o$; so $\ker \alpha$ can be identified with the set of $w \in W$ such that

i) $w \in \Lambda$

and

ii) $(o, w) \in \Gamma$

Suppose now that u is in the image of τ , i.e. $u = w_1 + \pi(v_2, w_2)$ with $(v_2, w_2) \varepsilon \Gamma$. Then $\Omega_W(w, w_1) = o$ by i) and $\Omega_W(w, w_2) = o$ by ii) so $\Omega_W(w, u) = o$. Since Γ and Λ are maximally isotropic, this argument works backward to show that $(\ker \alpha)^{\perp} = (\mathrm{Im}\ \tau)$ in **W**. It is easy now to show that the dimension of $\Lambda^{\#}$ is half the dimension of V using the exact sequences (2.1) and (2.2). We leave it for you to show as an easy exercise that $\Lambda^{\#}$ is isotropic, and thus Lagrangian, proving Lemma 1.

Lemma 2 Let $\alpha : F \to \Lambda^{\#}$ be the mapping defined by (2.2). Then there is a canonical mapping of half - densities

$$|\Lambda|^{\frac{1}{2}} \otimes |\Gamma|^{\frac{1}{2}} \overset{\cong}{\to} |\Lambda^{\#}|^{\frac{1}{2}} \otimes |\ker \alpha|$$

Proof From (2.1) we get an identification

$$|F|^{-\frac{1}{2}} \otimes |\Gamma|^{\frac{1}{2}} \otimes |\Lambda|^{\frac{1}{2}} \otimes |W|^{-\frac{1}{2}} \otimes |\mathrm{coker}\ \tau|^{\frac{1}{2}} \cong 1$$

or

$$|F|^{\frac{1}{2}} \otimes |W|^{\frac{1}{2}} \otimes |\mathrm{coker}|^{-\frac{1}{2}} \cong |\Gamma|^{\frac{1}{2}} \otimes |\Lambda|^{\frac{1}{2}}$$

From (2.2) we get

(2.3) (i) $|F|^{\frac{1}{2}} \cong |\Lambda^{\#}|^{\frac{1}{2}} \otimes |\ker\ \alpha|^{\frac{1}{2}}$

The sympletic structure on **W** gives us a trivialization

(2.3) (ii) $|W|^{\frac{1}{2}} \cong 1$

and finally the dual pairing of $\ker \alpha$ and $\mathrm{coker}\ \tau$ via the symplectic structure of V gives us a mapping

(2.3) (iii) $|\mathrm{coker}\ \tau|^{-\frac{1}{2}} \cong |\ker\ \alpha|^{\frac{1}{2}}$

Putting this all together we get the assertion of Lemma 2. Q.E.D.

Given manifolds X, Y, Z and maps $f : X \to Z$ and $g : Y \to Z$, f and g are said to intersect cleanly (See (1),) if the fiber product, F,

(2.4)
$$\begin{array}{ccc} X & \leftarrow & F \\ f\downarrow & & \downarrow \\ Z & \overset{g}{\leftarrow} & Y \end{array}$$

is a submanifold of X x Y, and in addition for each $p \varepsilon F$, $p = (x, y)$,

$$\begin{array}{ccc} T_xX & \leftarrow & T_pF \\ df_x \downarrow & & \downarrow \\ & dg_y & \\ T_zZ & \leftarrow & T_yY \end{array}$$

is a fiber product diagram. (For example, if f and g intersect transversally, they intersect cleanly.) We can associate to the diagram (2.4) a non-negative integer, e, called its excess :

$$e = \dim F + \dim Z - (\dim X + \dim Y)$$

Note that $e = o$ if and only if the clean diagram (2.4) is transversal.

Now let X and Y be compact manifolds and Γ and Λ closed, homogeneous Langrangian submanifolds of $T^*X \times T^*Y - o(X \times Y)$ and $T^*Y - o(Y)$ respectively. Let $\Gamma' = \{(x, \xi, y, \eta), (x, \xi, y, -\eta) \in \Gamma\}$; and let $\Lambda^{\#} = \Gamma' \circ \Lambda = \{(x, \xi), \exists (x, \xi, y, \eta) \in \Gamma', (y, \eta) \in \Lambda\}$. Let ρ and π be the projections of Γ on T^*X and T^*Y. Assume $\Lambda^{\#}$ contains no zero vectors, and Γ' no vectors of the form (x, ξ, y, o).

Lemma 3 If the fiber product diagram

$$(2.5) \qquad \begin{array}{ccc} \Gamma' & \leftarrow & F \\ \pi\downarrow & & \downarrow \\ T^*Y & \leftarrow & \Lambda \end{array}$$

is clean, then $\Lambda^{\#}$ is an (immersed) Lagrangian submanifold of T^*X, and the composite map, α ,

$$F \rightarrow \Gamma' \overset{\rho}{\rightarrow} \Lambda^{\#}$$

is a fiber mapping with compact fiber.

Proof The first part of the Lemma is just a rephrasing of Lemma 1. The fibers of $\alpha: F \rightarrow \Lambda^{\#}$ must be compact, for otherwise in view of the homogeneity of Γ' and Λ , $\Lambda^{\#}$ would have to contain zero vectors. Q.E.D.

Given half-densities on Γ and Λ , then by lemma 2 we get an object on F which is a half-density in the horizontal direction times a density in the fiber direction. Integrating this over the fibers, we get a half-density on $\Lambda^{\#}$ which we will denote by the composite of the given densities on Γ and Λ . It is easy to check that if σ and

τ are homogeneous densities, then so is their composite, $\sigma \circ \tau$, and

(2.6) $$\text{degree } \sigma \circ \tau = \text{degree } \sigma + \text{degree } \tau - \frac{(\dim Y - e)}{2},$$

e being the excess in the diagram (2.5). (The excess comes in because of the pairing (2.3) (iii), which is homogeneous of degree $\frac{e}{2}$.)

We can now state our second main theorem. For basic facts concerning oscillatory integrals and the Hörmander spaces, I_Λ, etc., see Hörmander, [8](*).

Theorem II Let κ be a generalized half-density on $X \times Y$ and K the operator associated with it. (K maps compactly supported half-densities on Y to generalized half-densities on X.) Let Γ and Λ be as sbove. If κ belongs to the Hörmander space I^m_Γ, then

$$K : I^s_\Lambda \to I^{s+m-\frac{(\dim Y - e)}{2}}_{\Lambda^\#}$$

and $\sigma(K\mu) = \sigma(\kappa) \circ \sigma(\mu)$, modulo Maslov factors.

Remark We have not attemped here to describe the Maslov aspect of the symbol formula above, as it is rather complicated. If one confines oneself to the "metalinear category" of Kostant-Sternberg (that is, manifolds for which the square of the first Stiefel-Whitney class is zero), then the Maslov factors can be disposed of by using half-forms everywhere in the above discussion instead of half-densities. See [7].

3. THE PROOF OF THEOREM I

Throughout this section we will adhere to the notation of section 1. Consider the mapping

(3.1) $$T^*X \times R \to T^*(X \times X \times R)$$

sending (x, ξ, t) to $(x, \xi, y, \eta, t, \tau)$ where $(y, -\eta) = \Phi_t(x, \xi)$ and $\tau = \rho(x, \xi)$. This imbeds $T^*X \times R$ as a Lagrangian submanifold, Λ, of $T^*(X \times X \times R)$. From (3.1) and the symplectic structure on T^*X, we get a canonical half-density, μ_{can}, on Λ.

(*) We will define the I^m_Λ a little differently from Hörmander, so that the symbol map preserves degree of homogeneity. Thus $I^m_\Lambda = I^{m-n/4}_\Lambda$ in Hörmander's definition, where $n = \dim \Lambda$.

Lemma 1 Let $e(x, y, t)$ be the Schwartz kernel of the operator $\exp \sqrt{-1}\, tP$, P being as in section 1. Then $e \in I^{n/2}(\Lambda)$ and $\sigma(e) = \mu_{can}$.

For the proof see Hörmander-Duistermaat [5].

Let $\Delta : X \times R \to X \times X \times R$ be the diagonal map. Given a half-density μ, on $X \times X \times R$, we can pull it back to the diagonal and multiply the two half-density factors in X to get an object $\Delta^*\mu$, which is a density in X times a half-density in R at each point of $X \times R$. This object can be integrated over X to get a half-density on R which we will denote by $\pi_*\Delta^*\mu$ ($\pi : X \times R \to R$ being the projection map). Since $\pi_*\Delta^*$ is an operator from half-densities on $X \times X \times R$ to half-densities on R, its Schwartz kernel, $K_{\pi_*\Delta^*}$, is a half-density on $X \times X \times R \times R$. In the following lemma, identify $X \times X \times R \times R$ with $(X \times R) \times (X \times R)$ via $(x, y, s, t) \to (x, s, y, t)$.

Lemma 2 $K_{\pi_*\Delta^*} = K_{Id}$, K_{Id} being the Schwartz kernel of the identity map.

Proof Denote by $\mathcal{D}^{\frac{1}{2}}(X)$, $\mathcal{D}^{\frac{1}{2}}(R)$ etc., the spaces of smooth half-densities associated to X, R, etc.. Both $K_{\pi_*\Delta^*}$ and K_{Id} can be viewed as multi-linear functionals on

$$\mathcal{D}^{\frac{1}{2}}(X) \times \mathcal{D}^{\frac{1}{2}}(X) \times \mathcal{D}^{\frac{1}{2}}(R) \times \mathcal{D}^{\frac{1}{2}}(R) .$$

We let you convince yourselves that they are identical. Q.E.D.

Let Γ be the normal bundle to the diagonal in $X \times X \times R \times R$. Identify

$$T^*(X \times R) \cong \Gamma \tag{3.2}$$

by $(x, t, \xi, \tau) \quad (x, t, \xi, \tau) \times (x, t, -\xi, -\tau)$.

Corollary $K_{\pi_*\Delta^*} \in I_{\Gamma}^{\frac{n+1}{2}}$ and its symbol, $\sigma = (\sigma K_{\pi_*\Delta^*})$ is just $|\Omega^{n+1}_{X \times R}|^{\frac{1}{2}}$ when we make the identification (3.2)

Proof Both these statements are well-known for K_{Id}. See for example [8].

Now let
$$\mathfrak{a}_1, \mathfrak{a}_2, \ldots, \mathfrak{a}_i \in \mathcal{D}^{\frac{1}{2}}(X)$$

be an orthonormal basis of eigenfunctions of P. in $L^2(X)$, and let λ_i be the eigenvalue associated with $\mathfrak{a}_i$. Then

$$e(x, y, t) = \sum e^{i\lambda_i t}\, \mathfrak{a}_i(x)\, \overline{\mathfrak{a}_i(y)} .$$

Setting x = y and integrating over **X** we get

$$\sum e^{i\lambda_i t} \langle \sigma_i , \sigma_i \rangle_{L^2} = \sum e^{i\lambda_i t}$$

so this proves :

<u>Lemma 3</u> $\sum e^{i\lambda_i t} = (\pi_* \Delta^*) e$.

We are now in position to apply the clean intersection theory of section 2 . To do so we need to know that the fiber product diagram

$$\begin{array}{ccc} \Gamma' & \leftarrow & F \\ \downarrow & & \downarrow \\ T^*(X \times X \times R) & \leftarrow & \Lambda \end{array}$$

is clean. Γ' injects into $T^*(X \times X \times R)$ and the image consists of the set of all points, $(x, \xi, x, -\xi, t, \tau)$. To apply theorem II we need to know that this intersects Λ cleanly. By (3, 1) the intersection can be identified with the set of all points (x, ξ, T) for which $\Phi_T(x, \xi) = (x, \xi)$, i.e. the fixed point set of Φ_T ; and this will be a clean intersection if and only if the fixed point set is clean.
Let i + 1 be the dimension of this fixed point set, (i.e. let i be the dimension of this fixed point set intersected with $\rho = 1$.). Then i + 1 is the excess in the diagram (2.5) , so by Theorem II and a simple dimension count, we get

$$(\pi_* \Delta^*) e \in I^{i/2}(\Lambda_T)$$

where

$$\Lambda_T = \{(T, \tau), \tau \in R^+\} .$$

By definition $I^{i/2}(\Lambda_T)$ consists of scalar multiples of the distribution

(3.3) $$\int_0^\infty s^{\frac{i-1}{2}} e^{\sqrt{-1}s(t-T)} ds$$

plus similar distributions of lower order. It is well-known (see, for example, Gelfand-Shilov (6) ,) that (3.3) has a pole of order $\frac{i+1}{2}$ at T. We get the residue at this pole from the symbol of $(\pi_* \Delta^*) e$ which is just $\sigma \circ \mu_{can}$ by Theorem II . This can be easily computed (for example, by comparing Lemma 1 of section 1 with Lemma 2 of section 2.) We let you check that the answer is as we stated it in Theorem I.

4. THE PROOF OF THEOREM II

We begin by considering oscillatory integrals of the type considered by Hörmander in [8], i.e. integrals of the form

(4.1) $$\mu(x) = \int a(x, \theta)\, e^{i\phi(x, \theta)}\, d\theta$$

where $a(x, \theta)$ and $\phi(x, \theta)$ are smooth functions on $X \times R^N$ with the properties

(1) $\phi(x, \theta)$ is homogeneous of degree one in θ and $d\phi \neq o$.

(2) $a(x, \theta) = o$ near $X \times \{0\}$ and is homogeneous of degree d for θ large.

Finally there is a third condition which Hörmander imposes on ϕ :

(3) If $\frac{\partial \phi}{\partial \theta}(x, \theta) = o$ then at (x, θ) the differentials

$d(\frac{\partial \phi}{\partial \theta_1}), \ldots, d(\frac{\partial \phi}{\partial \theta_N})$ are linearly independent .

Let C_ϕ be the set of points where $\frac{\partial \phi}{\partial \theta} = o$. Then condition (3) implies that C_ϕ is an n-dimensional submanifold of $X \times R^N$ and that the map

(4.2) $$C_\phi \longrightarrow T^*X, \quad (x, \theta) \longrightarrow \frac{\partial \phi}{\partial x}$$

immerses C_ϕ as a Lagrangian submanifold of T^*X. Hörmander defines $I_\Lambda^{d+\frac{N}{2}}$ to be the space of all distributions which have a local representation of the form (4.1). For our purposes we need to consider distributions of the type (4.1) satisfying (1) and (2), but with (3) replaced by a weaker "cleanness" condition, to wit :

(3') C_ϕ is a submanifold of $X \times R^N$ and at each point of C_ϕ the tangent space is the space of vectors annihilated by $d(\frac{\partial \phi}{\partial \theta_1}), \ldots, d(\frac{\partial \phi}{\partial \theta_N})$.

Suppose that the dimension of the space spanned by these differentials is $N - e$.

<u>Lemma 1</u> The map $C_\phi \longrightarrow T^*X$ defined by (4.2) has as its image an immersed Lagrangian manifold $\Lambda \subset T^*X$ and the map $C_\phi \longrightarrow \Lambda$ is a fiber mapping of fiber dimension e. Moreover the distribution $\mu(x)$ defined by (4.1) is in $I_\Lambda^{\frac{d+N+e}{2}}$.

<u>Proof</u> To see that the image of C_ϕ is a Lagrangian manifold, we apply lemma 3 of section 2 with Λ replaced by graph $d\phi$ and Γ replaced by the normal bundle to the graph of $\pi : X \times R^N \to X$. We leave for you to check that the fiber product diagram (2.5) of section 2 is clean if and only if (3)' holds. To prove the last assertion we write (4.1) in polar coordinate form. Set $\theta = s\omega$ with $\omega \in S^{N-1}$. Then

$$(4.3) \qquad \mu(x) = \int s^k a(x, \omega)\, e^{is\phi(x,\omega)}\, d\omega\, ds \quad \text{where} \quad k = d + N-1 \ .$$

We can assume that $a(x, \omega)$ has its support in a coordinate patch (on S^{N-1}), hence that when the integrand in (4.3) is non-zero, ω is in a compact subset of R^{N-1}. The "polar" critical set C'_ϕ in $X \times R^{N-1}$ is defined by the equations

$$(4.4) \qquad \phi = 0, \ \frac{\partial\phi}{\partial\omega_1} = 0, \ \frac{\partial\phi}{\partial\omega_2} = 0, \ \ldots, \ \frac{\partial\phi}{\partial\omega_{N-1}} = 0 \ .$$

By (3)' the differentials of these functions are the defining equations for the normal space at each point of C'_ϕ. By a change of coordinates we can assume that the first $N-e$ of these differentials are linearly independent, and that

$$d\phi\ ,\ d\left(\frac{\partial\phi}{\partial\omega_1}\right), \ldots, d\left(\frac{\partial\phi}{\partial\omega_{\ell-1}}\right),\ d\omega_\ell\ , \ldots, d\omega_{N-1}\ ,\quad \ell = N-e\ ,$$

are linearly independent. This implies that C'_ϕ is locally defined by the first ℓ equations of (4.4) and that C'_ϕ intersects the surface $\omega_\ell = \text{const}, \omega_{\ell+1} = \text{const}., \ldots, \omega_{N-1} = \text{const}.$ transversally. Let ω' denote the first $\ell -1$ coordinates and ω'' the remaining e coordinates. Then for $\omega'' = c$, the function $\phi(x, \omega'', c)$ is a non-degenerate phase function on $X \times R^{\ell-1}$ (in Hörmander's sense), and its critical set is just the intersection of C'_ϕ with $\omega'' = c$. So in particular its associated Lagrangian manifold in T^*X is Λ. Now write

$$(4.5) \qquad \mu(x) = \left(\int s^k\, a(x, \omega', \omega'')\, e^{i\phi(x,\omega',\omega'')}\, d\omega'\right) d\omega'' \ .$$

For fixed ω'' the inner integral is in $I_\Lambda^{d+\frac{N+e}{2}}$; therefore, so is μ itself.

<u>Remark</u> The symbol of $\mu(x)$ is the integral over ω'' of the symbol, $\sigma(\omega'')$, of the inner integral in (4.5)

We now proceed to the proof of Theorem II. With the notation of Theorem II,

With the notation of Theorem II, let $\mu \varepsilon I^k_\Lambda$. Represent μ and K by oscillatory integrals :

$$K(x, y) = \int a(x, y, \theta)\, e^{i\phi(x, y, \theta)}\, d\theta$$

$$\mu(y) = \int b(y, \xi)\, e^{i\psi(y, \xi)}\, d\xi .$$

Then $K \cdot \mu$ is represented by the oscillatory integral

$$K\mu(x) = \int a(x, y, \theta)\, b(y, \xi)\, e^{i(\phi(x, y, \theta) + \psi(y, \xi))}\, d\theta\, d\xi .$$

We let you check that the critical set of the phase function $\phi + \psi$ is just the fiber product of the critical sets of ϕ and $-\psi$ with respect to their natural projections on T^*Y and that $\phi + \psi$ is a clean phase function if and only if this fiber product is clean. Now apply Lemma 1 .

We omit the computation of the symbol of $K\mu$, which is rather messy. (The main idea in this computation, however, is to elaborate on the remark following Lemma 1 .)

BIBLIOGRAPHY

1. R. Bott, "On the iteration of closed geodesics and the Sturm intersection theory", Comm. Pure Appl. Math. 9 (1956) 176-206

2. J. Chazarain, "Formule de Poisson pour les variétés Riemanniennes", Invent. Math. 24, 65 - 82 (1974).

3. J. J. Duistermaat, "On the Morse index in variational calculus", to appear in Journal of Diff. Geom.

4. J. J. Duistermaat and V. Guillemin, "The spectrum of positive elliptic operators and periodic geodesics", Proc. AMS Summer institute on Diff. Geom. Stanford 1973 (to appear)

5. J. J. Duistermaat and L. Hörmander, "Fourier Integral Operators II", Acta Math. 128 (1972) 183-269

6. I. M. Gelfand and G. E. Shilov, Generalized Functions I, Academic Press, New York 1964

7. V. Guillemin and S. Sternberg, Geometric Asymptotics, AMS publications, (now in proof)

8. L. Hörmander, "Fourier Integral Operators", Acta Math. 127 (1971) 79-183.

NON-UNIQUENESS FOR THE CAUCHY PROBLEM

Lars Hörmander
University of Lund

1. Introduction. A series of counterexamples (Cohen [1], De Giorgi [2], Goorjian [3], Pliś [6-11]) has shown that uniqueness theorems for differential equations with non-analytic coefficients require much more restrictive conditions than those in Holmgren's uniqueness theorem. However, there is a considerable gap between these counterexamples and the uniqueness theorems available. In this paper we shall try to narrow the gap or at least make it well defined by making a systematic analysis of the scope of the constructions used in the counterexample

Let P(D) and Q(D) be two partial differential operators with constant coefficients in $\mathbb{R}^n$, $D = -i\,\partial/\partial x$ as usual, and let H_N be a half space

$$H_N = \{x \in \mathbb{R}^n;\ \langle x, N\rangle \geqq 0\}.$$

We shall study perturbations of P by the operator Q. The problem is to decide when there is a function a such that the equation

$$P(D)u + a\,Q(D)u = 0 \tag{1.1}$$

has a solution $u \in C^\infty(\mathbb{R}^n)$ with

$$\operatorname{supp} u = H_N. \tag{1.2}$$

We wish a to vanish when $\langle x, N\rangle = 0$ so that the operator P(D) is not perturbed there. The answer may of course depend on the conditions placed on a. We shall examine the cases where a is required to be analytic, C^∞ or C^j for some finite j. The main results are Theorems 2.2, 3.1, 3.7 and 4.1.

Most uniqueness theorems known for equations of the form (1.1) require that u vanishes outside a set with a strictly convex boundary. A direct comparison with the counterexamples proved here is therefore not possible.

For this and other reasons it would be interesting to modify the constructions with H_N replaced by a strictly convex set, compact sets being particularly important. However, we shall not consider this problem at all here.

2. Analytic perturbations. First we recall the situation for the unperturbed operator P:

Theorem 2.1. The equation $P(D)u = 0$ has a solution $u \in C^\infty(\mathbb{R}^n)$ with $\operatorname{supp} u = H_N$ if and only if $P_m(N) = 0$, where P_m is the principal part of P.

The necessity follows from Holmgren's uniqueness theorem (see Hörmander [4, Theorem 5.3.1]), and the sufficiency is proved by integrating suitable exponential solutions ([4, Theorem 5.2.2]). Holmgren's uniqueness theorem also gives the implication 2)$\Rightarrow$1) in the following

Theorem 2.2. The following conditions are equivalent if ∂H_N is non-characteristic with respect to P:

1) The order of P is smaller than the order of Q.
2) The equation (1.1) has a solution $u \in C^\infty(\mathbb{R}^n)$ satisfying (1.2) for some analytic a in $\mathbb{R}^n$ vanishing when $\langle x, N\rangle = 0$.
3) For any given integer k the equation (1.1) has a solution $u \in C^\infty(\mathbb{R}^n)$ satisfying (1.2) for some analytic a in $\mathbb{R}^n$ vanishing of order k when $\langle x, N\rangle = 0$.

Proof. Since 3)$\Rightarrow$2)$\Rightarrow$1) we just have to prove that 1)$\Rightarrow$3). Let m be the order of Q. If $Q_m(N) \neq 0$, that is, ∂H_N is non-characteristic with respect to Q, the proof is somewhat simpler so we consider this case first. Choosing coordinates with $\langle x, N\rangle = x_1$ and taking a and u as functions of x_1 only, we find that it is then sufficient to prove the theorem in the one-dimensional case. Thus we assume that $n = 1$ and set with a positive integer k and a large positive number A to be chosen later

$$U(x) = \exp A(x - x^{1-2k}),\ x \neq 0;\ u(x) = U(x),\ x > 0 \text{ and } u(x) = 0,\ x \leqq 0.$$

Then u is a classical example of a C^{∞} function with support equal to $\overline{\mathbb{R}}_+$. We have

$$dU/dx = A(1+(2k-1)x^{-2k})\, U\, .$$

If $Q = d^m/dx^m + \dots$ it follows that

$$QU = x^{-2km}\, G(x)\, U,$$

where

$$G(x) = (A(x^{2k} + 2k-1))^m + \dots$$

is a polynomial of order 2km where the coefficients of the terms indicated by dots are $O(A^{m-1})$. Hence $G(x) \neq 0$ for all real x if A is large enough. In the same way we obtain

$$PU = x^{-2k\mu} F(x) U$$

where $\mu < m$ is the order of P. Hence $Pu + aQu = 0$ if

$$a = -PU/QU = -x^{2k(m-\mu)} F(x)/G(x),$$

which has the properties required in 3).

If $Q_m(N) = 0$ it is clear in view of Holmgren's uniqueness theorem that we cannot work with functions of $\langle x, N\rangle$ only. In that case we choose a vector $\theta \in \mathbb{R}^n \setminus 0$ such that

$$(2.1) \qquad Q_m(\theta + itN) \neq 0,\ t \in \mathbb{R}.$$

This is possible, for if $Q_m(\theta_0) \neq 0$, then $\theta = \theta_0 + sN$ satisfies (2.1) if s is not the real part of one of the finitely many zeros of the equation $Q_m(\theta_0 + zN) = 0$.

Now choose the coordinates so that $x_1 = \langle x, \theta\rangle$ and $x_2 = \langle x, N\rangle$. Then (2.1) means that $Q_m(1, i\xi_2, 0, \dots, 0) \neq 0$ for real ξ_2. Taking a and u as functions of x_1 and x_2 only we may assume that $n = 2$ in what follows. Set

$$\Phi(x) = ix_1 + (1+x_2^{-2}),\ U(x) = \exp(-A\Phi(x)^k),\ x_2 \neq 0,$$

where k is a positive integer and A a large positive number. Then it is again clear that

$$u(x) = U(x),\ x_2 > 0;\ u(x) = 0,\ x_2 \leqq 0,$$

is a C^∞ function satisfying (1.2). It remains to show that

$$a = -P(D)U/Q(D)U$$

is analytic in R^n and vanishes of high order when $x_2 = 0$ if k and A are large enough.

Since

$$D_1U = -kA\Phi^{k-1}\,U,\quad D_2U = -2ik\ A\Phi^{k-1}\ U/x_2^3$$

it is clear that $(Q(D)U)/U$ is a polynomial in A with leading term

$$(-kA\Phi^{k-1})^m\ Q_m(1,\ 2i/x_2^3). \tag{2.2}$$

Here $|Q_m(1,\ 2i/x_2^3)|$ has a positive lower bound by (2.1). The other terms in $(Q(D)U)/U$ can be estimated by

$$C\ |A\Phi^{k-1}|^{m-1}\ (1 + |x_2|^{-3(m-1)}) \tag{2.3}$$

for some C. (Note that $|\Phi| \geqq 1$.) The quotient of (2.3) by (2.2) is bounded by

$$C(1+|x_2|^{-3(m-1)})\ |Q_m(1,\ 2i/x_2^3)|^{-1}\ A^{-1}\ |\Phi|^{1-k},$$

so it is $< 1/2$ for all x if $3(m-1) \leqq 2(k-1)$ and A is large enough. Hence

$$Q(D)U = GU,\ P(D)U = FU$$

where G and F are rational functions and F/G is analytic when $x_2 \neq 0$. It remains to examine the quotient when $x_2 = 0$.

Let $Q_m(1,\ 2it) = c\ t^\mu$ + terms of lower order. If $3(m-1) \leqq 2(k-1)$ then

$$x_2^{3\mu+2m(k-1)}\ G(x_1,\ x_2)$$

is a polynomial which is equal to $(-kA)^m c$ when $x_2 = 0$. Since P is of order m-1 the product

$$x_2^{(3+2(k-1))(m-1)} F(x)$$

is a polynomial because $x_2^{3+2(k-1)} \Phi^{k-1}/x_2^3$ is one. It follows that a is analytic and vanishes when $x_2 = 0$ of order

$$3\mu + 2m(k-1) - (3+2(k-1))(m-1) = 3\mu - 3(m-1) + 2(k-1),$$

which is arbitrarily large with k. This completes the proof.

For later reference we observe that the perturbation in Theorem 2.2 can be chosen so that it vanishes of infinite order when $\langle x, N\rangle = 0$ if one does not require that it shall be analytic:

Theorem 2.3. The conditions in Theorem 2.2 imply that the equation (1.1) has a solution $u \in C^\infty(\mathbb{R}^n)$ satisfying (1.2) for some $a \in C^\infty(\mathbb{R}^n)$ which vanishes of infinite order when $\langle x, N\rangle = 0$.

Proof. In the first part of the proof of Theorem 2.2 we replace $x - x^{1-2k}$ by $\Psi(x) = x - e^{x^{-\alpha}}$ where $\alpha > 0$. Then $\Psi'(x) = 1 + \alpha x^{-\alpha-1} e^{x^{-\alpha}}$ bounds any positive power of x^{-1}, and $|\Psi^{(k)}||\Psi'|^{-1-\varepsilon}$ is bounded for every $\varepsilon > 0$ if $k > 0$. If $U = e^{A\Psi}$ it follows for large A that

$$QU = U\,(A\Psi')^m(1+R)$$

where $R \in C^\infty$ and $|R| < 1/2$. A similar result is valid for PU, and since $1/\Psi'$ is a C^∞ function vanishing of infinite order when $x = 0$, the proof follows as before when $Q_m(N) \neq 0$. In the second part of the proof of Theorem 2.2 we replace the definition of U by $U(x) = \exp(-A \exp \Phi(x)^\alpha)$. Then

$$(D_1U, D_2U) = -A\, e^{\Phi^\alpha}\, \alpha \Phi^{\alpha-1}\,(1, 2i/x_2^3)\, U,$$

and $u \in C^\infty$ if $0 < \alpha < 1$. The leading term in $(QU)/U$ becomes

$$(-A\, e^{\Phi^\alpha}\, \alpha\Phi^{\alpha-1})^m\, Q_m(1, 2i/x_2^3),$$

and the proof can be completed as before.

3. Infinitely differentiable perturbations. When $P(D)$ is a differential operator with constant coefficients in $\mathbb{R}^n$ we shall write

$$\tilde{P}(\xi) = \Big(\sum_\alpha |P^{(\alpha)}(\xi)|^2\Big)^{1/2}.$$

This is a positive majorant of P. The quotient between $\tilde{P}(\xi)$ and

$$\sup_{|\eta|<R} |P(\xi+\eta)|, \quad R > 0,$$

is bounded from above and below by positive constants depending only on R, the dimension n and the degree m of P. (Here η may either be taken in $\mathbb{C}^n$ or be restricted to $\mathbb{R}^n$.) Another operator $Q(D)$ is said to be weaker than $P(D)$ if $\tilde{Q}/\tilde{P}$ is bounded in $\mathbb{R}^n$. (Cf. [4, section 3.3].) The order of Q is then at most equal to that of P but the converse is false if $n > 1$.

If $N \in \mathbb{R}^n \setminus 0$ we shall also use a smaller majorant of P defined by

$$\tilde{P}_N(\xi) = \Big(\sum_j |\langle D, N\rangle^j P(\xi)|^2\Big)^{1/2}.$$

If $\tilde{Q}_N(\xi) \leq C\tilde{P}_N(\xi)$ for all $\xi \in \mathbb{R}^n$, then

$$|Q(\xi)| \leq C' \sup_{|\tau|<1} |P(\xi+\tau N)|, \quad \xi \in \mathbb{R}^n,$$

where τ is chosen real, and this gives

$$\sup_{|\eta|<1} |Q(\xi+\eta)| \leq C' \sup_{|\tau|<1} \sup_{|\eta|<1} |P(\xi+\eta+\tau N)| \leq C' \sup_{|\eta|<|N|+1} |P(\xi+\eta)|.$$

Hence $\tilde{Q}(\xi)/\tilde{P}(\xi)$ is also bounded then. This shows that the following theorem implies its corollary.

Theorem 3.1. Let $P(D)$ and $Q(D)$ be two differential operators with constant coefficients, $N \in \mathbb{R}^n \setminus 0$, and assume that $\tilde{Q}_N(\xi)$ cannot be bounded by a constant times $\tilde{P}_N(\xi)$ when $\xi \in \mathbb{R}^n$. Then the equation (1.1) has a solution $u \in C^\infty(\mathbb{R}^n)$ satisfying (1.2) for some $a \in C^\infty(\mathbb{R}^n)$ vanishing in $\complement H_N$.

Corollary 3.2. Let $P(D)$ and $Q(D)$ be two differential operators with constant coefficients such that Q is not weaker than P. For any $N \in \mathbb{R}^n \setminus 0$ it is then possible to find $a \in C^\infty(\mathbb{R}^n)$ vanishing in $\complement H_N$ so that the equation (1.1) has a solution satisfying (1.2).

Theorem 3.1 follows from Theorems 2.1 and 2.3 unless

$$\deg Q \leqq \deg P = m, \quad P_m(N) \neq 0, \tag{3.1}$$

where P_m denotes the principal part of P. In what follows we assume (3.1).

<u>Lemma 3.3.</u> Let P and Q satisfy the hypotheses of Theorem 3.1 and (3.1). Then it is possible to find a Laurent series

$$\xi(t) = \sum_{-\infty}^{\mu} c_j t^j, \quad c_j \in \mathbb{R}^n,$$

converging for large t, and a positive integer $\varkappa$ such that when $t \to \infty$

$$P(\xi(t) + z\, t^{\varkappa} N)\, t^{-g(P)} \to c(P) z^{d(P)}, \quad Q(\xi(t) + z\, t^{\varkappa} N)\, t^{-g(Q)} \to c(Q)\, z^{d(Q)} \tag{3.2}$$

for some constants $c(P)$, $c(Q) \neq 0$ and integers $0 < g(P) < g(Q)$, $d(P) > d(Q) \geqq 0$.

<u>Proof.</u> $\tilde{Q}_N(\eta)/\tilde{P}_N(\eta)$ is continuous since $\tilde{P}_N(\eta) \geqq m!\,|P_m(N)|$. By the Tarski-Seidenberg theorem (see the appendix in [4]) the maximum when η is real and $|\eta| \leqq R$ is an algebraic function of R for large R. By hypothesis it tends to infinity with R, so it is increasing for large R. The maximum is then attained for some $\eta(R)$ with $|\eta(R)| = R$ and the Tarski-Seidenberg theorem shows that the lexicographically largest such $\eta(R)$ is an algebraic function of R for large R. From the Puiseux series expansion of $\eta(R)$ it follows that $\xi(t) = \eta(t^r)$ has a Laurent series expansion of the desired form for some positive integer r.

Writing $P^{(j)}(\xi) = \langle \partial, N\rangle^j P(\xi)$ we conclude that $P^{(j)}(\xi(t))$ is asymptotic to a constant times t^{μ_j} for some integer μ_j, or else $P^{(j)}(\xi(t))$ is identically 0 in which case we define $\mu_j = -\infty$. (3.1) gives $\mu_m = 0$, so

$$g_P(\varkappa) = \max_{j \geqq 0} (\varkappa j + \mu_j(P)), \quad \varkappa \geqq 0,$$

is a finite, convex, increasing, piecewise linear function. $\tilde{P}_N(\xi(t))\, t^{-g_P(0)}$ has a limit $\neq 0$ as $t \to \infty$, so $g_Q(0) > g_P(0)$ if g_Q is defined similarly. On the other hand, $g_P(\varkappa) = m\varkappa \geqq g_Q(\varkappa)$ when $\varkappa$ is large, in view of (3.1). It is therefore possible to find a rational number $\varkappa > 0$ such that

$$g_P(\varkappa) < g_Q(\varkappa)$$

and g_P, g_Q are both linear in a neighborhood of $\varkappa$, with slopes d_P and d_Q such that $d_P > d_Q$. In fact, g_P has to increase faster than g_Q some time before catching up. Now we have

$$\varkappa j + \mu_j(P) \leq g_P(\varkappa), \quad j \geq 0,$$

with equality for just one j, so $g_P(\sigma) = \sigma j + \mu_j(P)$ in a neighborhood of $\varkappa$. Hence $j = d_P$ and the first part of (3.2) is valid with $g_P(\varkappa) = g(P)$ and $d(P) = d_P$. The second part follows in the same way. Replacing t by an integral power of t we can make $\varkappa$ an integer, which completes the proof.

Proof of Theorem 3.1. To construct u we shall piece together functions u_ν which are appropriately chosen in the slabs

$$\{x;\ \langle x, N\rangle \in I_\nu\}, \quad I_\nu = (b_{\nu+1}, b_{\nu-1}).$$

Here we take $b_\nu = 1/\nu$, for example, and ν will be large. With $\xi(t)$ given by Lemma 3.3 and $t_\nu = 2^\nu$ we set

$$u_\nu(x) = \exp i(\langle x, \xi(t_\nu)\rangle + \varphi_\nu(\langle x,N\rangle)), \quad \langle x, N\rangle \in I_\nu, \tag{3.3}$$

where φ_ν will be chosen later. The plan is then to set for some large ν_0

$$u(x) = \sum_{\nu_0}^{\infty} \chi_\nu(\langle x,N\rangle)\, u_\nu(x), \quad x \in \mathbb{R}^n, \tag{3.4}$$

where $\chi_\nu \in C_0^\infty(I_\nu)$ is 1 in a neighborhood of $(b'_\nu, b'_{\nu-1})$; $b'_\nu = (b_\nu + b_{\nu+1})/2$. (The first term will be slightly modified.) Then $u(x) = u_\nu(x) + u_{\nu+1}(x)$ when $\langle x, N\rangle$ is close to b'_ν. When $\langle x, N\rangle$ is at some distance to the left (resp. right) of b'_ν, the function u_ν (resp. $u_{\nu+1}$) will then be cut off. To make $a = -P(D)u/Q(D)u$ smooth we have to make sure that $u_{\nu+1}$ (resp. u_ν) is much larger than u_ν (resp. $u_{\nu+1}$) then. When $\langle x, N\rangle$ is near b'_ν the two terms must therefore be of the same size and we shall arrange that both satisfy the same differential equation $P(D)u + aQ(D)u = 0$ with a constant a then. This can be done by choosing φ_ν and $\varphi_{\nu+1}$ as linear functions with suitable slopes there. Since (3.2) gives us information about the quotient of P and Q at $\xi(t) + zt^{\varkappa} N$ it is natural to choose the slope of φ_ν of the order of magnitude $t_\nu^{\varkappa}$.

After these motivations we set in I_ν

$$\varphi_\nu'(s) = t_\nu^{\varkappa} \psi_\nu(s).$$

We shall determine ψ_ν so that near b_ν' and $b'_{\nu-1}$, respectively, ψ_ν is equal to constants σ_ν^- and σ_ν^+ satisfying

$$(3.5) \qquad (P/Q)(\xi(t_\nu) + t_\nu^{\varkappa} \sigma_\nu^- N) = (P/Q)(\xi(t_{\nu+1}) + t_{\nu+1}^{\varkappa} \sigma_{\nu+1}^+ N).$$

With the notations of Lemma 3.3 the left hand side is asymptotically equal to

$$(c(P)/c(Q))\; t_\nu^{g(P)-g(Q)}\; (\sigma_\nu^-)^{d(P)-d(Q)},$$

so the equation (3.5) is closely approximated by

$$2^{g(Q)-g(P)}\; (\sigma_\nu^-)^{d(P)-d(Q)} = (\sigma_{\nu+1}^+)^{d(P)-d(Q)}.$$

If we choose $\sigma_\nu^- = i$, it follows that (3.5) for large ν has a solution $\sigma_{\nu+1}^+$ with

$$(3.6) \qquad \sigma_{\nu+1}^+ \to 2^{\gamma} i, \quad \nu \to \infty.$$

Here $\gamma = (g(Q) - g(P))/(d(P) - d(Q)) > 0$, so $2^{\gamma} > 1$. This means that the slope $-t_{\nu+1}^{\varkappa} \operatorname{Im} \sigma_{\nu+1}^+$ of $-\operatorname{Im} \varphi_{\nu+1}$ as a function of $\langle x, N\rangle$ will be far smaller than the slope $-t_\nu^{\varkappa} \operatorname{Im} \sigma_\nu^-$ of $-\operatorname{Im} \varphi_\nu$, so the ratio $|u_{\nu+1}/u_\nu|$ will decrease fast when $\langle x, N\rangle$ increases in a neighborhood of b_ν'.

Let $B \geq \operatorname{Im} \sigma_{\nu+1}^+$ for every ν, which implies that $B > 1$. For every $M > 0$ we can choose $\psi_\nu \in C^\infty(I_\nu)$ satisfying for large ν the following conditions for some constants $C > 0$, $C_k > 0$, and some compact set $K \subset \mathbb{C} \setminus \{0\}$

$$(3.7) \quad \psi_\nu(s) = \sigma_\nu^- \text{ when } |s-b_\nu'| < 1/4\nu^2, \quad \psi_\nu(s) = \sigma_\nu^+ \text{ when } |s-b'_{\nu-1}| < 1/4(\nu-1)^2;$$

$$\psi_\nu(s) \in K \text{ when } s \in I_\nu; \quad |\psi_\nu^{(k)}(s)| \leq C_k \nu^{2k} \text{ if } s \in I_\nu,\ k = 0, 1, \ldots;$$

$$\operatorname{Im} \psi_\nu(s) \leq B, \quad \int_{b_\nu'}^{b'_{\nu-1}} \operatorname{Im} \psi_\nu(s) < -M/\nu^2.$$

In fact, it suffices to piece together by a partition of unity the prescribed values σ_ν^- and σ_ν^+ near b'_ν and $b'_{\nu-1}$, and a value $1 - iA$ near b_ν, where A is chosen large. The derivatives of order k of the functions in the partition of unity (with three terms) are $O(\nu^{2k})$ since the length of I_ν is $2/(\nu^2-1)$.

Having chosen ψ_ν we determine φ_ν successively so that

$$\varphi'_\nu = t_\nu^{\varkappa}\psi_\nu \quad , \quad \varphi_\nu(b'_\nu) = \varphi_{\nu+1}(b'_\nu). \tag{3.8}$$

In view of (3.6) we have for some $C_1, C_2 > 0$ when ν is large

$$C_1\, 2^{\varkappa\nu} \leqq \operatorname{Im}\,(\varphi_{\nu+1}(s) - \varphi_\nu(s))/(s-b'_\nu) \leqq C_2\, 2^{\varkappa\nu}, \quad |s-b'_\nu| < 1/4\nu^2. \tag{3.9}$$

This gives the information about the relative sizes of u_ν and $u_{\nu+1}$ needed to prove that a is smooth. To prove that u itself is smooth we must also have some information about the absolute size of u_ν. From (3.7) it follows that

$$\operatorname{Im}\,(\varphi_\nu(b'_{\nu-1}) - \varphi_\nu(b'_\nu)) < -M\, 2^{\varkappa\nu}/\nu^2,$$

that is,

$$\operatorname{Im}\,\varphi_{\nu+1}(b'_\nu) > \operatorname{Im}\,\varphi_\nu(b'_{\nu-1}) + M\, 2^{\varkappa\nu}/\nu^2.$$

Hence

$$\operatorname{Im}\,\varphi_\nu(b'_{\nu-1}) > M\, 2^{\varkappa(\nu-1)}/\nu^2$$

if we make sure that this is true at the beginning of the construction. Since $\operatorname{Im}\,\varphi'_\nu \leqq 2^{\varkappa\nu}B$, it follows that

$$\operatorname{Im}\,\varphi_\nu(s) \geqq 2^{\varkappa\nu}/\nu^2, \quad b'_\nu \leqq s \leqq b'_{\nu-1} \tag{3.10}$$

provided that

$$M\, 2^{\varkappa(\nu-1)}/\nu^2 - 2^{\varkappa\nu}B(b'_{\nu-1} - b'_\nu) \geqq 2^{\varkappa\nu}/\nu^2,$$

which is true for large ν if $M > (B+1)\, 2^{\varkappa}$. In view of (3.9) the estimate (3.10) is also valid with ν replaced by $\nu+1$ or $\nu-1$ in the right hand side when $b'_\nu - 1/4\nu^2 < s < b'_\nu$ or $b'_{\nu-1} < s < b'_{\nu-1} + 1/4(\nu-1)^2$.

Now choose $\chi_\nu \in C_0^\infty(b'_\nu - 1/4\nu^2,\ b'_{\nu-1} + 1/4(\nu-1)^2)$ so that $\chi_\nu = 1$ in $(b'_\nu - 1/8\nu^2,\ b'_{\nu-1} + 1/8(\nu-1)^2)$ and

(3.11) $$|\chi_\nu^{(k)}(s)| \leq C_k \nu^{2k},\quad k \geq 0.$$

With some ν_0 so large that the preceding estimates and some later ones are valid for $\nu \geq \nu_0$ we define u by (3.4). (The first term will be modified later on.) In view of (3.7), (3.10) and (3.11) derivatives of order k of the νth term can be estimated by

$$C_k \exp(-2^{\varkappa(\nu-1)}/\nu^2)\quad (t_\nu^\varkappa + |\xi(t_\nu)|)^k,$$

which converges to 0 very rapidly when $\nu \to \infty$. Since only two terms in (3.4) are simultaneously different from 0, it follows that $u \in C^\infty$.

Next we consider $a = -P(D)u/Q(D)u$, defined as 0 when $\langle x, N\rangle \leq 0$. In the slab where

$$|\langle x, N\rangle - b'_\nu| < 1/8\nu^2$$

the construction has been made so that $u = u_\nu + u_{\nu+1}$ and $-a$ is the constant (3.5), which tends to 0 when $\nu \to \infty$ in view of (3.2) where $g(P) < g(Q)$. The derivatives of a are all 0 in this set. Passing to the set where

$$b'_\nu + 1/8\nu^2 < \langle x, N\rangle < b'_\nu + 1/4\nu^2,$$

we write $u = u_\nu(1 + R_\nu)$, where

$$R_\nu(x) = \chi_{\nu+1}(\langle x,N\rangle) \exp i(\langle x, \xi(t_{\nu+1}) - \xi(t_\nu)\rangle + \varphi_{\nu+1}(\langle x,N\rangle) - \varphi_\nu(\langle x,N\rangle)).$$

In view of (3.9) it is clear that if $\sigma \geq \varkappa$ and $\xi(t) = O(t^\sigma)$ then

(3.12) $$|D^\alpha R_\nu(x)| \leq C_\alpha 2^{|\alpha|\sigma\nu} \exp(-c\, 2^{\varkappa\nu}/\nu^2)$$

for some positive constants C_α and c. Since u_ν is a pure exponential, we have

$$(Q(D)u)/u_\nu = Q(\xi_\nu + i\sigma_\nu^- t_\nu^\varkappa N + D)(1 + R_\nu) = Q(\xi_\nu + i\sigma_\nu^- t_\nu^\varkappa N)(1+S_\nu)$$

where for some constant C

(3.12)' $$|D^\alpha S_\nu(x)| \leq C_\alpha 2^{|\alpha|\sigma\nu} \exp(C\nu - c\, 2^{\varkappa\nu}/\nu^2).$$

In particular, $|S_\nu| < 1/2$ when ν is large. Replacing Q by P we also obtain

$$(P(D)u)/u_\nu = P(\xi_\nu + i\sigma_\nu^- t_\nu^\varkappa N)(1 + T_\nu)$$

where T_ν also satisfies estimates of the form (3.12)'. Thus

$$a = -(P/Q)(\xi_\nu + i\sigma_\nu^- t_\nu^\varkappa N)(1+T_\nu)/(1+S_\nu),$$

so it follows from (3.2) and (3.12)' that all derivatives have bounds converging to 0 as $\nu \to \infty$. The same argument is applicable in the slab

$$b'_{\nu-1} - 1/4(\nu-1)^2 < \langle x, N\rangle < b'_{\nu-1} - 1/8(\nu-1)^2.$$

It remains to consider the set where

$$b'_\nu + 1/4\nu^2 < \langle x, N\rangle < b'_{\nu-1} - 1/4(\nu-1)^2. \tag{3.13}$$

There we have $u = u_\nu$, and

$$Q(D)u = c(Q)\, t_\nu^{g(Q)}\, f_{Q\nu}(\langle x, N\rangle)\, u,$$

where

$$f_{Q\nu}(s) = e^{-i\varphi_\nu(s)}\, q_\nu(D/t_\nu^\varkappa)\, e^{i\varphi_\nu(s)},$$

$$q_\nu(z) = Q(\xi(t_\nu) + z\, t_\nu^\varkappa N)/t_\nu^{g(Q)} c(Q).$$

From (3.2) we know that $q_\nu(z) - z^{d(Q)}$ is a polynomial with coefficients which are $O(t_\nu^{-1})$. Recalling that $\varphi_\nu'/t_\nu^\varkappa = \psi_\nu$ we obtain

$$f_{Q\nu}(s) = \psi_\nu(s)^{d(Q)}(1 + r_{Q\nu}(s))$$

where $r_{Q\nu}(s)$ is a polynomial in ψ_ν, the derivatives of ψ_ν, and ψ_ν^{-1} with coefficients $O(1/t_\nu)$. Since t_ν grows exponentially it follows from (3.7) that $r_{Q\nu}(s)$ and any one of its derivatives $\to 0$ as $\nu \to \infty$. If $r_{P\nu}$ is defined in the same way, we have in the set defined by (3.13)

$$a = -P(D)u/Q(D)u = -(c(P)/c(Q))\, t_\nu^{g(P)-g(Q)}\psi_\nu^{d(P)-d(Q)}(1 + r_{P\nu})/(1 + r_{Q\nu})$$

where the argument is x on the left and $s = \langle x, N\rangle$ on the right. Since $g(P) < g(Q)$ it follows that the right hand side and all its derivatives tend to 0 exponentially when $\nu \to \infty$. Thus $a \in C^\infty$ when $\langle x, N\rangle < b'_{\nu_0 - 1}$ if ν_0 is chosen large enough.

Now change the definitions of χ_{ν_0} and φ_{ν_0} when $\langle x, N\rangle > b'_{\nu_0-1}$ so that $\chi_{\nu_0} = 1$ and φ_{ν_0} is linear in $(b'_{\nu_0-1} - 1/4(\nu_0-1)^2)$. It is then clear that $u \in C^\infty(\mathbf{R}^n)$, and $a \in C^\infty(\mathbf{R}^n)$ since a is constant when $\langle x,N\rangle > b'_{\nu_0-1} - 1/4(\nu_0-1)^2$. The proof is complete.

Remark. The proof shows that sup $|a|$ is as small as we please if ν_0 is chosen large. Thus the perturbation of P in (1.1) can be taken small in the whole space. For any compact interval $I \subset \mathbf{R}$ we can construct u and a satisfying (1.1) and

$$\operatorname{supp} u = \operatorname{supp} a = \{x;\ \langle x, N\rangle \in I\}$$

by making the same choices as above near the two boundaries of the slab.

Before proceeding we shall discuss some examples. The first is essentially (the first) Theorem 1 of Cohen [1]. (After the proof of that theorem Cohen made the following remark which might be interpreted as a hint of Theorem 3.1: "For a more general choice of the operator Q_1 which can be chosen, it would seem that the condition of Hörmander [ ] theorem is very likely a sufficient one. However this would be quite complicated and we prefer not to discuss it here.")

Example 3.4. If P(D) is homogeneous of degree m and $k \geqq 0$, then the hypotheses of Theorem 3.1 are fulfilled for some homogeneous Q of order m-k unless

$$\xi \in \mathbf{R}^n,\ \langle D, N\rangle^j P(\xi) = 0,\ j \leqq k \Rightarrow \xi = 0.$$

The hypotheses of Corollary 3.2 are fulfilled for some such Q unless

$$\xi \in \mathbf{R}^n,\ D^\alpha P(\xi) = 0,\ |\alpha| \leqq k \Rightarrow \xi = 0.$$

In particular, Example 3.4 with k = 0 is applicable to every non-elliptic P. When k > 0 the lower order terms may become important:

Example 3.5. The hypotheses of Corollary 3.2 are fulfilled by some Q of strictly lower order than $P = P_m + P_{m-1} + \dots$ unless every $\xi \in \mathbb{R}^n \setminus 0$ with $P_m'(\xi) = 0$ has a neighborhood U such that $P_{m-1}(\xi)$ is not in the closed conic hull of $\{P_m(\eta), \eta \in U\}$.

Indeed, Corollary 3.2 is applicable for some such Q unless

$$(1+|\theta|)^{m-1} \leq C \sum_{\alpha} |P^{(\alpha)}(\theta)|, \quad \theta \in \mathbb{R}^n.$$

For large $|\theta|$ this implies that

$$|\theta|^{m-1} \leq 2C(|P_m(\theta) + P_{m-1}(\theta)| + \sum_{1}^{n} |P_m^{(j)}(\theta)|).$$

Taking $\theta = -t\eta$ where $t \in \mathbb{R}$ is large and η is in a small neighborhood of ξ, we obtain

$$1 \leq 3C|tP_m(\eta) - P_{m-1}(\xi)|$$

which proves the assertion.

The following example contains the second Theorem 1 in Cohen [1] and also Theorem 8.9.2 in [4].

Example 3.6. The hypotheses of Corollary 3.2 are fulfilled if for some $\eta \in \mathbb{R}^n$ the polynomial $P(\xi+z\eta)$ is independent of z while $Q(\xi+z\eta)$ is not.

If P is hyperbolic with respect to N and Q is weaker than P, it is easy to show that $P + aQ$ is hyperbolic provided that sup $|a|$ is sufficiently small. In fact, if E_0 is a fundamental solution of P then QE_0 is bounded for an appropriate norm (see the proof of Theorem 5.5.7 in [4]), and $E_0(I + aQE_0)^{-1}$ is a fundamental solution of $P + aQ$. In particular, if $(P+aQ)u = 0$ and u vanishes outside H_N then u must vanish identically. Theorem 3.1 can therefore not be improved when P is hyperbolic.

On the other hand, Theorem 3.1 fails completely when P is elliptic (unless Q is of higher degree than P in which case Theorem 2.3 is already applicable).

The only examples of non-uniqueness for C^∞ perturbations of elliptic operators have been given by Pliś [8]. His constructions are much more delicate than those in the proof of Theorem 3.1. There we could choose the functions (3.3) rather arbitrarily, just making sure that Im φ_ν and φ_ν' were of the right order of magnitude and exercising some care at the points b_ν' where a switch was made from u_ν to $u_{\nu+1}$. In the case of complex characteristics, on the other hand, the analogue of the function φ_ν has to be chosen so that it fits an initial analytic perturbation of the operator. We shall do so by means of the asymptotic expansions of geometrical optics which seem to have greater generality than the arguments used by Pliś [8].

Theorem 3.7. Assume that there exists a sequence $\zeta_\nu \in \Sigma_N$,

$$\Sigma_N = \{ \xi + isN;\ \xi \in \mathbb{R}^n,\ s \leq 0 \},$$

sequences T_ν, K_ν of positive numbers and a_ν, b_ν of complex numbers such that

(3.14) $$a_\nu P(\zeta_\nu + T_\nu zN) \to q(z), \quad b_\nu Q(\zeta_\nu + T_\nu zN) \to q(z), \quad z \in \mathbb{C},$$

(3.15) $$b_\nu / a_\nu \to 0,$$

(3.16) $$K_\nu (a_\nu P(\zeta_\nu + T_\nu zN) - b_\nu Q(\zeta_\nu + T_\nu zN)) \to r(z),$$

(3.17) $$T_\nu / K_\nu \to \infty, \quad T_\nu K_\nu / (1 + |\mathrm{Im}\ \zeta_\nu|) \to \infty.$$

Here q and r are polynomials in one variable which are not identically 0, and we assume that

(3.18) $r(z)/q(z)$ is not a first order polynomial $az+b$ with $\mathrm{Im}\ a \geq 0$.

Then one can find $a \in C^\infty(\mathbb{R}^n)$ vanishing of a given, arbitrarily high order when $\langle x, N\rangle = 0$, such that (1.1) has a solution $u \in C^\infty(\mathbb{R}^n)$ satisfying (1.2).

Note that (3.17) implies that

(3.17)' $$T_\nu^2 / (1 + |\mathrm{Im}\ \zeta_\nu|) \to \infty, \text{ thus } T_\nu \to \infty.$$

This is reminiscent of the sufficient conditions for uniqueness obtained in Hörmander [5]. If the conclusion of Theorem 3.7 is not already contained in Theorem 3.1 we have

$$\tilde{Q}_N(\xi) \leqq C\,\tilde{P}_N(\xi), \quad \xi \in \mathbb{R}^n,$$

and by [4, Theorem 3.3.2] this implies that

$$\sum_j |\langle D, N\rangle^j Q(\zeta)|\,T^j \leqq C' \sum_j |\langle D, N\rangle^j P(\zeta)|\,T^j, \quad \zeta \in \sum_N, T \geqq |\operatorname{Im}\zeta|+1.$$

Thus the condition (3.15) requires that $\operatorname{Im}\zeta_\nu \to \infty$ and that

$$T_\nu / |\operatorname{Im}\zeta_\nu| \to 0,$$

so T_ν will have to be somewhere between $|\operatorname{Im}\zeta_\nu|^{1/2}$ and $|\operatorname{Im}\zeta_\nu|$.

The condition (3.18) will be used in an equivalent and perhaps more illuminating version,

(3.18)' $\quad \operatorname{Im} d(r(z)/q(z)) < 0$ for some z with $q(z) \neq 0$.

To prove the equivalence we assume that (3.18)' is not valid and set

$$f(z) = r(z)/q(z).$$

Then $\operatorname{Im} f'(z) \geqq 0$ for every z which is not a pole. Thus the main term of f' at a finite or infinite pole has to have non-negative imaginary part, which shows that f' has no poles so that f' is a constant a. Thus $f(z) = az + b$ and $\operatorname{Im} a \geqq 0$, which is precisely the case excluded by (3.18).

In the hypotheses of Theorem 3.7 we may replace b_ν by $b_\nu(1 + c/K_\nu)$. Since $K_\nu \to \infty$ by (3.16) the only consequence is that r is replaced by $r - cq$. Since c may be any complex number we may therefore strengthen (3.18)' to

(3.18)" $\quad r(z) = 0$ and $\operatorname{Im} r'(z)/q(z) < 0$ for some z with $q(z) \neq 0$.

The length of the proof of Theorem 3.7 is in proportion to the complexity of the statement so we shall postpone it until we have discussed some examples.

Example 3.8. Set $P(\xi) = (\xi_1^2 + \xi_2^2 + \xi_3^2)^a - \xi_1^{2a}/2$ and $Q(\xi) = \xi_2^b$, where $a > 1$ and $b \leqq 2a$. Choose $\zeta_\nu = (\xi_{\nu 1}, \xi_{\nu 2}, -i\nu)$ where $\xi_{\nu 1}, \xi_{\nu 2}$ are real and $\xi_{\nu 1}^2 + \xi_{\nu 2}^2 = \nu^2$. Then we obtain for $N = (0, 0, 1)$

$$P(\zeta_\nu + T_\nu zN) = (T_\nu^2 z^2 - 2i\nu T_\nu z)^a - \xi_{\nu 1}^{2a}/2, \quad Q(\zeta_\nu + T_\nu zN) = \xi_{\nu 2}^b.$$

With $a_\nu = -2\xi_{\nu 1}^{-2a}$, $b_\nu = \xi_{\nu 2}^{-b}$ and $q = 1$ we have (3.14), (3.15) if

$$\xi_{\nu 1}^{2a} \xi_{\nu 2}^{-b} \to 0, \quad \nu T_\nu / \xi_{\nu 1}^2 \to 0.$$

Since $b \leqq 2a$ the first condition implies $\xi_{\nu 1}/\nu \to 0$. Defining K_ν by

$$K_\nu (\nu T_\nu)^a = \xi_{\nu 1}^{2a}$$

we obtain $r(z) = C z^a$. Condition (3.17) becomes

$$(\nu T_\nu)^a T_\nu / \xi_{\nu 1}^{2a} \to \infty, \quad \xi_{\nu 1}^{2a} T_\nu / \nu (\nu T_\nu)^a \to \infty.$$

Summing up, we have the conditions

$$\xi_{\nu 1}^{2a}/\nu^b \to 0, \quad \nu T_\nu/\xi_{\nu 1}^2 \to 0, \quad T_\nu^{1+a} \xi_{\nu 1}^{-2a} \nu^a \to \infty, \quad T_\nu^{a-1} \xi_{\nu 1}^{-2a} \nu^{1+a} \to 0.$$

Since $|\xi_{\nu 1}| < \nu$, the last condition implies the second. The last two conditions can be satisfied by an appropriate choice of T_ν if and only if

$$(\xi_{\nu 1}^{2a} \nu^{-1-a})^{a+1} (\xi_{\nu 1}^{-2a} \nu^a)^{a-1} \to \infty,$$

so our remaining conditions are

$$\xi_{\nu 1} \nu^{-b/2a} \to 0, \quad \xi_{\nu 1} \nu^{-(3a+1)/4a} \to \infty.$$

These are compatible if and only if $(3a+1)/2 < b$, that is,

$$b > 2a - (a-1)/2.$$

The case $a = 2$, $b = 4$ is essentially Theorem 1 of Pliś [8]. Starting with $a = 4$ we can choose Q of lower order than P. Theorem 1 of Pliś [8] is also contained in the following

Corollary 3.9. Let P be a homogeneous polynomial, and assume that $P(\zeta) = 0$ for some $\zeta \neq 0$ with $\operatorname{Im} \zeta$ proportional to N. Unless P has a polynomial factorization

$$P = P_1^j P_2 \; ; \; \langle D, N \rangle P_1(\zeta) \neq 0, \; P_2(\zeta) \neq 0,$$

one can for every homogeneous Q with $\deg Q = \deg P$, $Q(\xi+zN) = Q(\xi)$ and $Q(\zeta) \neq 0$ find $u \in C^\infty(\mathbb{R}^n)$ and $a \in C^\infty(\mathbb{R}^n)$ satisfying (1.1), (1.2) so that a vanishes of any prescribed order when $\langle x, N\rangle = 0$.

Proof. We choose the coordinates so that $N = (0, \ldots, 0, 1)$ and set $\zeta = (\xi_0', \lambda_0)$, $\xi_0' \in \mathbb{R}^{n-1}$. In view of Theorem 2.1 and Example 3.4 we may assume that $\xi_0' \neq 0$ and that $\operatorname{Im} \lambda_0 \neq 0$, hence that $\operatorname{Im} \lambda_0 < 0$, for may be replaced by $-\zeta$. Let λ_0 be a zero of multiplicity μ of $P(\xi_0', \lambda)$ as a polynomial in λ. Then the equation $P(\xi', \lambda) = 0$ has μ roots, and $\partial P(\xi', \lambda)/\partial\lambda = 0$ has $\mu - 1$ roots close to λ_0 if ξ' is close to ξ_0'. If $P = 0$ at all such zeros of $\partial P/\partial\lambda$, it follows that there can only be one, for a zero of P is a zero of $\partial P/\partial\lambda$ of multiplicity decreased by one. If we consider the decomposition of P in a product of irreducible polynomials it follows easily that $P = P_1{}^j P_2$ with $P_2 \neq 0$ and $D_n P_1 \neq 0$ at (ξ_0', λ_0).

Thus we have $\partial P(\xi', \lambda)/\partial\lambda = 0$ but $P(\xi', \lambda) \neq 0$ for a sequence (ξ_ν', λ_ν) in Σ_N converging to (ξ_0', λ_0). With $P^{(j)}(\xi', \lambda) = \partial^j P(\xi', \lambda)/\partial\lambda^j$ we have

$$P(\xi_\nu', \lambda_\nu + S_\nu z) = \sum_{j \neq 1} P^{(j)}(\xi_\nu', \lambda_\nu)\, (S_\nu z)^j / j!\,.$$

If S_ν is sufficiently small, then

$$P(\xi_\nu', \lambda_\nu + S_\nu z)/P(\xi_\nu', \lambda_\nu) - 1 = O(S_\nu \nu^{-1}(|z|^2 + \ldots + |z|^m)).$$

Now we set $\zeta_\nu = \rho_\nu(\xi_\nu', \lambda_\nu)$, $T_\nu = \rho_\nu S_\nu$ where $\rho_\nu \to +\infty$. Then

$$P(\zeta_\nu + T_\nu zN) = \rho_\nu^{\,m} P(\xi_\nu', \lambda_\nu + S_\nu z).$$

We set $a_\nu = \rho_\nu^{-m}/P(\xi_\nu', \lambda_\nu)$ and $b_\nu = 1/Q(\zeta_\nu) = \rho_\nu^{-m}/Q(\xi_\nu')$. Then we have (3.14), (3.15), and (3.16) follows for some r of degree ≥ 2 and $K_\nu \geq \nu/S_\nu$. The conditions (3.17) can be written

$$\rho_\nu S_\nu / K_\nu \to \infty, \quad S_\nu K_\nu \to \infty.$$

The first is valid if ρ_ν is sufficiently rapidly increasing, and the second follows since $S_\nu K_\nu \geq \nu$. The proof is complete.

The case excluded in Corollary 3.9 is considered in the following example which is Theorem 4 of Pliś [8].

Example 3.10. Let $P(\xi) = (\xi_1 - i\xi_2)^a - \xi_1^{b-1}$, $Q(\xi) = \xi_1^b$, $N = (0, 1)$. With $\zeta_\nu = (\nu, -i\nu)$ we have

$$P(\zeta_\nu + T_\nu zN) = (-iT_\nu z)^a - \nu^{b-1}, \quad Q(\zeta_\nu + T_\nu zN) = \nu^b.$$

(3.14) is valid with $q = 1$ if $a_\nu = -\nu^{1-b}$, $b_\nu = \nu^{-b}$ and $T_\nu^a \nu^{1-b} \to 0$. Then we obtain

$$a_\nu P(\zeta_\nu + T_\nu zN) - b_\nu Q(\zeta_\nu + T_\nu zN) = -(-iT_\nu z)^a \nu^{1-b},$$

so we put $K_\nu = \nu^{b-1} T_\nu^{-a}$. Then we have (3.16) with $r(z) = C z^a$ so (3.18) is valid if $a > 1$. (3.15) is fulfilled so all the required conditions are

$$T_\nu \nu^{(1-b)/a} \to 0, \quad T_\nu \nu^{-(b-1)/(a+1)} \to \infty, \quad T_\nu \nu^{-(b-2)/(a-1)} \to 0.$$

These are compatible if and only if $(b-1)/(a+1) < (b-2)/(a-1)$, that is, $a < 2b-3$ or

$$b > a - (a-3)/2.$$

Note that the multiplicity a has to be two units higher than in Example 3.8 if the order of Q shall be (strictly) smaller than that of P. There is of course no difficulty in extending the example to the general exceptional case in Corollary 3.9.

We shall now pass to the proof of Theorem 3.7. The first step is to rephrase the hypothesis using the Tarski-Seidenberg theorem.

Lemma 3.11. Assume that the hypotheses of Theorem 3.7 are fulfilled. Then there exist Laurent polynomials $\zeta(\varepsilon)$, $T(\varepsilon)$, $K(\varepsilon)$, $a(\varepsilon)$, $b(\varepsilon)$ such that $\zeta(\varepsilon) \in \Sigma_N$, $T(\varepsilon) > 0$, $K(\varepsilon) > 0$ for small $\varepsilon > 0$, and for $\varepsilon \to 0$

(3.14)' $\quad a(\varepsilon)P(\zeta(\varepsilon)+T(\varepsilon)zN) - q(z) = O(\varepsilon)$, $b(\varepsilon)(Q(\zeta(\varepsilon)+T(\varepsilon)zN) - q(z) = O(\varepsilon)$,

(3.15)' $\quad b(\varepsilon)/a(\varepsilon) = O(\varepsilon)$,

(3.16)' $\quad K(\varepsilon)(a(\varepsilon)P(\zeta(\varepsilon)+T(\varepsilon)zN) - b(\varepsilon)Q(\zeta(\varepsilon)+T(\varepsilon)zN)) = r(z) + O(\varepsilon)$,

(3.17)' $\quad K(\varepsilon)/T(\varepsilon) = O(\varepsilon)$, $\quad (1+|\operatorname{Im}\zeta(\varepsilon)|)/T(\varepsilon)K(\varepsilon) = O(\varepsilon)$.

In (3.14)', (3.16)' the notation $O(\varepsilon)$ indicates a polynomial in z with coefficients $O(\varepsilon)$. If the hypothesis of Theorem 3.1 is not fulfilled, we also have

$$(1+T(\varepsilon))/|\operatorname{Im}\zeta(\varepsilon)| = O(\varepsilon).$$

$b(\varepsilon)$ and $a(\varepsilon)$ can be chosen so that $b(\varepsilon)/a(\varepsilon)$ is analytic on $\mathbb{R}$.

Proof. Let E be the set of all $(\zeta, T, K, a, b, \varepsilon)$ with $\zeta \in \Sigma_N$, $T > 0$, $K > 0$, $\varepsilon > 0$, $a, b \in \mathbb{C}$, such that the coefficients of the polynomials in z

$$aP(\zeta+TzN) - q(z), bQ(\zeta+TzN) - q(z), \quad K(aP(\zeta+TzN) - bQ(\zeta+TzN)) - r(z)$$

have modulus $\leqq \varepsilon$, $K \leqq \varepsilon T$ and $1 + |\operatorname{Im}\zeta|^2 \leqq (\varepsilon TK)^2$. This is a closed semi-algebraic set, and the hypothesis of Theorem 3.7 shows that E contains points with arbitrarily small $\varepsilon > 0$. Without giving up this property we can shrink E by imposing further minimality conditions on the other variables until they are uniquely determined by ε. From the Tarski-Seidenberg theorem it follows then that they are algebraic functions of ε for small $\varepsilon > 0$. Thus they have Laurent series expansions in $\varepsilon^{1/k}$ for some integer $k > 0$. If we replace ε by ε^k and take sufficiently high partial sums of these series, we obtain all the required properties except that $b(\varepsilon)/a(\varepsilon)$ may have a finite number of poles on $\mathbb{R} \setminus 0$. However, if ν is sufficiently large we can replace $a(\varepsilon)$ by $a(\varepsilon+h\varepsilon^{\nu})$. If h is purely imaginary it is clear that $a(\varepsilon+h\varepsilon^{\nu})$ cannot have a real zero $\varepsilon \neq 0$ except for finitely many values of h so this permits us to choose b/a analytic.

To avoid an interruption of the proof later on we give a version of the expansions of geometrical optics (the WKB method) which will be important.

Lemma 3.12. Let $I \subset \mathbb{R}$ be a compact interval and let

$$G_\delta(S, D_S) = \sum_0^m g_j(S, \delta)\, D_S{}^j, \quad D_S = -i\, d/dS,$$

be an ordinary differential operator with C^∞ coefficients when $S \in I$ and $\delta \in \mathbb{R}$ is small. Assume that there exist positive integers m_0 and m_1 such that

$$\delta^{m_1} G_\delta(S, \delta^{-m_0} z) = H_\delta(S, z)$$

also has C^∞ coefficients and $H_0 \not\equiv 0$. Assume further that

$$H_0(S, z) = 0, \ \partial H_0(S, z)/\partial z \neq 0 \text{ when } z = \varphi_0'(S),$$

where $\varphi_0 : I \to \mathbb{C}$ is a C^∞ function. Then there exist C^∞ functions $\varphi(S, \delta)$ and $W(S, \delta)$ when $S \in I$ and $|\delta|$ is small, such that

$$\varphi(S, 0) = \varphi_0(S), \quad W(S, 0) \neq 0 \ ; \ S \in I; \tag{3.19}$$

$$\exp(-i\varphi(S,\delta)\,\delta^{-m_0})\, G_\delta(S, D_S)\,(W(S,\delta)\exp(i\varphi(S,\delta)\,\delta^{-m_0})) = R(S, \delta) \tag{3.20}$$

where $R(S, \delta)$ is a C^∞ function vanishing of infinite order when $\delta = 0$. If Γ_1 and Γ_2 are curves in the (S, δ) plane intersecting $I \times \{0\}$ transversally, and if $g_m(S, \delta)$ does not vanish of infinite order on any one of them when $\delta = 0$, then W can be chosen so that R vanishes of infinite order on Γ_1 and Γ_2 also.

<u>Proof.</u> Since $H_\delta(S, z)$ is a polynomial in z, thus analytic in z, we can use the implicit function theorem to determine $\psi \in C^\infty$ with $H_\delta(S, \psi(S, \delta)) = 0$ when $S \in I$, δ is small, and so that $\psi(S, 0) = \varphi_0'(S)$. Choose φ with $\partial\varphi(S, \delta)/\partial S = \psi(S, \delta)$ and $\varphi(S, 0) = \varphi_0(S)$.

The equation (3.20) can be written

$$\delta^{m_1} G_\delta(S, D_S + \varphi_S'(S,\delta)\,\delta^{-m_0})\, W(S, \delta) = \delta^{m_1} R(S, \delta)$$

or

$$\delta^{-m_0}(H_\delta(S, \delta^{m_0} D_S + \varphi_S'(S, \delta)) - H_\delta(S, \varphi_S'(S, \delta)))\, W(S, \delta) = \delta^{m_1 - m_0} R(S, \delta). \tag{3.21}$$

When $\delta = 0$ the left hand side reduces to $L\, W(S, 0)$ where

$$L = \partial H_0(S, z)/\partial z\, D_S + B, \quad z = \varphi_0'(S),$$

for some $B \in C^\infty$. By hypothesis the coefficient of D_S has no zero in I. Introducing the formal Taylor expansion $W(S, \delta) \sim \sum_0^\infty W^{(j)}(S, 0)\, \delta^j/j!$ in (3.21), we find that (3.21) for some R vanishing of infinite order when $\delta = 0$

is equivalent to a system of equations

$$L\,W(S, 0) = 0, \ldots, L\,W^{(j)}(S, 0) + E_j = 0, \ldots$$

where E_j is determined by $W, \ldots, W^{(j-1)}$. These can be solved successively, and $W(S, 0)$ can be chosen with no zero in I. By a classical theorem of E.Borel there exists a C^∞ function $W(S, \delta)$ with these derivatives when $\delta = 0$, which proves the first part of the lemma.

To prove the last assertion we have to find a function $V \in C^\infty$ vanishing of infinite order when $\delta = 0$ such that (3.20) is valid with W replaced by V apart from an error vanishing of infinite order on the curves Γ_j. The difference $W - V$ will then have the required properties. The condition on V is that

$$\delta^{m_1} G_\delta(S, D_S + \varphi_S'(S, \delta)\, \delta^{-m_0})\, V(S, \delta) - R(S, \delta)\, \delta^{m_1}$$

shall vanish of infinite order on the curves Γ_j. The differential operator has C^∞ coefficients and the coefficient of the highest derivative is $\delta^{m_1} g_m(S, \delta)$, which vanishes at most as a power of δ on Γ_j when $\delta = 0$. Now we require that derivatives of order $< m$ of V with respect to S shall vanish on Γ_j. Using the equation we can then compute $D_S^m V$ on Γ_j, which yields a function vanishing of infinite order when $\delta \to 0$. Repeating the argument we find that V has the desired properties if the derivatives with respect to S on Γ_j are certain C^∞ functions vanishing of infinite order when $\delta \to 0$. By Whitney's extension theorem [12] it is possible to find V so that V vanishes of infinite order when $\delta = 0$ and has these derivatives with respect to S on the transversal curves Γ_j. This completes the proof.

<u>Remark.</u> It is obvious that the lemma remains valid if the coefficients g_j are singular when $\delta = 0$ but $\delta^N g_j \in C^\infty$ for some integer $N > 0$. We may also replace δ^{m_0} by a C^∞ function of δ vanishing precisely of order m_0 when $\delta = 0$.

Proof of Theorem 3.7. With the notations of Lemma 3.11 we introduce

$$A(\varepsilon) = b(\varepsilon)/a(\varepsilon),$$

which is a rational function of ε vanishing at 0 and with no poles on $\mathbb{R}$. With a positive integer ρ to be chosen later we shall take the coefficient a in (1.1) as $-A(\langle x, N\rangle^{\rho})$ apart from a term vanishing in $\complement H_N$. For small $\delta > 0$ we set in analogy to (3.3)

$$u_\delta(x) = v_\delta(\langle x, N\rangle) \exp i\langle x, \zeta(\delta^\rho)\rangle .$$

The differential equation $(P(D) - A(\langle x, N\rangle^\rho)Q(D))\, u_\delta = 0$ can then be written

$$(3.22) \qquad (P(\zeta(\delta^\rho)+D_sN) - A(s^\rho)Q(\zeta(\delta^\rho)+ D_sN))\, v_\delta(s) = 0,$$

and we shall solve it approximately using Lemma 3.12. To be able to use (3.14)', (3.16)' we multiply by $a(\delta^\rho)$ and obtain the equivalent equation

$$(3.22)' \qquad (a(\delta^\rho)P(\zeta(\delta^\rho) +D_sN) -b(\delta^\rho)\, Q(\zeta(\delta^\rho) +D_sN)) +$$
$$+ (1 -A(s^\rho)/A(\delta^\rho))\, b(\delta^\rho)\, Q(\zeta(\delta^\rho) +D_sN))\, v_\delta(s) = 0.$$

The difference $1 - A(s^\rho)/A(\delta^\rho)$ vanishes when $s = \delta$ and the first order term in the Taylor expansion at $s = \delta$ is

$$- A'(\delta^\rho)/A(\delta^\rho)\, \rho\, \delta^{\rho-1}(s-\delta) = - \rho j_0(s -\delta)/\delta$$

if A has a zero of order j_0 at 0. To balance the two terms in (3.22)' we therefore want $(s -\delta)/\delta$ and $1/K(\delta^\rho)$ to be of the same order of magnitude. Since $K(\varepsilon)$ may be replaced by the leading term in the Laurent expansion at $\varepsilon = 0$ we may assume that $K(\varepsilon) = \varepsilon^{-\varkappa}$ where $\varkappa$ is a positive integer. Thus we wish $s-\delta$ to be of the order of magnitude $\delta^{1+\varkappa\rho}$. To be able to apply Lemma 3.12 in a fixed interval we must now introduce a new variable S through

$$s = \delta + S\, \delta^{\varkappa\rho+1}.$$

With the notation $v_\delta(s) = V_\delta(S)$ the equation (3.22)' becomes

$$(3.22)'' \qquad (K(\delta^\rho)(a(\delta^\rho)P -b(\delta^\rho)Q)(\zeta(\delta^\rho) +\delta^{-\varkappa\rho -1}D_SN) +$$
$$+ C(S, \delta)\, b(\delta^\rho)\, Q(\zeta(\delta^\rho) +\delta^{-\varkappa\rho -1}D_SN))\, V_\delta(S) = 0.$$

Here

$$C(S, \delta) = K(\delta^{\rho})(1 - A(s^{\rho}))/A(\delta^{\rho}) \to -\rho j_0 S, \; \delta \to 0.$$

It is clear that $C(S, \delta)$ is analytic for small δ.

The coefficients of (3.22)" become smooth after multiplication by some power of δ. If D_S is replaced by $\delta^{\varkappa\rho+1} T(\delta^{\rho}) z$ we obtain a polynomial converging to

$$(3.23) \qquad r(z) - \rho j_0 S \, q(z)$$

when $\delta \to 0$. Note that

$$(3.24) \qquad \delta^{\varkappa\rho+1} T(\delta^{\rho}) = \delta \, T(\delta^{\rho})/K(\delta^{\rho}) \to \infty, \; \delta \to 0,$$

by the first part of (3.17)' if $\rho > 1$, as we assume from now on. This allows us to apply Lemma 3.12 (and the remark following its proof). In doing so we may assume that (3.18)" is fulfilled for some z_0. By the implicit function theorem the polynomial (3.23) has a unique C^{∞} zero $z(S)$ with $z(0) = z_0$ defined for S in a neighborhood of 0. Since

$$r'(z_0) \, dz/dS = \rho j_0 \, q(z_0) \text{ when } S = 0,$$

it follows from (3.18)" that

$$\operatorname{Im} dz/dS = \rho j_0 \operatorname{Im} q(z_0)/r'(z_0) > 0 \text{ when } S = 0.$$

Summing up, there is a symmetric interval $I \subset \mathbb{R}$ such that (3.23) has a simple root $z(S)$ when $S \in I$, $q(z(S)) \neq 0$ and

$$(3.25) \qquad \operatorname{Im} dz/dS > 0, \; S \in I.$$

We can now choose φ and W by applying Lemma 3.12 to (3.22)" with $\varphi_0'(S) = z(S)$. From (3.25) it follows that for small δ

$$(3.25)' \qquad \partial^2 \operatorname{Im} \varphi(S, \delta)/\partial S^2 \geqq c > 0, \; S \in I.$$

The choice of the curves Γ_j in Lemma 3.12 is left open for the moment.

Returning to the original variables we define

$$u_\delta(x) = \exp i(\langle x, \zeta(\delta^\varrho)\rangle + \varphi(S, \delta)\, \delta^{\varkappa\varrho+1}\, T(\delta^\varrho))\, W(S, \delta),$$

$$r_\delta(x) = R(S, \delta)/(K(\delta^\varrho)a(\delta^\varrho)W(S, \delta)) = R_1(S, \delta),$$

where $S = (\langle x, N\rangle - \delta)\, \delta^{-\varkappa\varrho-1} \in I$. Here R_1 is also a C^∞ function vanishing of infinite order when $\delta = 0$ and on the curves Γ_j. We have

$$(3.26)\qquad (P(D) - A(\langle x, N\rangle^\varrho)Q(D))\, u_\delta = r_\delta\, u_\delta \text{ if } (\langle x, N\rangle - \delta)\, \delta^{-\varkappa\varrho-1} \in I.$$

Furthermore,

$$(3.27)\qquad Q(D)\, u_\delta = M_\delta\, u_\delta$$

where

$$M_\delta(x) = b(\delta^\varrho)^{-1} m(S, \delta)\ ;\ m \in C^\infty \text{ and } m(S, 0) = q(\varphi'_S(S, 0)) \neq 0.$$

Following the proof of Theorem 3.1 we shall now piece together u by means of the functions u_δ.

For $\nu = 1, 2, \ldots$ we put $\delta_\nu = c_0 \nu^{-\gamma}$ where $c_0 > 0$ and $\gamma > 0$. For the interval

$$I_\nu = \{ s \in \mathbb{R};\ (s - \delta_\nu)\, \delta_\nu^{-\varkappa\varrho-1} \in I \}$$

the length $|I_\nu|$ is $|I|\, (c_0 \nu^{-\gamma})^{\varkappa\varrho+1}$, and the distance $\delta_\nu - \delta_{\nu+1}$ between the centers of I_ν and $I_{\nu+1}$ is asymptotically $c_0 \gamma \nu^{-\gamma-1}$. Choose γ so that

$$\gamma(\varkappa\varrho + 1) = \gamma + 1, \text{ that is, } \gamma = 1/\varkappa\varrho,$$

and choose c_0 so that $c_0\gamma = |I|\, c_0^{\varkappa\varrho+1}/2$. For large ν the end points of I_ν are then close to the centers of $I_{\nu+1}$ and of $I_{\nu-1}$. We shall switch from one u_{δ_ν} to the next when $\langle x, N\rangle$ is near the center of the interval where they are both defined.

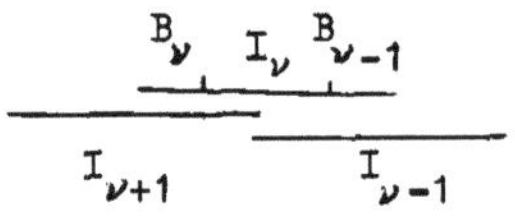

Let $I = (-2B, 2B)$. The center $(-B, 0)$ of the left half corresponds in I_ν to

$$B_\nu = \delta_\nu - B\, \delta_\nu^{\varkappa\varrho+1}.$$

For large ν the point $B_{\nu-1}$ is close to the center of the right half of I_ν. To confirm this we set $B_{\nu-1} = \delta_\nu + S\,\delta_\nu^{\varkappa\varrho+1}$ and obtain

$$S = (\delta_{\nu-1} - \delta_\nu)/\delta_\nu^{\varkappa\varrho+1} - B(\delta_{\nu-1}/\delta_\nu)^{\varkappa\varrho+1} = B + \dots$$

where dots indicate a convergent power series in $1/\nu$ or in $\delta_\nu^{\varkappa\varrho}$, with no constant term. Thus $S = B + f(\delta_\nu)$ where $f(\delta)$ is an analytic function of δ vanishing at least to the second order when $\delta = 0$. We choose the curves Γ_j in Lemma 3.12 to be $S = -B$ and $S = B + f(\delta)$. This guarantees that the right hand side of (3.26) vanishes of infinite order where the switch over occurs.

When $\langle x, N\rangle \in I_\nu$ we set

$$U_\nu(x) = C_\nu\, u_{\delta_\nu}(x)$$

where $C_\nu > 0$ is determined successively so that with the notation in (3.27)

(3.28) $\qquad |M_{\delta_\nu}(x)\, U_\nu(x)| = |M_{\delta_{\nu-1}}(x)\, U_{\nu-1}(x)|$ when $\langle x, N\rangle = B_{\nu-1}$.

Note that the two sides are constant in this plane. Choose $\chi \in C_0^\infty(-3B/2, 3B/2)$ equal to 1 in $(-5B/4, 5B/4)$, and set $\chi_\nu(x) = \chi((\langle x,N\rangle - \delta_\nu)\,\delta_\nu^{-\varkappa\varrho-1})$,

$$u(x) = \sum_{\nu_0}^{\infty} U_\nu(x)\,\chi_\nu(x).$$

When ν_0 is large and the first term is appropriately modified we shall see that $u \in C^\infty(\mathbb{R}^n)$ and that (1.1) is satisfied with a C^∞ function a such that $a(x) + A(\langle x, N\rangle^\varrho)$ vanishes outside H_N.

The first step is to study $F_\nu = |M_{\delta_\nu}(x)\, U_\nu(x)|$ as a function of $s = \langle x, N\rangle$. Apart from a constant term $\log F_\nu(s)$ is equal to

$$\log|m(S, \delta_\nu)W(S, \delta_\nu)| + s\,|\mathrm{Im}\,\zeta(\delta_\nu^\varrho)| - \mathrm{Im}\,\varphi(S, \delta_\nu)\delta_\nu^{\varkappa\varrho+1}\, T(\delta_\nu^\varrho).$$

Here $S = (s - \delta_\nu)\delta_\nu^{-\varkappa\varrho-1}$. By Lemma 3.11 we may assume that

$$T(\delta_\nu^\varrho)/|\mathrm{Im}\,\zeta(\delta_\nu^\varrho)| = O(\delta_\nu^\varrho),$$

and in view of (3.24) this implies that $d(\log F_\nu(s))/ds$ is asymptotically equal to $|\mathrm{Im}\,\zeta(\delta_\nu^\varrho)|$. Moreover, since

$$K(\varepsilon)/|\mathrm{Im}\,\zeta(\varepsilon)| = (K(\varepsilon)/T(\varepsilon))(T(\varepsilon)/|\mathrm{Im}\,\zeta(\varepsilon)| = O(\varepsilon^2)$$

we have

$$\delta_\nu^{\varkappa\rho+1}\ |\operatorname{Im}\zeta(\delta_\nu^\rho)| \geqq c\ \delta_\nu^{\varkappa\rho+1-(\varkappa+2)\rho} = c\ \delta^{1-2\rho} \geqq c_1\nu^\gamma$$

so for large ν it follows that

$$F_\nu(B_\nu)/F_\nu(B_{\nu-1}) < C\ \exp(-c\nu^\gamma).$$

Since $F_\nu(B_\nu) = F_{\nu+1}(B_\nu)$ by (3.28) it follows that

$$F_\nu(B_{\nu-1}) < C_1\ \exp(-c_1\nu^{\gamma+1}).$$

Hence

$$F_\nu(s) < C_2\ \exp(-c_2\nu^{\gamma+1}),\ B_\nu \leqq s \leqq B_{\nu-1},$$

and (3.29) below will show that the same estimate is valid in supp χ_ν. It follows immediately that all derivatives of u have bounds converging to 0 when $\langle x, N\rangle \to 0$ so $u \in C^\infty(\mathbb{R}^n)$.

To show that $a = -P(D)u/Q(D)u$ is in C^∞ when $\langle x, N\rangle < B_{\nu_0-1}$ we first prove

(3.29) $\quad c_1T(\delta_\nu^\rho) \leqq (s-B_\nu)^{-1}\log(F_\nu(s)/F_{\nu+1}(s)) \leqq c_2T(\delta_\nu^\rho),\ s \in I_\nu \cap I_{\nu+1}.$

Since $F_\nu(s)/F_{\nu+1}(s) = 1$ when $s = B_\nu$ we only have to examine the derivative of the logarithm. Define $S_j = (s-\delta_j)\ \delta_j^{-\varkappa\rho-1}$, $j = \nu, \nu+1$. Then

$$S_{\nu+1} - S_\nu - 2B \to 0$$

uniformly when $\nu \to \infty$, so $S_{\nu+1} > S_\nu + B$ if ν is large. As we saw above, the derivative of $\log(F_\nu(s)/F_{\nu+1}(s))$ with respect to s is

(3.30) $\quad O(\delta_\nu^{-\varkappa\rho-1}) + |\operatorname{Im}\zeta(\delta_\nu^\rho)| - |\operatorname{Im}\zeta(\delta_{\nu+1}^\rho)| +$

$$+\ T(\delta_{\nu+1}^\rho)\ \operatorname{Im}\varphi'(S_{\nu+1}, \delta_{\nu+1}) - T(\delta_\nu^\rho)\ \operatorname{Im}\varphi'(S_\nu, \delta_\nu).$$

Here $\delta_\nu^{-\varkappa\rho-1}/T(\delta_\nu^\rho) \to 0$ by (3.24). Since

$$T(\varepsilon)\ \varepsilon^{-\varkappa}/|\operatorname{Im}\zeta(\varepsilon)| \to \infty,\ \varepsilon \to 0,$$

by the second part of (3.17)' and since $\operatorname{Im}\zeta(\varepsilon)$ has a pole when $\varepsilon = 0$, we have

$$T(\varepsilon)\ \varepsilon^{-\varkappa-1}/|\operatorname{Im}\zeta'(\varepsilon)| \to \infty,\ \varepsilon \to 0.$$

Hence

$$|\operatorname{Im}\zeta(\delta_{\nu+1}^\rho)| - |\operatorname{Im}\zeta(\delta_\nu^\rho)| = o(T(\delta_\nu^\rho)),\ \nu \to \infty,$$

for $\delta_\nu^{\varkappa\rho}$ and $\delta_\nu^{-\rho}(\delta_\nu^\rho - \delta_{\nu+1}^\rho)$ are asymptotic to constants times $1/\nu$ so $\delta_\nu^{-\varkappa\rho-\rho}(\delta_\nu^\rho - \delta_{\nu+1}^\rho)$ has a finite limit when $\nu \to \infty$. The last two terms in (3.30) are therefore dominating. The lower bound for $S_{\nu+1} - S_\nu$ and (3.25)' now shows that (3.30) can be bounded from above and below by constants times $T(\delta_\nu^\rho)$, which proves (3.29).

Let us now study a in the neighborhood of B_ν where $u = U_\nu + U_{\nu+1}$. There we have

$$-a = (P(D)U_\nu + P(D)U_{\nu+1})/(Q(D)U_\nu + Q(D)U_{\nu+1}) =$$

$$= A(\langle x, N\rangle^\rho) + (r_{\delta_\nu} U_\nu + r_{\delta_{\nu+1}} U_{\nu+1})/(M_{\delta_\nu} U_\nu + M_{\delta_{\nu+1}} U_{\nu+1}).$$

U_ν dominates when $S = (s - B_\nu)\delta_\nu^{-\varkappa\rho-1} > 0$, so then we divide by $M_{\delta_\nu} U_\nu$. In view of (3.29) we have for some constant $c > 0$

$$|1 + M_{\delta_{\nu+1}} U_{\nu+1}/M_{\delta_\nu} U_\nu| \geq c \min (S\, \delta_\nu^{\varkappa\rho+1}\, T(\delta_\nu^\rho), 1),$$

for $(1 - e^{-t}) \geq (1 - e^{-1}) \min (1, t)$, $t > 0$. On the other hand, r_{δ_ν} and $r_{\delta_{\nu+1}}$ can be estimated by any desired power of δ_ν or S. It follows that

$$a(x) + A(\langle x, N\rangle^\rho)$$

in the set now considered can be estimated by any power of $1/\nu$. Since the derivatives of r_δ have estimates similar to those we have used for r_{δ_ν}, the same is true of all the derivatives.

In the part of the left half of I_ν where $\chi_{\nu+1}$ is cutting off $U_{\nu+1}$, we know by (3.29) that $F_{\nu+1}(s)/F_\nu(s)$ is exponentially small, so similar estimates are immediately obtained there. The argument is even simpler than the corresponding point in the proof of Theorem 3.1 so we omit the details.

In the middle of I_ν where $u = U_\nu$ we have

$$-a = A(\langle x, N\rangle^\rho) + r_{\delta_\nu}/M_{\delta_\nu},$$

and all derivatives of the second term have bounds converging to 0 as $\nu \to \infty$.

This proves the smoothness of a for $\langle x, N\rangle \leq B_{\nu_0 - 1}$. We can continue a as a function of $\langle x, N\rangle$ for $\langle x, N\rangle > B_{\nu_0 - 1}$ and obtain u there by just solving an ordinary differential equation. This completes the proof.

In the final part of the proof we can replace Q by any other operator $R \neq 0$ with $\deg R \leq \deg P$. In fact, if h is sufficiently large and we set $\zeta_1(\varepsilon) = \zeta(\varepsilon) + (\varepsilon^{h}, \varepsilon^{h^2}, \dots, \varepsilon^{h^n})$ then

$$R(\zeta_1(\varepsilon)) \not\equiv 0 \text{ if } \deg R \leq \deg P,\ R \neq 0.$$

This follows from Taylor's formula since there is a uniform bound for the order of 0 as zero or pole of $P^{(\alpha)}(\zeta(\varepsilon))$. As in the proof of Lemma 3.11 we can modify $\zeta_1(\varepsilon)$ to another function $\zeta_2(\varepsilon)$ which also has all properties stated there.

Now we have an analogue of (3.27) with Q replaced by R, and we use the corresponding M_δ in the following normalizations. The result is as follows:

Theorem 3.13. Let the hypotheses be as in Theorem 3.7. Then there is a rational function a_1 of $\langle x, N\rangle$ with no real pole, vanishing at 0 of given order, such that for every $R \neq 0$ with $\deg R \leq \deg P$ one can find $a_2 \in C^\infty(\mathbb{R}^n)$ vanishing in $\complement H_N$ and $u \in C^\infty(\mathbb{R}^n)$ with supp $u = H_N$ such that

$$(P(D) + a_1 Q(D) + a_2 R(D))\, u = 0.$$

4. Hölder continuous perturbations. A function a defined in $\mathbb{R}^n$ is said to be Hölder continuous of order $\lambda \in (0, 1]$ and one writes $a \in H^\lambda(\mathbb{R}^n)$ if

$$|a(x) - a(y)| \leq C_K |x - y|^\lambda \ ; \ x, y \in K;$$

for every compact set $K \subset \mathbb{R}^n$. When $\lambda \in (j, j+1]$, j a positive integer, then H^λ is the set of all $a \in C^j$ such that $D^\alpha a \in H^{\lambda - j}$ when $|\alpha| = j$. We shall now discuss the equation (1.1) when a is merely Hölder continuous.

If P(D) and Q(D) are two differential operators with constant coefficients and $N \in \mathbb{R}^n \setminus 0$, we set

$$\tilde{P}_N(\zeta, T) = (\sum_j |\langle D, TN\rangle^j P(\zeta)|^2)^{1/2}$$

and define $\tilde{Q}_N$ similarly. If Theorem 3.1 does not give examples of non-uniqueness for an operator (1.1) with $a \in C^\infty$, then

$$\tilde{Q}_N(\xi, 1) \leq C\, \tilde{P}_N(\xi, 1),\ \xi \in \mathbb{R}^n, \tag{4.1}$$

which implies in view of [4, Theorem 3.3.2] that

$$\tilde{Q}_N(\xi, T) \leq C_1\, \tilde{P}_N(\xi, T);\ T \geq 1,\ \xi \in \mathbb{R}^n.$$

With Σ_N defined as in Theorem 3.7 Taylor's formula gives

$$\tilde{Q}_N(\zeta, T) \leq C_2\, \tilde{P}_N(\zeta, T);\ T \geq |\mathrm{Im}\,\zeta| + 1,\ \zeta \in \Sigma_N. \tag{4.2}$$

We shall now consider the quotient $\tilde{Q}_N/\tilde{P}_N$ when T is smaller than $|\mathrm{Im}\,\zeta| + 1$ but larger than $(1 + |\mathrm{Im}\,\zeta|)^{1/2}$. The reason for the latter restriction will become clear later on. Note that this quotient with $T = (1 + |\mathrm{Im}\,\zeta|)^{1/2}$ occurs in the uniqueness theorems of [5].

Let s > 0 and form

$$f(s) = \sup \left\{ \tilde{Q}_N(\zeta, T)/\tilde{P}_N(\zeta, T);\ \zeta \in \Sigma_N;\ |\mathrm{Im}\,\zeta| \leq sT,\ T \geq \sqrt{1 + |\mathrm{Im}\,\zeta|} \right\}.$$

In view of (4.2) the increasing function f(s) is bounded by a power of s, and it follows from the Tarski-Seidenberg theorem that f(s) is an algebraic function of s for large s. Hence

$$f(s)\, s^{-\lambda} \to c > 0,\ s \to \infty,$$

for some rational $\lambda \geq 0$. Writing the condition $|\mathrm{Im}\,\zeta| \leq sT$ in the form $1 + |\mathrm{Im}\,\zeta|/T \leq s + 1$, we conclude that

$$\tilde{Q}_N(\zeta, T) \leq C\, \tilde{P}_N(\zeta, T)\, (1 + |\mathrm{Im}\,\zeta|/T)^\lambda,\ T \geq \sqrt{1 + |\mathrm{Im}\,\zeta|},\ \zeta \in \Sigma_N, \tag{4.3}$$

and that no such estimate is valid for a smaller value of λ. After these preliminaries we can state

<u>Theorem 4.1.</u> Assume that (4.1) is valid but that $\lambda > 1$ for the smallest λ such that (4.3) is fulfilled. Then one can find $a \in \bigcap_{\mu<\lambda} H^{\mu}(\mathbb{R}^n)$ such that (1.1) has a solution $u \in C^{\infty}(\mathbb{R}^n)$ satisfying (1.2).

Examples given later on in the section show that the statement is sometimes but not always true also when $0 < \lambda \leqq 1$. Our first aim now is to prove an analogue of Lemma 3.3. Since

$$\{(\zeta, T, s) \in \Sigma_N \times \mathbb{R}_+ \times \mathbb{R}_+ ;\ |\operatorname{Im} \zeta| \leqq sT,\ T^2 \geqq 1 + |\operatorname{Im} \zeta|,$$
$$\tilde{Q}_N(\zeta, T) \geqq \tilde{P}_N(\zeta, T)\, f(s)/2\}$$

is closed, semi-algebraic and contains points with arbitrary s, we can use the Tarski-Seidenberg theorem as in the proof of Lemma 3.3 to select an algebraic curve $\zeta(s)$, $T(s)$ in the set, $s \to \infty$. Then we have

$$\tilde{Q}_N(\zeta(s), T(s)) \geqq C_1\, \tilde{P}_N(\zeta(s), T(s))\, (1 + |\operatorname{Im} \zeta(s)|/T(s))^{\lambda}, \tag{4.4}$$
$$T(s)^2 \geqq 1 + |\operatorname{Im} \zeta(s)|,\quad \zeta(s) \in \Sigma_N.$$

When $f(s)/2 > f(s')$ we have $|\operatorname{Im} \zeta(s)|/T(s) > s'$, which shows that $|\operatorname{Im} \zeta(s)|/sT(s)$ is bounded from above and below as $s \to \infty$, so a limit $\neq 0$ exists.

When $s \to \infty$ we have $T(s)\, s^{-\varkappa} \to c$ for some $c, \varkappa > 0$, and (4.4) remains valid for another C_1 if we replace $T(s)$ by $s^{\varkappa}$. Since $|\operatorname{Im} \zeta(s)|\, s^{-\varkappa-1}$ has a limit $\neq 0$ as $s \to \infty$, it follows that $2\varkappa \geqq \varkappa + 1$, that is, $\varkappa \geqq 1$.

As in the proof of Lemma 3.3 we obtain increasing, convex, piecewise linear functions g_P and g_Q, with integer slopes, such that $\tilde{P}_N(\zeta(s), s^{\mu})$ is asymptotic to a constant times $s^{g_P(\mu)}$ and similarly for Q. By (4.4)

$$g_Q(\varkappa) \geqq g_P(\varkappa) + \lambda .$$

In view of (4.3) we have on the other hand

$$g_Q(\mu) \leqq g_P(\mu) + \lambda(\varkappa + 1 - \mu),\quad 2\mu \geqq \varkappa + 1.$$

When $\mu = \varkappa$ we conclude that $g_Q(\varkappa) = g_P(\varkappa) + \lambda$, hence

$$g_Q(\mu) - g_Q(\varkappa) - (g_P(\mu) - g_P(\varkappa)) \leqq \lambda(\varkappa - \mu), \mu \geqq \varkappa .$$

Since $\lambda > 1$ it follows that the integer slope of $g_Q - g_P$ must be $\leqq -2$ at $\varkappa + 0$. Thus

$$P(\zeta(s) + z\, s^{\varkappa} N)\, s^{-g_P(\varkappa)} \to p(z)$$

and similarly for Q, where p and q are polynomials with deg p $\geqq$ deg q + 2. (This is the important information obtained from the hypothesis $\lambda > 1$. It may sometimes be true although $\lambda \leqq 1$ and all that follows is applicable then.) Writing $s = t^j$ for some large j we have now proved

<u>Lemma 4.2.</u> Assume that the hypotheses of Theorem 4.1 are fulfilled. Then it is possible to find a Laurent series $\zeta(t)$, converging to a point in Σ_N for large $t > 0$, and positive integers $\varkappa, \sigma$ such that

(4.5) $\quad |\operatorname{Im} \zeta(t)|\, t^{-\sigma} \to c \neq 0, \quad \varkappa < \sigma \leqq 2\varkappa,$

(4.6) $\quad P(\zeta(t) + zt^{\varkappa} N)\, t^{-g(P)} \to p(z), \quad Q(\zeta(t) + zt^{\varkappa} N)\, t^{-g(Q)} \to q(z),$

where p and q are polynomials $\not\equiv 0$ with deg p $\geqq$ deg q + 2 and

$$g(Q) - g(P) = \lambda(\sigma - \varkappa). \tag{4.7}$$

<u>Proof of Theorem 4.1.</u> Following Pliś [10] we choose a sequence b_ν decreasing to 0 more slowly than any power of ν^{-1}, for example

$$b_\nu = 1/(\log \nu).$$

(Choosing $b_\nu = \nu^{-\gamma}$ with small $\gamma > 0$ we could get $a \in H^{\mu}$ for a fixed $\mu < \lambda$.) Then

$$\Delta b_\nu = b_\nu - b_{\nu+1} \sim 1/\nu(\log \nu)^2$$

is only slightly smaller than $1/\nu$. With $\rho = 1/(\sigma - \varkappa)$ we set

$$t_\nu = \nu^\rho, \quad \zeta_\nu = \zeta(t_\nu), \quad T_\nu = t_\nu^{\varkappa}(\log \nu)^4 .$$

Since $2\varkappa \geqq \sigma$, we have $\varkappa\rho = \varkappa/(\sigma - \varkappa) \geqq 1$ and

$$T_\nu \Delta b_\nu \geqq C\, (\log \nu)^2 \to \infty . \tag{4.8}$$

Furthermore,

$$|\mathrm{Im}\,(\zeta_{\nu+1} - \zeta_\nu)|/T_\nu < C_1(t_{\nu+1} - t_\nu)\, t_\nu^{\sigma-1}/T_\nu < C_2\, t_\nu^{\sigma}/\nu\, T_\nu$$

which gives

$$|\mathrm{Im}\,(\zeta_{\nu+1} - \zeta_\nu)|/T_\nu < C_2(\log \nu)^{-4} \to 0,\ \nu \to \infty. \tag{4.9}$$

The condition (4.9) did not appear in the proof of Theorem 3.1 since we were only using real frequencies then. The restriction to $T^2 \geqq 1 + |\mathrm{Im}\,\zeta|$ in (4.3) was made precisely to guarantee (4.9). We shall now see that (4.8), (4.9) permit us to repeat the proof of Theorem 3.1 with only minor changes.

From (4.6) it follows that the coefficients of

$$P(\zeta(t) + z\, t^{\varkappa} N)\; t^{-g(P)} - p(z)$$

as a polynomial in z are $O(1/t)$. If p is of order $d(P)$ with leading coefficient $c(P)$, we conclude with $t = t_\nu$ and z replaced by $z\, T_\nu\, t_\nu^{-\varkappa} = z(\log \nu)^4$ that

$$P(\zeta_\nu + zT_\nu N)\; t_\nu^{-g(P)}\; (\log \nu)^{-4d(P)} \to c(P)\; z^{d(P)},\ \nu \to \infty.$$

If $d = d(P) - d(Q)$, which is an integer larger than 1, we set $z^+ = i$ and choose z^- so that

$$z^{+\,d} = z^{-\,d},\ z^- \neq z^+,\ \text{thus } \mathrm{Im}\, z^- < \mathrm{Im}\, z^+. \tag{4.10}$$

We set $z_\nu^+ = z^+$ and determine a sequence $z_\nu^- \to z^-,\ \nu \to \infty$, so that

$$(P/Q)(\zeta_\nu + z_\nu^- T_\nu\, N) = (P/Q)(\zeta_{\nu+1} + z_{\nu+1}^+ T_{\nu+1} N). \tag{4.11}$$

This is possible since the equation converges to $z^{+\,d} = z_\nu^{-\,d}$ after multiplication by $t^{g(Q) - g(P)}(\log \nu)^{-4d}$.

Next we choose $\psi_\nu \in C^\infty(I_\nu)$, $I_\nu = (b_{\nu+1}, b_{\nu-1})$ so that with $b'_\nu = (b_\nu + b_{\nu+1})/2$ the following conditions are fulfilled for large ν

(4.12) $\psi_\nu(s) = z_\nu^-$ when $|s-b'_\nu| < (\Delta b_\nu)/4$; $\psi_\nu(s) = z_\nu^+$ when $|s-b'_{\nu-1}| <$
$< (\Delta b_{\nu-1})/4$.

$$\psi_\nu(s) \in K,\ s \in I_\nu;\ |\psi_\nu^{(k)}(s)| < C_k(\Delta b_\nu)^{-k} \text{ if } s \in I_\nu \text{ and } k = 0, 1, 2, \ldots$$

$$\operatorname{Im} \psi_\nu(s) \leqq 1, \quad \int_{b'_\nu}^{b'_{\nu-1}} \operatorname{Im} \psi_\nu(s) < -\Delta b_\nu.$$

(Compare with (3.7).) Here C_k are constants and K a compact set $\subset \mathbb{C} \setminus \{0\}$. Having chosen ψ_ν we determine φ_ν inductively so that

(4.13) $\varphi'_\nu = T_\nu \psi_\nu$; $\Phi_\nu(b'_\nu) = \Phi_{\nu+1}(b')$ if $\Phi_\nu(s) = \varphi_\nu(s) + s\,\zeta_{\nu,n}$.

Using (4.9), (4.10) we conclude that for some C_1, $C_2 > 0$

(4.14) $C_1 T_\nu \leqq \operatorname{Im}(\Phi_{\nu+1}(s) - \Phi_\nu(s))/(s-b'_\nu) \leqq C_2 T_\nu$, $|s-b'_\nu| < (\Delta b_\nu)/4$.

This gives the information about the relative sizes of u_ν and $u_{\nu+1}$ required to estimate the derivatives of a in the interval where we switch over from one to the other. From (4.12) we also obtain

$$\operatorname{Im}(\Phi_\nu(b'_{\nu-1}) - \Phi_\nu(b'_\nu)) < -T_\nu \Delta b_\nu,$$

that is,

$$\operatorname{Im} \Phi_{\nu+1}(b') > \operatorname{Im} \Phi_\nu(b'_{\nu-1}) + T_\nu \Delta b_\nu.$$

Since $T_\nu \Delta b_\nu \geqq t_\nu^\varkappa (\log \nu)^2/2\nu$ it follows that for some $C > 0$

$$\operatorname{Im} \Phi_\nu(b'_{\nu-1}) > C\, t_\nu^\varkappa (\log \nu)^2.$$

Noting that $\operatorname{Im} \Phi'_\nu \leqq T_\nu$ and that $T_\nu \Delta b_\nu = O((\log \nu)^2\, t_\nu^\varkappa/\nu)$, we obtain first in $(b'_\nu, b'_{\nu-1})$ and then using (4.14) outside this interval that

(4.15) $\operatorname{Im} \Phi_\nu(s) > C\, t_\nu^\varkappa (\log \nu)^2$ in $(b'_\nu - 4^{-1}\Delta b_\nu,\ b'_{\nu-1} + 4^{-1}\Delta b_{\nu-1})$.

Now choose $\chi_\nu \in C_0^\infty(b'_\nu - 4^{-1}\Delta b_\nu,\ b'_{\nu-1} + 4^{-1}\Delta b_{\nu-1})$ so that $\chi_\nu = 1$ in $(b'_\nu - 8^{-1}\Delta b_\nu,\ b'_{\nu-1} + 8^{-1}\Delta b_{\nu-1})$ and

(4.16) $$|\chi_\nu^{(k)}(s)| \leqq C_k (\Delta b_\nu)^{-k},\ k \geqq 0.$$

With ν_0 so large that the preceding conditions and a few later ones are fulfilled when $\nu \geqq \nu_0$ we define u by (3.4) where

$$u_\nu(x) = \exp i(\langle x, \zeta_\nu\rangle + \varphi_\nu(\langle x, N\rangle)).$$

Since $|u_\nu(x)| < \exp(-C\nu(\log \nu)^2)$ by (4.15), it is immediately seen that

$u \in C^\infty$.

Finally we have to consider $a = -P(D)u/Q(D)u$. The construction has been made so that a is a constant converging to 0 with ν^{-1} in the set where

$$b'_\nu - 8^{-1}\Delta b_\nu < \langle x, N\rangle < b'_\nu + 8^{-1}\Delta b_\nu .$$

In the set where

$$b'_\nu + 8^{-1}\Delta b_\nu < \langle x, N\rangle < b'_\nu + 4^{-1}\Delta b_\nu$$

the estimate (4.14) and the fact that $T_\nu \Delta b_\nu \geq C(\log \nu)^2$ is much larger than $\log \nu$ makes it possible to argue exactly as in the corresponding part of the proof of Theorem 3.1. In this set the derivatives of a therefore have bounds converging to 0 rapidly as $\nu \to \infty$. The only new point occurs in the set where

$$b'_\nu + 4^{-1}\Delta b_\nu < \langle x, N\rangle < b'_{\nu-1} - 4^{-1}\Delta b_{\nu-1}.$$

There we have $u = u_\nu$, and

(4.17) $Q(D)u = t_\nu^{g(Q)} f_Q(\langle x, N\rangle)u, \quad f_Q(s) = e^{-i\varphi_\nu(s)}\, q_\nu(D/t_\nu^\varkappa)\, e^{i\varphi_\nu(s)},$

where we have written

$$q_\nu(z) = Q(\zeta_\nu + z t_\nu^\varkappa N)\, t_\nu^{-g(Q)}.$$

This is a polynomial and the coefficients of $q_\nu(z) - q(z)$ are $O(t_\nu^{-1})$.

Recall now that $\varphi'_\nu = T_\nu \psi_\nu$ and that $T_\nu/t_\nu^\varkappa = (\log \nu)^4$. This gives

$$f_Q(s) = (\log \nu)^{4d(Q)}\, \psi_\nu^{d(Q)}\, c(Q)\, (1 + r_{Q\nu}(s))$$

where $r_{Q\nu}$ is a sum of products of powers of $1/\psi_\nu$ and $(\Delta b_\nu)^k \psi_\nu^{(k)}$ with coefficients converging to 0 with $1/\nu$. Here it is important that (4.8) shows that the factor T_ν brought out by differentiation of the exponential in (4.17) exceeds the deterioration of the bounds by $1/\Delta b_\nu$ which occurs when ψ_ν is differentiated (see (4.12)). Using (4.7) we therefore obtain in the considered interval

$$|D^\alpha a(x)| \leq C_\alpha\, t_\nu^{-\lambda(\sigma-\varkappa)}\, (\log \nu)^{4d}\, (\Delta b_\nu)^{-|\alpha|}.$$

For any $\varepsilon > 0$ it follows that

$$|D^\alpha a(x)| < C_\varepsilon \nu^{\varepsilon+|\alpha|-\varepsilon},$$

so we have proved

$$(4.18) \qquad |D^\alpha a(x)| \leq C_{\varepsilon,\alpha} \langle x, N\rangle^{\lambda - |\alpha| - \varepsilon}, \quad \langle x, N\rangle < b'_{\nu_0 - 1}.$$

The proof is now completed by a modification of the first term in the definition of u as in the proof of Theorem 3.1, and an application of the following elementary lemma to show that (4.18) implies $a \in H^\mu$ for every $\mu < \lambda$.

Lemma 4.3. Let $f \in C^\infty$ in $\mathbb{R}^n$ when $\langle x, N\rangle \neq 0$ and assume that

$$|f(x)| \leq C|\langle x, N\rangle|^a, \quad |\text{grad } f(x)| \leq C|\langle x,N\rangle|^{-b}, \quad x \in \mathbb{R}^n.$$

Then it follows that $f \in H^{a/(a+b)}$.

Proof. If $\langle x, N\rangle \leq \langle y, N\rangle \leq 2\langle x, N\rangle$ we have the estimates

$$|f(y) - f(x)| \leq C|x-y|\langle x,N\rangle^{-b}, \quad |f(y) - f(x)| < C \langle x, N\rangle^a.$$

Raising the estimates to the powers $a/(a+b)$ and $b/(a+b)$ and multiplying, we obtain $|f(y) - f(x)| < C|x-y|^{a/(a+b)}$. On the other hand, if $0 < 2\langle x, N\rangle < \langle y, N\rangle < 1$, then the same estimate follows if we use the triangle inequality after inserting on the segment between x and y a sequence of points with $\langle x_k, N\rangle = 2^k\langle x, N\rangle$, $k = 1, 2, \ldots$ The details are left for the reader.

Example 4.4. Let $P(\xi) = (\xi_1+i\xi_2)^p$, $Q(\xi) = \xi_1{}^q$ where $q \leq p$. If we take $N = (0, 1)$, $\zeta = (t, it)$ and $T^2 = 1-t$, $t < 0$, the inequality (4.3) implies

$$|t|^{q - p/2} \leq C\ |t|^{\lambda/2}, \quad t < -1,$$

hence $\lambda \geq 2q - p$. If $2q - p > 1$ Theorem 4.1 gives non-uniqueness for (1.1) with some $a \in \bigcap_{\mu < 2q-p} H^\mu$. The proof is in fact applicable unless $p = q = 1$. In that case well known results on quasiconformal mappings give uniqueness for measurable a in any open set with $\sup |a| < 1$.

The preceding example indicates that the question on the validity of Theorem 4.1 does not have an obvious answer when $0 < \lambda \leq 1$. A case where Theorem 4.1 remains valid for such λ is given by the arguments of Pliś [10] which are applicable unless $|\mathrm{Im}\,\zeta|$ is bounded when $\zeta \in \Sigma_N$ and $P(\zeta) = Q(\zeta) = 0$.

References

1. P. Cohen, The non-uniqueness of the Cauchy problem. O.N.R.Techn.Report 93, Stanford 1960.
2. E. De Giorgi, Un esempio di non-unicitá della soluzione del problema di Cauchy relativo ad una equazione differenziale lineare a derivate parziale di tipo parabolico. Rend.di Mat.e Appl. 14(1955),382-387.
3. P.M.Goorjian, The uniqueness of the Cauchy problem for partial differential equations which may have multiple characteristics. Trans. Amer. Math. Soc. 149(1969), 493-509.
4. L. Hörmander, Linear partial differential operators. Springer Verlag, Berlin-Göttingen-Heidelberg 1963.
5. - , On the uniqueness of the Cauchy problem. Math. Scand. 6(1958), 213-225.
6. A. Pliś, The problem of uniqueness for the solution of a system of partial differential equations. Bull. Acad. Pol. Sci. 2(1954), 55-57.
7. - , Non-uniqueness in Cauchy's problem for differential equations of elliptic type. J. Math. Mech. 9(1960), 557-562.
8. - , A smooth linear elliptic differential equation without any solution in a sphere. Comm. Pure Appl. Math. 14(1961), 599-617.
9. - , Unique continuation theorems for solutions of partial differential equations. Proc. Int. Congr. Math. Stockholm 1962, 397-402.
10. - , On non-uniqueness in Cauchy problem for an elliptic second order differential equation. Bull. Acad. Pol. Sci. 11(1963), 95-100.
11. A. Pliś, Homogeneous partial differential equations posessing solutions with arbitrary small supports. Bull. Acad. Polon. Sci. 12(1964), 205-206.
12. H. Whitney, Analytic extensions of differentiable functions defined in closed sets. Trans. Am. Math. Soc. 36(1934), 63-89.

SOLUTIONS ASYMPTOTIQUES ET GROUPE SYMPLECTIQUE

par Jean LERAY

Collège de France, Paris 05

INTRODUCTION.

Il est nécessaire d'expliciter et de justifier la notion, due à V.P. Maslov [3], de solution asymptotique ; je l'ai fait, par exemple à Rome en décembre 1972 [1].

Le Traité de V.P. Maslov et mon exposé emploient un choix particulier de coordonnées pour construire des notions, qui se révèlent finalement indépendantes de ce choix. Le présent exposé libère cette théorie d'un tel choix, en employant -au lieu du groupe fini engendré par les transformations de Fourier opérant chacune sur l'une des coordonnées- une représentation unitaire Sp_2 du revêtement à deux feuillets du groupe symplectique Sp .

Cette représentation Sp_2 fut employée par D. Shale [5] et V.C. Bouslaev [3], qui développaient tous deux des notions introduites en théorie quantique par I. Segal [4]. Cette représentation Sp_2 est l'un des groupes algébriques d'opérateurs unitaires qu'A. Weil [6] relie aux travaux de théorie des nombres de K. Siegel. Mais aucun de ces auteurs n'énonce les propriétés de Sp_2 qu'emploie la théorie des solutions asymptotiques.

§ 1. LE REVÊTEMENT $Sp_2(\ell)$ DU GROUPE SYMPLECTIQUE $Sp(\ell)$.

1. LE GROUPE MÉTAPLECTIQUE. - Notons : $X = R^\ell$;

$\mathcal{S}(X)$ l'espace des fonctions $X \to C$ dont toutes les dérivées sont à décroissance rapide ;

$\mathcal{S}'(X)$ l'espace des distributions tempérées sur X (L. Schwartz) ;

$\mathcal{H}(X)$ l'espace de Hilbert des fonctions $X \to C$ de carré sommable ;

$X^* = R^\ell$ le dual de X ; $< p,x > \in R$ la valeur en $x \in X$ de $p \in X^*$;

$Z(\ell)$ l'espace vectoriel $X \oplus X^*$, muni de la structure symplectique $[.,.]$ que voici : soient z et $z' \in Z(\ell)$; soient x, x', p et p' tels que

$$z = x+p \ , \quad z' = x'+p' \ , \quad x \text{ et } x' \in X \ , \quad p \text{ et } p' \in X^* \ ;$$

alors

(1.1) $$[z,z'] = < p,x' > - < p',x > \ ;$$

ν un nombre imaginaire pur, non nul : $\nu \in i\dot{\mathbb{R}}$;

$\frac{1}{\nu}\frac{\partial}{\partial x}$ et le produit par x seront donc, sur $\mathcal{H}(X)$, des opérateurs self-adjoints.

Tout $a \in Z(\ell)$ définit une fonction linéaire de

$z = x+p \in Z(\ell)$ $(x \in X \ , \ p \in X^*)$ valant en z , par définition :

(1.2) $$a(z) = a(x,p) = [a,z] \ ;$$

elle définit donc un opérateur différentiel $a(x,\frac{1}{\nu}\frac{\partial}{\partial x})$, linéaire en $(x,\frac{1}{\nu}\frac{\partial}{\partial x})$; cet opérateur est un endomorphisme de $\mathcal{S}'(X)$; il est self-adjoint sur X .

Tout automorphisme S de $\mathcal{S}'(X)$ le transforme en un endomorphisme SaS^{-1} de $\mathcal{S}'(X)$.

<u>Définition</u>.- Nous notons $G(\ell)$ le groupe de ceux des automorphismes S de $\mathcal{S}'(X)$ qui transforment tous les opérateurs différentiels $a(x,\frac{1}{\nu}\frac{\partial}{\partial x})$ linéaires en x et $\frac{1}{\nu}\frac{\partial}{\partial x}$ en opérateurs du même type.

<u>Propriétés</u>.- Tout S de $G(\ell)$ induit donc un endomorphisme

$$s : a \mapsto SaS^{-1}$$

de $Z(\ell)$; mais l'opérateur différentiel

$$a(x,\frac{1}{\nu}\frac{\partial}{\partial x})\, b(x,\frac{1}{\nu}\frac{\partial}{\partial x}) - b(x,\frac{1}{\nu}\frac{\partial}{\partial x})\, a(x,\frac{1}{\nu}\frac{\partial}{\partial x})$$

est la multiplication par $\frac{1}{\nu}[a,b] \in \mathbb{C}$; l'endomorphisme s de $Z(\ell)$ doit donc laisser $[.,.]$ invariant, c'est-à-dire être un automorphisme symplectique de $Z(\ell)$; le groupe de ces automorphismes est noté $Sp(\ell)$.

L'application $S \to s$ est donc un morphisme naturel :

(1.3) $$G(\ell) \to Sp(\ell) \ ;$$

la valeur de $sa = SaS^{-1}$ en z est donc

$$(sa)(z) = [sa, z] = [a, s^{-1}z] = a(s^{-1}z) = (a \circ s^{-1})(z) \quad ;$$

autrement dit l'endomorphisme sa de $Z(\ell)$ est l'application composée

$$sa = a \circ s^{-1} \quad . \tag{1.4}$$

Le noyau du morphisme naturel (1.3) est l'ensemble des automorphismes S de $\mathcal{S}'(X)$ commutant à x et $\frac{\partial}{\partial x}$; $S1$ est donc une constante $c \neq 0$; si P est un polynome, $SP = cP$; or les polynomes sont denses dans $\mathcal{S}'(X)$; donc S est la multiplication par c . Le noyau de (1.3) est donc le sous-groupe $\dot{\mathbb{C}}$ du centre de $G(\ell)$ ayant pour éléments les multiplications cE par les nombres complexes $c \neq 0$. Notons $\dot{\mathbb{C}} = \mathbb{S}^1 \times \mathbb{R}_+$, $\mathbb{S}^1$ étant le groupe multiplicatif des nombres complexes de module 1 et $\mathbb{R}_+$ celui des nombres réels > 0 .

Le morphisme naturel (1.3) est un épimorphisme. Pour le prouver, notons A la donnée d'un nombre réel ou imaginaire pur $\Delta(A) \neq 0$ et d'une forme quadratique sur $X \oplus X^*$, à valeurs réelles :

$$X \oplus X^* \ni (x, x') \longmapsto A(x, x') = \frac{1}{2} < Px, x > - < Lx, x' > + \frac{1}{2} < Qx', x' > \in \mathbb{R} \quad , \tag{1.5}$$

où $\quad L \, , \; P = {}^tP, \; Q = {}^tQ : X \to X^*, \;$ dét $L = \Delta^2(A) \quad ;$

notons ν un nombre imaginaire pur et $i^{\ell/2} = e^{\pi \ell i/4}$; pour tout $u' \in \mathcal{S}(X)$, définissons $u \in \mathcal{S}(X)$ par l'intégrale

$$u(x) = \left(\frac{|\nu|}{2\pi i}\right)^{\ell/2} \Delta(A) \int_X e^{\nu A(x,x')} u'(x') \, d^\ell x' \quad ; \tag{1.6}$$

$$S_A : u' \longmapsto u$$

est un automorphisme unitaire, puisqu'il est le composé des quatre automorphismes unitaires de $\mathcal{S}(X)$ que voici :

(1.7) une multiplication de v et v' par e^{iq} et $e^{iq'}$, q et q' étant des formes quadratiques $X \to \mathbb{R}$;

(1.8) une transformation de Fourier ;

(1.9) un automorphisme de $\mathcal{S}(X)$ qui applique $v \in \mathcal{S}(X)$ sur u , valant

$$u(x) = \sqrt{\text{dét } T}\ \ v(Tx) \quad ,$$

où T est un automorphisme de X .

Ces automorphismes, donc leur produit S_A , se prolongent en automorphismes unitaires de $\mathcal{H}(X)$ et en automorphismes de $\mathcal{S}''(X)$.

D'autre part, la dérivation en x de la relation (1.5) montre que $S_A \in G(\ell)$; l'image s_A de S_A par (1.3) dans $Sp(\ell)$ est l'automorphisme symplectique

$$s_A : Z(\ell) \ni (x',p') \mapsto (x,p) \in Z(\ell)$$

défini par les relations

(1.10) $$p = A_x(x,x') \ , \ p' = - A_{x'}(x,x') \quad ,$$

c'est-à-dire par les relations, où L est inversible :

(1.11) $$p = Px - {}^tLx' \ , \ p' = Lx - Qx' \ .$$

<u>Le groupe métaplectique</u> $Mp(\ell)$ est le sous-groupe de $G(\ell)$ dont les éléments sont les éléments de $G(\ell)$ ayant une restriction à $\mathcal{H}(\ell)$ qui soit unitaire ; nous venons de voir que $S_A \in Mp(\ell)$; il en résulte que la restriction à $Mp(\ell)$ du morphisme (1.3) est un épimorphisme $Mp(\ell) \longrightarrow Sp(\ell)$; donc

(1.12) $$G(\ell) = Mp(\ell) \times \mathbb{R}_+ \ .$$

Puisque le noyau de $G(\ell) \to Sp(\ell)$ est $\dot{\mathbb{C}}$, le noyau de $Mp(\ell) \to Sp(\ell)$ est $\mathbb{S}^1$; donc

(1.13) $$Mp(\ell)/\mathbb{S}^1 = Sp(\ell) \ .$$

2. LE GROUPE UNITAIRE $Sp_2(\ell)$.-

On déduit aisément de la définition (1.6) de S_A que,

(2.1) si $$s_A\, s_{A'}\, s_{A''} = E \ , \ \text{alors}\ S_A\, S_{A'}\, S_{A''} = \pm E \ .$$

Il en résulte que le sous-groupe de $Mp(\ell)$ engendré par les S_A est l'ensemble des produits d'un couple d'éléments de S_A . La restriction à ce sous-groupe du morphisme canonique (1.3) est donc un épimorphisme. On déduit de la définition (1.6) de S_A que

(2.2) si $s_A \, s_{A'} = E$, alors $S_A \, S_{A'} = \pm E$.

Il en résulte que le noyau de cet épimorphisme est le sous-groupe à deux éléments $S^o = \{E, -E\}$; rappelons que $-E : v \mapsto -v$ (produit par -1 de $v \in \mathcal{S}'(X)$) . Donc: <u>les automorphismes</u> S_A (de $\mathcal{S}(X)$, de $\mathcal{S}'(X)$ et de $\mathcal{H}(X)$) <u>engendrent un revêtement à 2 feuillets,</u> $Sp_2(\ell)$ <u>de</u> $Sp(\ell)$.

Leurs restrictions à $\mathcal{H}(X)$ constituent <u>une représentation unitaire de</u> $Sp_2(\ell)$.

La projection naturelle de $Sp_2(\ell)$ sur $Sp(\ell)$ est : $\pm S \mapsto s$.

On prouve que ce revêtement est connexe, donc n'est pas trivial.

On prouve enfin que tout élément S de $Sp_2(\ell)$ est encore le produit des quatre automorphismes (1.7), (1.8) et (1.9), si l'on permet à la transformation de Fourier de n'opérer que sur certaines des variables indépendantes.

Nous noterons $\Sigma(\ell)$ [et $\Sigma_2(\ell)$] l'ensemble des $s \in Sp(\ell)$ [et des $S \in Sp_2(\ell)$] qui ne sont pas du type s_A [et S_A] ; $\Sigma(\ell)$ [et $\Sigma_2(\ell)$] sont des hypersurfaces de $Sp(\ell)$ [et de $Sp_2(\ell)$] ; la projection naturelle de $Sp_2(\ell)$ sur $Sp(\ell)$ applique $\Sigma_2(\ell)$ et $Sp_2(\ell) \setminus \Sigma_2(\ell)$ sur $\Sigma(\ell)$ et $Sp(\ell) \setminus \Sigma(\ell)$.

<u>Note</u> . -

(2.3) $s \notin \Sigma(\ell)$ signifie : X^* et sX^* sont transverses.

3. INERTIE ; INDICE DE MASLOV, mod. 4. - La preuve de (2.1) emploie la formule (3.2) que voici.

Définissons A' (et A") comme A l'a été, par la donnée de $\Delta(A')$, L', P', Q'; la condition

(3.1) $s_A \; s_{A'} \, s_{A''} = E$ s'énonce

$$P'' + Q' = L'(P' + Q)^{-1}\,{}^tL' \quad , \quad P + Q'' = {}^tL(P' + Q)^{-1}\,L \ , \tag{3.2}$$
$$L'' = -{}^tL(P' + Q)^{-1}\,{}^tL'$$

Cette formule (3.1) prouve que les formes quadratiques définies par les morphismes symétriques et inversibles

$$P' + Q \quad , \quad P'' + Q' \quad , \quad P + Q''$$

ont le même indice d'inertie[1] ;

$$\text{Inert}(P'+Q) = \text{Inert}(P''+Q') = \text{Inert}(P+Q'')$$

nous le nommons <u>inertie</u> de $(s_A\ ,\ s_{A'}, s_{A''})$ et le notons

$$\text{Inert}(s_A, s_{A'}, s_{A''}) \quad \text{ou} \quad \text{Inert}(S_A, S_{A'}, S_{A''}) \ .$$

Ainsi

Inert(...) est une fonction, à valeurs $\in \{0,\ldots,\ell\}$, de (S,S',S'') , <u>définie pour</u>

$$S, S' \text{ et } S'' \notin \Sigma \ , \quad S\,S'\,S'' = \pm E \ ;$$

<u>elle ne dépend que des projections</u> s,s',s'' <u>de</u> S,S',S'' <u>dans</u> $Sp(\ell)$; nous avons :

(3.3) $\text{Inert}(S,S',S'') = \text{Inert}(S',S'',S) = \text{Inert}(S'',S,S') = \ell - \text{Inert}(S''^{-1},S'^{-1},S^{-1})$

Supposons (3.1) vérifié ; (2.1) peut être précisé comme suit. Notons

(3.4) $$m(S_A) \equiv \frac{2}{\pi} \arg \Delta(A) \mod 4 \ ;$$

alors on a

$$S_A\ S_{A'} S_{A''} = E$$

quand

$$\text{Inert}(S_A, S_{A'}, S_{A''}) \equiv m(S_A) - m(S_{A'}^{-1}) + m(S_{A''}) \mod. 4 \ ,$$

Ainsi :

m <u>est une fonction localement constante, à valeurs dans</u> $\mathbb{Z}_4$, <u>de</u> S , <u>définie sur</u> $Sp_2(\ell) \setminus \Sigma_2(\ell)$; <u>elle vérifie</u>

(3.5) $$\text{Inert}(S,S',S'') \equiv m(S) - m(S'^{-1}) + m(S'') \mod. 4$$

Ces deux propriétés la caractérisent évidemment.

Elle possède les propriétés suivantes :

(3.6) $$m(S^{-1}) \equiv \ell - m(S) \ , \quad m(-S) \equiv m(S)+2 \mod. 4 \ .$$

---

[1] Un morphisme $q : X \to X^*$ inversible et symétrique, c'est-à-dire tel que ${}^tq = q$, définit une forme quadratique :

$$x \mapsto \langle qx,x \rangle \ ; \quad \langle qx,x \rangle = -\sum_j L_j^2(x) + \sum_k L_k^2(x),$$

les L_j et L_k étant ℓ formes linéaires indépendantes ; Inert(q) est le nombre des j

Note.- m est donc définie mod. 2 sur $Sp(\ell) - \Sigma(\ell)$:

$$m(s_A) \equiv m(\pm S_A) \equiv \text{signe (dét } L) \quad \text{mod. } 2$$

Note.- Le § 3 définira m comme étant une fonction à valeurs dans $\mathbb{Z}$ et non plus dans $\mathbb{Z}_4$.

§ 2. OPÉRATEURS DIFFÉRENTIELS À COEFFICIENTS POLYNOMIAUX.

4. Sur X , soit a un opérateur différentiel à coefficients polynomiaux, dépendant de la variable imaginaire pure ν déjà introduite dans (1.6).

Il a deux formes canoniques :

$$(4.1) \qquad a^+(\nu,x,\frac{1}{\nu}\frac{\partial}{\partial x}). = \sum_\alpha a^+_\alpha(\nu,x)(\frac{1}{\nu}\frac{\partial}{\partial x})^\alpha. \quad ;$$

$$(4.2) \qquad a^-(\nu,\frac{1}{\nu}\frac{\partial}{\partial x},x). = \sum_\alpha (\frac{1}{\nu}\frac{\partial}{\partial x})^\alpha [a^-_\alpha(\nu,x).] \quad .$$

A ces deux formes faisons correspondre deux polynomes, valant :

$$a^+(\nu,x,p) = \sum_\alpha a^+_\alpha(\nu,x)p^\alpha \quad ; \quad a^-(\nu,x,p) = \sum_\alpha a^-_\alpha(\nu,x)p^\alpha \quad .$$

On prouve aisément ceci :

<u>A cet opérateur différentiel</u> a <u>est associé un polynome</u> $a^o: (x,p) \mapsto a^o(x,p)$ <u>tel que</u> :

$$(4.3) \qquad a^o(\nu,x,p) = e^{-\frac{1}{2\nu}\frac{\partial^2}{\partial x.\partial p}} a^+(\nu,x,p) = e^{\frac{1}{2\nu}\frac{\partial^2}{\partial x.\partial p}} a^-(\nu,x,p) \quad ;$$

on a noté

$$\frac{\partial^2}{\partial x.\partial p} = \sum_j \frac{\partial^2}{\partial x_j\, \partial p_j} \quad , \quad \{x_j\} \text{ et } \{p_j\} \text{ étant des coordonnées duales de } X \text{ et } X^*;$$

$$e^{\frac{1}{2\nu}\frac{\partial^2}{\partial x.\partial p}} = \sum_{n\in\mathbb{N}} \frac{1}{n!}\left(\frac{1}{2\nu}\frac{\partial^2}{\partial x.\partial p}\right)^n$$

opère évidemment sur les polynomes de (x,p) .

Si le polynome a^o est associé à l'opérateur différentiel a alors le polynome

$$a^o \circ s^{-1} : (x,p) \to a^o(\nu, s^{-1}(x,p)) \ ;$$

est associé à l'opérateur différentiel SaS^{-1} . s désigne la projection sur $Sp(\ell)$ de $S \in Sp_2(\ell)$.

§ 3. AUTRES INDICES D'INERTIES DE MASLOV SUR $Z(\ell)$.

La topologie algébrique permet d'établir les résultats suivants : il suffit d'employer des méthodes dues à I. Arnold [3] , comme je l'ai fait à Rome en janvier 1973 [2].

5. LE GROUPE FONDAMENTAL DE $Sp(\ell)$. - Ce groupe fondamental est (cf. E. Cartan):

(5.1) $$\pi_1 \,[Sp(\ell)] \simeq \mathbb{Z} \ .$$

Le groupe $Sp(\ell)$ possède donc un seul revêtement non trivial $Sp_q(\ell)$ d'ordre q ($q \in \mathbb{N}$ ou $q = +\infty$) ; nous avons identifié $Sp_2(\ell)$ à un groupe unitaire opérant sur $\mathcal{H}(X)$.

Notons $\pi_1 \,[Sp(\ell)]$ mod. q l'image de $\pi_1 \,[Sp(\ell)]$ isomorphe à $\mathbb{Z}_q$;

(5.2) $\pi_1 \,[Sp(\ell)]$ mod. q s'identifie à un sous-groupe du centre de $Sp_q(\ell)$.

6. LA GRASSMANNIENNE LAGRANGIENNE $\Lambda(\ell)$. - On nomme sous-espace lagrangien de Z tout sous-espace de Z sur lequel la forme symplectique $[\cdot,\cdot]$ s'annule identiquement.

L'ensemble $\Lambda(\ell)$ des ℓ-sous-espaces lagrangiens est un espace homogène ; on peut l'identifier à $U(\ell)/O(\ell)$, quotient du groupe unitaire $U(\ell)$ par le groupe orthogonal $O(\ell)$. I. Arnold [3] a prouvé que le groupe fondamental de $\Lambda(\ell)$ est

(6.1) $$\pi_1 \,[\Lambda(\ell] \simeq \mathbb{Z} \ .$$

Cette grassmannienne $\Lambda(\ell)$ possède donc un seul revêtement non trivial d'ordre q ($q \in \mathbb{N}$ ou $q = +\infty$) : $\Lambda_q(\ell)$; $\pi_1 \,[\Lambda(\ell)]$ opère sur $\Lambda_q(\ell)$; si β est le générateur de $\pi_1 \,[\Lambda(\ell)]$ et si $\lambda_q \in \Lambda_q(\ell)$, alors

(6.2) $\beta^p \lambda_q = \lambda_q$ <u>si et seulement si</u> $p = 0$ mod. q .

D'autre part, $Sp(\ell)$ opère effectivement et transitivement sur $\Lambda(\ell)$; il en résulte que $Sp_\infty(\ell)$ opère effectivement et transitivement sur $\Lambda_\infty(\ell)$; en choisissant de façon cohérente <u>les générateurs</u> α de $\pi_1[Sp(\ell)]$ et β de $\pi_1[\Lambda(\ell)]$, on obtient la formule :

(6.3) $$\alpha\,\lambda_\infty = \beta^2\,\lambda_\infty \quad , \quad \text{où} \quad \lambda_\infty \in \Lambda_\infty(\ell)$$

Vu (6.2), il en résulte que $Sp_q(\ell)$ opère sur $\Lambda_{2q}(\ell)$, l'image α_q de α dans $Sp_q(\ell)$ opérant comme suit :

(6.4) $$\alpha_q\,\lambda_{2q} = \beta^2\,\lambda_{2q} \quad .$$

<u>En particulier</u> : $Sp_2(\ell)$ opère sur $\Lambda_4(\ell)$;
l'élément $-E$ de $Sp_2(\ell)$ et l'élément β^2 de $\pi_1[\Lambda(\ell)]$ définissent le même homéomorphisme de $\Lambda_4(\ell)$.

7. L'INERTIE D'UN TRIPLET DE ℓ-PLANS LAGRANGIENS. - Soient trois ℓ- sous-espaces lagrangiens de Z , <u>deux à deux transverses</u> : $\lambda, \lambda', \lambda''$; nous avons donc

$$Z = \lambda \oplus \lambda' = \lambda' \oplus \lambda'' = \lambda'' \oplus \lambda \quad .$$

Les conditions

(7.1) $$z \in \lambda \quad , \quad z' \in \lambda' \quad , \quad z'' \in \lambda'' \quad , \quad z+z'+z'' = 0$$

définissent évidemment trois isomorphismes

(7.2) 

dont le produit est l'identité et tels que

(7.3) $$[z,z'] = [z',z''] = [z'',z]$$

est la valeur d'une forme quadratique de $z \in \lambda$, d'une forme de $z' \in \lambda'$ et d'une forme de $z'' \in \lambda''$. Ces trois formes sont les transformées l'une de l'autre par les isomorphismes (7.2) ; elles ont donc le même indice d'inertie ; elles sont de rang maximum.

C'est l'indice d'inertie de la forme opposée que nous emploierons.

Définition.- Etant donné le triplet $\lambda,\lambda',\lambda''$ d'éléments de $\Lambda(\ell)$, deux à deux transverses, la condition

$$z \in \lambda\, ,\, z' \in \lambda' \quad ,\quad z - z' \in \lambda''$$

définit un isomorphisme

$$z \mapsto z' \quad ,\quad \lambda \rightarrow \lambda' \ ;$$

$[z,z']$ est donc une forme quadratique de z , dont l'indice d'inertie sera noté $\mathrm{Inert}(\lambda,\lambda',\lambda'')$.

Evidemment

(7.4) $\mathrm{Inert}(\lambda,\lambda',\lambda'') = \mathrm{Inert}(\lambda',\lambda'',\lambda) = \mathrm{Inert}(\lambda'',\lambda,\lambda') = \ell - \mathrm{Inert}(\lambda,\lambda'',\lambda') = \dots$

Inert est une fonction localement constante, à valeurs dans $\{0,\dots,\ell\}$.

Soient $\lambda_q,\lambda'_q,\lambda''_q \in \Lambda_q(\ell)$; supposons-les deux à deux transverses, c'est-à-dire leurs projections naturelles $\lambda,\lambda',\lambda''$ sur $\Lambda(\ell)$ deux à deux transverses ; nous définirons :

$$\mathrm{Inert}(\lambda_q,\lambda'_q,\lambda''_q) = \mathrm{Inert}(\lambda,\lambda',\lambda'') \quad .$$

8. L'INDICE DE MASLOV D'UN COUPLE D'ÉLÉMENTS DE $\Lambda_\infty(\ell)$. On peut construire une fonction, évidemment unique, appelée indice de Maslov et notée m , qui a les trois propriétés suivantes :

- elle est définie sur tout couple d'éléments transverses de $\Lambda_\infty(\ell)$ et est à valeurs entières :

$$m(\lambda'_\infty,\lambda_\infty) \in \mathbf{Z} \ ;$$

- elle est localement constante (en tout point de son domaine de définition) ;
- elle vérifie la relation

(8.1) $$\mathrm{Inert}(\lambda',\lambda'',\lambda) = m(\lambda''_\infty,\lambda_\infty) - m(\lambda'_\infty,\lambda_\infty) + m(\lambda'_\infty,\lambda''_\infty) \quad .$$

Note.- Cette relation (8.1) définit le cobord en topologie algébrique, sous des hypothèses différentes des précédentes.

Note.- Cette relation (8.1) prouve la suivante :

(8.2) $\mathrm{Inert}(\lambda,\lambda'\lambda'') - \mathrm{Inert}(\lambda,\lambda',\lambda''') + \mathrm{Inert}(\lambda,\lambda'',\lambda''') - \mathrm{Inert}(\lambda',\lambda'',\lambda''') = 0$.

Voici les propriétés de cet indice de Maslov : il est invariant par $Sp_\infty(\ell)$, c'est-à-dire :

(8.3) $$m(S_\infty\lambda'_\infty,S_\infty\lambda_\infty) = m(\lambda'_\infty,\lambda_\infty) \quad , \quad \text{où } S_\infty \in Sp_\infty(\ell) \ ;$$

(8.4) $$m(\lambda'_\infty,\lambda_\infty) + m(\lambda_\infty,\lambda'_\infty) = \ell \ ;$$

(8.5) $$m(\beta^{p'}\lambda'_\infty,\beta^p\lambda_\infty) - m(\lambda'_\infty,\lambda_\infty) = p-p' \ ,$$

à condition de choisir convenablement le générateur β de $\pi_1[\Lambda(\ell)]$.

Note.- La relation (8.5) prouve que m est défini mod. q sur les couples λ'_q,λ_q d'éléments de $\Lambda_q(\ell)$, les relations précédentes valant alors mod. q .

Par exemple : m est défini mod. 2 sur $\Lambda_2(\ell)$, qui est l'ensemble des ℓ-sous-espaces lagrangiens de Z orientés (au sens euclidien du terme) ; les orientations de λ_2 et λ'_2 sont compatibles avec la dualité de λ et λ' définie par la fonction bilinéaire de $z \in \lambda$ et $z' \in \lambda'$ valant

$$(-1)^{m(\lambda'_2,\lambda_2)}[z,z'] \ ;$$

$m(X^*_2,\lambda_2) \equiv 0$ mod. 2 signifie que la projection parallèle à X^* projette λ_2 orienté sur X_2 orienté.

9. L'INDICE DE MASLOV de $s_\infty \in Sp_\infty(\ell)$.- L'inertie d'un triplet d'éléments, deux à deux transverses, de $\Lambda(\ell)$ (n° 7) et celle d'un triplet d'éléments de $Sp(\ell) \setminus \Sigma(\ell)$, dont le produit est l'identité,(n° 3) sont liées comme suit :

Soient $s,s',s'' \in Sp(\ell) \setminus \Sigma(\ell)$ (voir (2.3)) tels que $s\,s's'' = E$; on a

(9.1) $$\mathrm{Inert}(s,s',s'') = \mathrm{Inert}(s^{-1}X^*,X^*,s'X^*) = \mathrm{Inert}(s''^{-1}X^*,X^*,sX^*) = \dots$$

Définissons sur $Sp_\infty(\ell)\setminus \Sigma_\infty(\ell)$ une fonction m par la relation :

(9.2) $$m(s_\infty) = m(X^*_\infty,s_\infty X^*_\infty)$$

$(\lambda_\infty \in \Sigma_\infty(\ell)$ signifie $\lambda \in \Sigma(\ell)$; nous choisissons $X^*_\infty \in \Lambda_\infty(\ell)$ de projection X^*)
Cette fonction m a donc les propriétés suivantes, qui la caractérisent :

- elle est à valeurs entières ;
- elle est localement constante ;
- elle vérifie la relation, où $s_\infty s'_\infty s''_\infty = E$ (dans $Sp_\infty(\ell)$)

(9.3) $$\mathrm{Inert}(s,s',s'') = m(s_\infty) - m(s'^{-1}_\infty) + m(s''_\infty) \quad .$$

Elle possède en outre les propriétés que voici

(9.4) $$m(s_\infty) + m(s_\infty^{-1}) = \ell$$

(9.5) $$m(\alpha^q s_\infty) - m(s_\infty) = 2q \quad ,$$

à condition de choisir convenablement le générateur α de $\pi_1[Sp(\ell)]$.

<u>Note</u>.- La relation (9.5) prouve que m est défini mod. 2q sur

$$Sp_q(\ell) \setminus \Sigma_q(\ell) \quad ;$$

$Sp_q(\ell)$ opère sur $\Lambda_{2q}(\ell)$; les relations précédentes valent mod. 2q quand on y remplace Sp_∞, Λ_∞, X^*_∞ par Sp_q , Λ_{2q} , X^*_{2q} .

L'unicité de m prouve que, sur $Sp_2(\ell)$, m mod. 4 est l'indice de Maslov défini mod. 4 par (3.4).

10. UN INDICE D'INERTIE MIXTE est l'indice d'inertie qu'emploiera l'étude des variétés lagrangiennes (§ 4, n° 12).

<u>Définitions</u>.- Soient $s \in Sp(\ell) \setminus \Sigma(\ell)$, λ et $\lambda' \in \Lambda(\ell)$; supposons λ et λ' transverses à X^* et tels que

$$\lambda = s\lambda' \quad ;$$

nous définissons alors

(10.1) $$\mathrm{Inert}(s,\lambda,\lambda') = \mathrm{Inert}(s^{-1}X^*, X^*, \lambda') = \mathrm{Inert}(X^*, sX^*, \lambda)$$

<u>Les propriétés</u> de cet indice d'inertie sont évidentes :

(10.2) $$\mathrm{Inert}(s,\lambda,\lambda') = \ell - \mathrm{Inert}(s^{-1},\lambda',\lambda) \quad ;$$

(10.3) $$\text{Inert}(s,\lambda,\lambda') = m(s_q) - m(X^*_{2q},\lambda_{2q}) + m(X^*_{2q},\lambda'_{2q}) \mod. 2q$$

$$\text{si} \quad \lambda_{2q} = s_q\, \lambda'_{2q} \quad .$$

Note.- C'est le cas $q = 2$ qu'emploie la théorie des solutions asymptotiques.

§ 4. VARIÉTÉS LAGRANGIENNES DANS $Z(\ell)$.

11. DÉFINITION D'UNE VARIÉTÉ LAGRANGIENNE. - Une variété $V(\ell)$ de $Z(\ell)$ est dite lagrangienne quand

(11.1) $$\dim V(\ell) = 1 \quad , \; d\,p \wedge dx = 0 \quad \text{sur} \quad V(\ell) \quad ,$$

en notant $dp \wedge dx = d < p,dx >$.

Phase.- Puisque la forme $< p,dx >$ est régulière et fermée sur $V(\ell)$, l'équation

(11.2) $$d\varphi = < p,dx > = \frac{1}{2}\,[z,dz] + \frac{1}{2}\, d < p,x >$$

définit, à une constante additive près, sur le revêtement universel $\check{V}(\ell)$ une fonction

(11.3) $$\varphi = \check{V}(\ell) \rightarrow \mathbb{R} \quad ;$$

on dit que φ est la phase associée à $V(\ell)$.

12. GROUPE SYMPLECTIQUE ET VARIÉTÉS LAGRANGIENNES. - Tout élément s de $Sp(\ell)$ transforme évidemment une variété lagrangienne $V'(\ell)$ en une variété lagrangienne $V(\ell) = sV'(\ell)$; si on choisit de façon cohérente les constantes additives de leurs phases φ et φ' , alors

(12.1) $$\varphi \circ s - \varphi'$$

est la restriction à $V'(\ell)$ de la forme quadratique valant en (x',p') :

$$\frac{1}{2} < p,x > - \frac{1}{2} < p',x' > \quad , \quad \text{où} \quad (x,p) = s(x',p') \quad .$$

Supposons les ℓ-plans tangents à $V(\ell)$ et à $V'(\ell)$ transverses à X^* : on peut prendre pour coordonnée locale x sur $V(\ell)$ et x' sur $V'(\ell)$; les conditions

$$(x,p) \in V(\ell) \quad , \quad (x',p') \in V'(\ell)$$

s'énoncent respectivement

(12.2) $$p = \varphi_x \quad , \quad p' = \varphi'_{x'} \quad (\varphi_x = \frac{\partial\varphi(x)}{\partial x})$$

L'automorphisme s de $Z(\ell)$ a pour restriction à $V'(\ell)$ une application

(12.3) $$s : V'(\ell) \rightarrow V(\ell) \quad ,$$

que nous noterons, en coordonnées locales,

$$x' \rightarrow x(x') \quad .$$

Supposons $s \notin \Sigma(\ell)$, c'est-à-dire s du type s_A (n° 1) ; alors (12.2) s'explicite comme suit, vu (1.10) :

(12.4) $$\varphi_x = A_x(x,x') \quad , \quad \varphi'_{x'} = - A_{x'}(x,x') \quad ; \quad \text{où} \quad x = x(x') \quad ;$$

puisque $\text{dét}(A_{xx'}) = \text{dét}\, L \neq 0$, chacune des deux solutions (12.2) définit l'application $x' \mapsto x(x')$.

Bien entendu, (12.3) donne, conformément à (12.1) :

$$\varphi(x) - \varphi'(x') = A(x,x')$$

c'est-à-dire

(12.5) $$\varphi(x) - \frac{1}{2} < p,x > = \varphi'(x') - \frac{1}{2} < p',x' > \quad , \quad \text{vu (12.2)} \quad .$$

Les deux définitions équivalentes (12.3) de l'application $x \mapsto x(x')$ permettent de calculer deux expressions équivalentes que voici de son déterminant fonctionnel :

(12.6) $$\frac{d^\ell x}{d^\ell x'} = \frac{\text{Hess}_{x'} \, [\varphi'(x') + A(o,x')]}{\Delta^2(A)} = \frac{\Delta^2(A)}{\text{Hess}_x \, [\varphi(x) + A(x,o)]}$$

Ce calcul montre en outre ceci : <u>les deux hessiens figurant</u> dans (12.6) <u>ont le même indice d'inertie</u>.

(12.7) $$\text{Inert Hess}_{x'} \, [\varphi'(x') + A(o,x')] = \text{Inert Hess}_x \, [\varphi(x) + A(x,o)]$$

$$= \text{Inert}(s,\lambda'(x'),\lambda(x)) \quad ,$$

où $\lambda'(x')$ est la direction du ℓ-plan tangent à $V'(\ell)$ en x' ,

$\lambda(x)$ est la direction du ℓ-plan tangent à $V(\ell)$ en x ;

(12.8) $$s\lambda'(x') = \lambda(x) \quad \text{quand} \quad x = x(x') \ .$$

13. UNE q-ORIENTATION DE $V(\ell)$ est une application continue

$$V_q(\ell) \rightarrow \Lambda_q(\ell)$$

dont la composée avec la projection naturelle

$$\Lambda_q(\ell) \rightarrow \Lambda(\ell) \ .$$

est l'application $V(\ell) \rightarrow \Lambda(\ell)$ appliquant chaque point $z \in V(\ell)$ sur la direction $\lambda \in \Lambda(\ell)$ de son ℓ-plan tangent.

Soit $s_q \in Sp_q(\ell)$, de projection naturelle $s \in Sp(\ell)$. Puisque s_q opère sur $\Lambda_{2q}(\ell)$, s_q applique une 2q-orientation de $V'(\ell)$ sur une 2q-orientation de $V(\ell) = sV'(\ell)$.

Reprenons les formules (12.6) et (12.7) ; supposons $s \notin \Sigma(\ell)$, donc $s = s_A$; définissons, si Q est une forme quadratique de rang maximum,

(13.1) $$\arg \text{Hess}(Q) = \pi \ \text{Inert}(Q) \ .$$

<u>L'argument du déterminant fonctionnel</u> $d^\ell x/d^\ell x'$ <u>peut être défini comme suit mod.</u>$2q\pi$, compte-tenu de (3.4) :

(13.2) $$\arg \frac{d^\ell x}{d^\ell x'} \equiv \pi\Big[\text{Inert. Hess}_{x'}[\varphi'(x') + A(o,x')] - m(s_q)\Big] \ \text{mod. } 2q\pi$$

$$\equiv \pi[\text{Inert}(s,\lambda'(x'),\lambda(x)) - m(s_q)] \ \text{mod. } 2q\pi$$

$$\equiv \pi[m(X^*_{2q},\lambda'_{2q}(x')) - m(X^*_{2q},\lambda_{2q}(x)] \ \text{mod. } 2q\pi \ ;$$

cette dernière expression emploie (10.3), suppose $x = x(x')$ et note $\lambda'_{2q}(x')$ l'image dans $\Lambda_{2q}(\ell)$ du point d'abcisse x' de $V'(\ell)$ 2q-orientée ;

(13.3) $$\lambda_{2q}(x) = s_q \ \lambda'_{2q}(x')$$

est l'image dans $\Lambda_{2q}(\ell)$ du point d'abcisse x de $V(\ell)$ 2q-orientée.

<u>Note</u>.- Le § 6 emploiera pour $q = 2$ ce résultat, qui définit alors une déter-

mination de

$$\sqrt{\frac{d^\ell x}{d^\ell x'}} .$$

§ 5. LES ESPACES q-SYMPLECTIQUES.

14. L'ESPACE Z_q ET SES REPÈRES. - Soit Z l'espace vectoriel symplectique de dimension 2ℓ , c'est-à-dire $\mathbb{R}^{2\ell}$ muni d'une forme bilinéaire alternée $[.,.]$ de rang maximum ; notons Z_q la donnée de Z et de $q \in \{1,2,\dots,\infty\}$; (nous ne définirons et n'utiliserons dans Z_q que des 2q-orientations).

Notons $\Lambda(Z)$ la grassmannienne lagrangienne de Z , c'est-à-dire l'ensemble de ses ℓ-sous-espaces lagrangiens ; soit $\Lambda_{2q}(Z)$ son revêtement connexe à 2q feuillets ; q est un entier ≥ 1 donné.

Notons $X = \mathbb{R}^\ell$, X^* son dual ; soit $Z(\ell)$ l'espace symplectique défini par $X \oplus X^*$ et la forme valant

$$[z,z'] = \langle p,x' \rangle - \langle p',x \rangle$$

pour $z = x+p$, $z' = x'+p'$, x et $x' \in X$, p et $p' \in X^*$.

Soit $\Lambda(\ell)$ la grassmannienne lagrangienne de $Z(\ell)$. $Sp(\ell)$ est un groupe d'automorphismes de $Z(\ell)$; il induit un groupe d'homéomorphismes de $\Lambda(\ell)$; $Sp_q(\ell)$ induit un groupe d'homéomorphismes de $\Lambda_{2q}(\ell)$.

Un q-<u>repère</u> R de Z_q est constitué par :

- un isomorphisme $j_R : Z \to Z(\ell)$, compatible avec la structure symplectique ;
- un homéomorphisme $h_R : \Lambda_{2q}(Z) \to \Lambda_{2q}(\ell)$ ayant pour projection naturelle l'homéomorphisme $\Lambda(Z) \to \Lambda(\ell)$ induit par j_R .

Soient deux repères de Z_q ;

$$R = j_R \times h_R \quad ; \quad R' = j_{R'} \times h_{R'} .$$

Evidemment :

$$j_R\, j_{R'}^{-1} \in Sp(\ell) \quad ;$$

$$h_R\, h_{R'}^{-1} \; : \; \Lambda_{2q}(\ell) \to \Lambda_{2q}(\ell)$$

a pour projection l'homéomorphisme $\Lambda(\ell) \longrightarrow \Lambda(\ell)$ induit par $j_R\, j_{R'}^{-1}$. Il est évident que l'homéomorphisme $h_R\, h_{R'}^{-1}$ est induit par un élément $s_R^{R'}$ de $Sp_q(\ell)$; cet élément est unique ; nous pouvons donc l'identifier à

$$R\, R'^{-1} = j_R\, j_{R'}^{-1} \times h_R\, h_{R'}^{-1} \quad ,$$

donc écrire :

(14.1) $$R = s_R^{R'}\, R' \quad , \quad \text{où} \quad s_R^{R'} \in Sp_q(\ell) \quad .$$

$s_R^{R'}$ est défini par la donnée de R et R' ; c'est le changement de repères ; évidemment :

$$s_R^{R'}\, s_{R'}^{R''}\, s_{R''}^{R} = E \quad ; \quad s_R^{R'} = E \ \text{ si et seulement si } \ R = R' \quad .$$

Nous écrirons désormais R pour j_R , h_R ou $j_R \times h_R$.

Un automorphisme s_q de Z_q est constitué par un automorphisme s de Z et un homéomorphisme de $\Lambda_{2q}(Z)$, dont la projection sur $\Lambda(Z)$ soit induit par s . Son image dans R et dans R' est $Rs_q\, R^{-1} \in Sp_q(\ell)$, $R'\, s_q\, R'^{-1} \in Sp_q(\ell)$, liés par la relation

$$R\, s_q\, R^{-1} = s_R^{R'}\, R'\, s_q\, R'^{-1}\, s_{R'}^{R} \quad .$$

Le groupe $Sp_q(Z)$ des automorphismes de Z_q est donc isomorphe à $Sp_q(\ell)$; chaque repère R' définit un isomorphisme $R' : Sp_q(Z) \longrightarrow Sp_q(\ell)$; $s_R^{R'}$ le transforme en $R : Sp_q(Z) \longrightarrow Sp_q(\ell)$.

La notion d'inertie et celle d'indice de Maslov mod. q ont évidemment, sur Z_q , un sens invariant par $Sp_q(Z)$, puisque dans chaque repère elles ont un sens indépendant du choix de ce repère.

15. VARIÉTÉ LAGRANGIENNE de Z_q .- Dans l'espace q-symplectique Z_q , les notions suivantes on évidemment un sens : variété lagrangienne V ; 2q-orientation de V .

Tout q-repère R définit, à une constante additive près, sur le revêtement universel $\check{V}$ de V , une <u>phase</u> $\varphi_R : \check{V} \to \mathbb{R}$ par la relation

$$(15.1) \qquad d\varphi_R = < p, dx > \quad , \quad \text{où} \quad Rz = x+p \quad , \quad x \in X \; , \quad p \in X^* \; .$$

Vu (12.5)

$$(15.2) \qquad \varphi(z) = \varphi_R(z) - \frac{1}{2} < p,x >$$

est la valeur d'une fonction $\varphi : \check{V} \to \mathbb{R}$, indépendante de R , définie par

$$(15.3) \qquad d\varphi = \frac{1}{2}\,[z,dz] \quad .$$

Evidemment, x est une coordonnée locale de V au voisinage de tout point z de V tel que $\lambda(z)$ soit transverse à $R^{-1} X^*$; nous noterons $V \setminus \Sigma_R$ l'ensemble de ces points z ; Σ_R ne dépend que de $R^{-1} X^*$; Σ_R est <u>le contour apparent</u> de V relativement à R ou, plus précisément, à $R^{-1} X^*$. Soient x et x' les coordonnées locales définies au voisinage de $z \in V \setminus \Sigma_R \cup \Sigma_{R'}$ par R et R' ; soit $d^{\ell}x \,/\, d^{\ell}x'$ le déterminant fonctionnel de la bijection $x' \longmapsto x(x')$; la formule (12.6) donne une expression de $d^{\ell}x \,/\, d^{\ell}x'$; la formule (13.2) définit son argument mod. $2q\pi$.

Voici l'une des expressions de cet argument : soit $\lambda_{2q}(z)$ la direction en z du plan tangent à la variété V munie d'une $2q$-orientation ; soit $X^*_{2q} \subset \Lambda_{2q}(\ell)$, de projection X^* sur $\Lambda(\ell)$; notons

$$(15.4) \qquad m_R(z) = m(R^{-1} X^*_{2q} \, , \, \lambda_{2q}) \quad ;$$

alors

$$(15.5) \qquad \arg \frac{d^{\ell}x}{d^{\ell}x'} \equiv \pi\,[m_{R'}(z) - m_R(z)] \mod. 2q\pi \quad .$$

La valeur de m_R dépend du choix de la $2q$-orientation de V et du choix de X^*_{2q} ; mais la valeur de $m_{R'} - m_R$ en est indépendante.

<u>Note</u>.- La formule (15.5) est compatible avec la définition suivante :

$$(15.6) \qquad \arg.\, d^{\ell}x \equiv -\pi\, m_R(z) \mod. 2q\pi \quad .$$

§ 6. SOLUTIONS LAGRANGIENNES ET ASYMPTOTIQUES.

Nous supposerons désormais $q = 2$. Rappelons (§ 1) que $Sp_2(\ell)$ est un groupe d'automorphismes S de $\mathcal{S}(X)$, $\mathcal{H}(X)$, $\mathcal{S}'(X)$, unitaires sur $\mathcal{H}(X)$.

La projection naturelle de S sur $Sp(\ell)$ est notée s .

16. FONCTIONS LAGRANGIENNES. - Donnons-nous dans l'espace symplectique Z_2 une variété lagrangienne lisse V de phase φ :

$$d\varphi = \frac{1}{2}[z,dz] .$$

Soit R' un 2-repère de Z_2 ; définissons

$$\varphi_{R'} : \check{V} \to \mathbb{R} \quad \text{par} \quad \varphi_{R'}(z) = \varphi(z) + \frac{1}{2} < p',x' > ,$$

où $z \in V$, $R'z = x'+p'$, $x' \in X$, $p' \in X^*$; donc $d\varphi_{R'}(z) = < p',dx' >$.

Notons $U_{R'}$ une fonction de $z \in V \setminus \Sigma_{R'}$, fonction formelle de $\nu \in i\dot{\mathbb{R}}$, du type :

$$(16.1) \qquad U_{R'}(\nu,z) = \alpha_{R'}(\nu,z)\, e^{\nu\varphi_{R'}(z)}$$

où $\alpha_{R'}$ est la série formelle, à coefficients indéfiniment différentiables :

$$\alpha_{R'}(\nu,z) = \sum_j \nu^{-j}\, \alpha_{jR'}(z) .$$

Soit R un autre 2-repère de Z_2 ; le changement de repère est

$$S_R^{R'} \in Sp_2(\ell) .$$

Soit f' une fonction de (ν,x') admettant, pour ν tendant vers $i\infty$, le développement asymptotique

$$(16.2) \qquad u'_{R'}(\nu,x') = \sum_{\{z | R'z \in x'+X^*\}} U_{R'}(\nu,z) ;$$

$S_R^{R'} f'$ est une fonction f de (ν,x) ; la méthode de la phase stationnaire montre que f admet un développement asymptotique

$$u_R(\nu,x) = \sum_{\{z | Rz \in x+X^*\}} U_R(\nu,z)$$

où U_R est défini comme $U_{R'}$ l'est par (16.1) et est unique ; nous écrirons

$$(16.3) \qquad u_R = S_R^{R'} u_{R'} \quad , \quad U_R = S_R^{R'} U_{R'} \quad ;$$

$S_R^{R'}$ <u>opère localement sur</u> $u_{R'}$ <u>et</u> $U_{R'}$, <u>en conservant le support de</u> $U_{R'}$:

$$\mathrm{Supp}\, U_{R'} = \bigcup_j \mathrm{Supp}\, \alpha_{jR'} \subset V \ .$$

Nous dirons que U_R <u>et</u> u_R <u>sont des fonctions</u> ν-<u>formelles</u> définies respectivement sur V et X ; u_R sera nommée : <u>projection</u> de U_R .

La donnée <u>sur</u> $V \setminus \Sigma_R$ <u>pour chaque</u> 2-<u>repère</u> R <u>d'une fonction</u> ν-<u>formelle</u> U_R <u>telle que</u>

$$(16.4) \qquad U_R = S_R^{R'}\, U_{R'}$$

constituera <u>une fonction lagrangienne</u> $U = \{U_R\}$ définie sur V ; U_R sera son <u>expression</u> dans le repère R et u_R sa projection sur X dans ce repère.

L'allure au voisinage de Σ_R de l'expression U_R de U peut être précisée : au voisinage d'un point z de Σ_R n'appartenant pas à $\Sigma_{R'}$ on calcule

$$U_R = S_R^{R'}\, U_{R'}$$

au moyen de $U_{R'}$ par la méthode de la phase stationnaire ; elle introduit l'indice d'inertie d'un hessien ; cet indice s'identifie à

$$\mathrm{Inert}\, (S_R^{R'} \ , \ R\lambda_4 \ , \ R'\lambda_4) \ ,$$

λ_4 étant le plan tangent à $\check{V}$ en z ; plus précisément, vu (3.4), cette méthode introduit

$$\mathrm{Inert}\, (S_R^{R'} \ , \ R\lambda_4 \ , \ R\lambda_4) - m(S_R^{R'}) \ \mathrm{mod.}\ 4$$

c'est-à-dire, vu (10.3) et la définition (15.5), où $q = 2$,

$$\arg \sqrt{\frac{d^\ell x}{d^\ell x'}} \ \mathrm{mod.}\ 2\pi \ .$$

On obtient ainsi <u>la structure des expressions</u> U_R <u>des fonctions lagrangiennes</u> U :

(16.5) $$U_R(\nu,z) = \sum_{j=0}^{\infty} \left(\frac{1}{\nu}\frac{\eta}{d^{\ell}x}\right)^{j+\frac{1}{2}} \beta_{Rj}(z)\, e^{\nu\varphi_R(z)} \quad ,$$

où : η est une <u>mesure régulière</u> > 0 <u>sur</u> V ;

$\sqrt{d^{\ell}x}$ est une demi-mesure, définie sur $\check{V}$ par (15.6), où $q = 2$;

les β_{Rj} sont des fonctions $\check{V} \to \mathbf{C}$, <u>indéfiniment différentiables</u> ;

β_{R0} <u>est indépendante</u> de j et est notée β_0 .

Puisque $U_R(\nu,z)$ est une fonction ν-formelle sur $V \setminus \Sigma_R$, chacun des termes de (16.5) doit être une fonction définie (donc uniforme) sur $V \setminus \Sigma_R$; autrement dit :

(16.6) $$\beta_{Rj}\, e^{\nu\varphi + (j+\frac{1}{2})\pi i\, m_R} \quad \text{est uniforme sur } V \setminus \Sigma_R \ ;$$

<u>si</u> V <u>est orientable</u> (au sens enclidien) (16.6) s'énonce :

(16.7) $$\beta_{Rj}\, e^{\nu\varphi + \frac{\pi}{2} i\, m_R} \quad \text{est uniforme sur } V \setminus \Sigma_R \ .$$

17. OPÉRATEURS PSEUDO-DIFFÉRENTIELS. - Soit a^o une fonction ν-formelle , définie sur Z et de phase nulle :

(17.1) $$a^o(\nu,z) = \sum_j \nu^{-j} a^o{}_j(z) \quad \text{(série formelle)} .$$

Soit R un repère de Z ; notons $a^o{}_R$ la fonction ν-formelle définie sur $Z(\ell)$ par

(17.2) $$a^o{}_R(\nu,x,p) = a^o(\nu,R^{-1}(x+p)) \quad ;$$

donc

(17.3) $$a^o{}_R = a^o{}_{R'} \circ s^R_{R'} \quad ;$$

soit

$$a_R^+(\nu,x,p) = e^{\frac{1}{2\nu} < \frac{\partial}{\partial x}, \frac{\partial}{\partial p} >} a^o(\nu,x,p) \quad ;$$

si a^o est un polynome en (ν^{-1},x,p) , alors $a_R = a_R^+ (\nu,x,\frac{1}{\nu}\frac{\partial}{\partial x})$ est un opérateur différentiel à coefficients polynomiaux ; l'application (§2, n° 4)

$$a^o \mapsto a_R = a_R^+(\nu,x,\frac{1}{\nu}\frac{\partial}{\partial x})$$

se prolonge par complétion en une application de l'ensemble des a^o sur un ensemble d'opérateurs $a_R = a_R^+ (\nu,x,\frac{1}{\nu}\frac{\partial}{\partial x})$; a_R est un endomorphisme de l'ensemble des u_R et de l'ensemble des U_R ; a_R opère localement :

$$\text{Supp } u_R \subset \text{Supp } a_R\, u_R \quad ; \quad \text{Supp } U_R \subset \text{Supp } a_R\, U_R \quad .$$

a_R est le transformé de $a_{R'}$ par $S_R^{R'}$:

(17.4) $$a_R = S_R^{R'}\, a_{R'}\, S_{R'}^{R} \quad ;$$

donc $$a_R\, U_R = S_R^{R'}(a_{R'}\, U_{R'}) \quad \text{si} \quad U_R = S_R^{R'}\, U_{R'} \quad .$$

Etant donnée une fonction lagrangienne, $U = \{U_R\}$, il existe donc une fonction lagrangienne $aU = \{a_R\, U_R\}$.

L'opérateur

(17.5) $$a = \{a_R\} : U \mapsto a\, U$$

est nommé <u>opérateur pseudo-différentiel</u> de Z ; a_R est son expression dans le repère R .

$a\, U$ ne dépend que de U , qui est défini sur V , et du germe de a^o sur V , c'est-à-dire des valeurs sur V de a^o et de toutes ses dérivées.

Nous nommerons <u>solution lagrangienne</u> de l'équation pseudo-différentielle

$$a\, u = 0$$

toute fonction lagrangienne U vérifiant cette équation ; en général cette solution n'existera que pour certaines valeurs particulières de ν .

<u>Note</u>.- Un cas important, où a est <u>self-adjoint</u>, est le suivant :

a^o est indépendant de ν , est à valeurs réelles et

(17.6) $$< \frac{\partial}{\partial x} , \frac{\partial}{\partial p} > a^o(x,p) = 0 \quad , \quad (\text{donc } a^+ = a^o) \quad .$$

18. SOLUTIONS ASYMPTOTIQUES. - Soit, sur X , un opérateur différentiel $a_R(\nu,x,\frac{1}{\nu} , \frac{\partial}{\partial x})$; il est évidemment l'expression dans R d'un opérateur <u>pseudo-différentiel de</u> Z <u>unique</u> : a .

Soit $u_R(\nu,x)$ une solution asymptotique de l'équation

(18.1) $$a_R(\nu,s,\frac{1}{\nu} \frac{\partial}{\partial x}) \; u_R(\nu,x) = 0 \quad ;$$

c'est, par définition, une fonction ν-formelle sur X vérifiant (18.1). Un calcul classique donne la phase φ_R de u_R par résolution d'une équation aux dérivées partielles du premier ordre [c'est-à-dire par construction d'une variété lagrangienne V de $Z(\ell)$ appartenant à une hypersurface donnée de $Z(\ell)$] et l'amplitude α_R de u_R par intégrations le long des caractéristiques de cette équation : u_R est la projection d'une solution ν-formelle U_R sur $V \setminus \Sigma_R$ de l'équation

(18.2) $$a_R \, U_R = 0 \quad .$$

La théorie précédente montre que localement, de chaque côté de Σ_R , U_R a la structure (16.5), β_{Rj} pouvant donc faire un saut à la traversée de Σ_R ; en général U_R est <u>indéterminé</u>.

On lève cette indétermination <u>en imposant à</u> U_R <u>d'avoir la structure</u> (16.5) qui implique (16.6) ou (16.7) ; c'est imposer à U_R d'être <u>l'expression dans</u> R <u>d'une fonction lagrangienne sur</u> V , U , <u>qui est évidemment solution lagrangienne de l'équation</u>

(18.3) $$a \, U = 0 \quad .$$

On peut dire que c'est imposer à U_R de vérifier, en un certain sens, (18.1) même sur Σ_R .

C'est la condition que Maslov impose aux solutions asymptotiques(sans la justifier, puisqu'il n'emploie pas la notion d'opérateur pseudo-différentiel).

Note.- Dans le cas particulier ou a^o vérifie (17.6), β_o est constant et cette condition s'énonce

(18.4) $$e^{\nu\varphi + \frac{\pi}{2} i\, m_R} \text{ et les } \beta_{Rj} \text{ sont uniformes sur } V .$$

19. LES APPLICATIONS DE CETTE THÉORIE semblent limitées à des équations très particulières. Voir [7].

L'une d'elles est l'équation relativiste stationnaire de Schrödinger, avec champ magnétique non nul ; cette équation vérifie (17.6). Elle dépend d'un paramètre : "l'énergie" ; l'ensemble des valeurs de l'énergie pour lesquelles elle possède une solution, d'ailleurs unique, est "le spectre". Ce spectre se trouve être rigoureusement le même, qu'on impose aux solutions d'être des fonctions de carré sommable ou d'être des solutions asymptotiques, c'est-à-dire des fonctions ν-formelles (ici, $\nu = \frac{i}{\hbar}$ où $2\pi\hbar$ = constante de Planck).

C'est également vrai de l'équation de Dirac.

Le spectre est repéré par des entiers : les nombres quantiques ; c'est seulement quand ces nombres sont grands que la solution fonction de carré sommable est approchée par la solution asymptotique. Celle-ci est toujours définie en première approximation par une trajectoire et une densité d'électrons relativistes. La notion de solution asymptotique donne donc une formalisation de la première théorie des quanta qui diffère de la mécanique ondulatoire, qui emploie cependant les équations de Schrödinger et de Dirac sans altérer leurs spectres.

BIBLIOGRAPHIE

[1] LERAY, J. Solutions asymtotiques des équations aux dérivées partielles ; (une adaptation du traité de V.P. Maslov). Convegno internaziole Metodi valutativi nelle fisicamatematica ; Accad. Naz. dei Lincei, Roma, 1972 (sous presse).

[2] LERAY, J. Complément à la théorie d'Arnold de l'indice de Maslov. Convegno di Geometrica simplettica e Fisica matematica, Istituto di Alta Matematica, Roma, 1973 (sous presse).

[3] MASLOV, V.P. Théorie des perturbations et méthodes asymptotiques (M.G.U., Moscou, 1965).

ARNOLD, V.I. Une classe caractéristique intervenant dans les conditions de quantification, Analyse fonctionnelle (en russe), 1 (1967) 1-14.

BOUSLAEV, V.C. Intégrale génératrice et opérateur canonique de Maslov par la méthode W.K.B.

Traduits par LASCOUX, J. et SENEOR, R. (Dunod 1972)

[4] SEGAL, I.E. Foundations of the theory of dynamical systems of infinitely many degrees of freedom (I). Mat-Fys. Medd. Danske Vid. Selsk. 31, n° 12 (1959) 1-39.

[5] SHALE, D. Linear symmetrics of free boson fields, Trans. Amer. Math. Soc. 103 (1962), 149-167.

[6] WEIL, A. Sur certains groupes d'opérateurs unitaires, Acta math. 111 (1964) 143-211.

En préparation

[7] LERAY, J. Exposé au Colloque d'Aix en Provence, Géométrie symplectique et physique mathématique (Juin 1974).

LE POLYNOME DE BERNSTEIN D'UNE SINGULARITE ISOLEE

B. Malgrange

1. INTRODUCTION

Dans [7], j'avais essayé d'examiner une question posée par différents auteurs (notamment I.N. Bernstein et M. Sato), à savoir la rationalité lité des zéros des polynômes que I.N. Bernstein et J.E. Björk attachent à une singularité. Le résultat de [7], quoique très partiel, semble indiquer une étroite relation entre ces zéros et la monodromie de ladite singularité.

Dans cet article, je reprends la question pour une singularité isolée, cas dans lequel on obtient une réponse complète. L'idée essentielle est travailler avec la "cohomologie de de Rham" de certains modules à gauche sur l'anneau des opérateurs différentiels, modules qui s'introduisent spontanément dans l'étude des "polynômes de Bernstein", et de montrer que cette cohomologie est, à très peu près, la "cohomologie de de Rham relative" des spécialistes de la monodromie. Ceci, joint à un résultat de Kashiwara sur les $\mathcal{D}$-modules de support d'origine permet de traiter le cas des singularités isolées. Ceci devrait peut-être aussi permettre d'étudier le cas des singularités quelconques, si l'on avait à sa disposition des variantes locales (analytique et formelle) du "théorème de régularité" de la connexion de Gauss-Manin, dû à Griffiths et Nilsson.

L'auteur souhaite attirer l'attention du public sur les remarques (3.6) et (4.4), concluant à l'intérêt qu'il y aurait à avoir une bonne extension à n variables de la théorie des "points singuliers réguliers" de Fuchs, dans le cadre des $\mathcal{D}$-modules. Cette idée rejoint une idée de F. Pham (non publiée) sur les "hyperfonctions de classe de Nilsson". Pour ce faire, la théorie des "connexions régulières" au sens de Deligne [4] est un ingrédient nécessaire, mais elle n'est pas suffisante, du moins en l'état actuel des choses.

Comme on le constatera, une partie des idées de cet article sont dûes à M. Kashiwara ; je le remercie très vivement de me les avoir communiquées. Je signale d'autre part qu'il avait obtenu antérieurement le théorème (5.4) pour une singularité isolée quasi-homogène, cas beaucoup plus simple que le cas général. A ma connaissance, ce résultat non plus n'est pas publié.

2. PRELIMINAIRES SUR LES $\mathcal{D}$-MODULES

Notations. $\mathcal{O} = \mathbb{C}\{x_1,\ldots,x_n\}$, l'anneau des séries convergentes de n variables ;

$\hat{\mathcal{O}} = \mathbb{C}[[x_1,\ldots,x_n]]$, l'anneau des séries formelles de n variables ;

$\mathcal{D}$ (resp. $\hat{\mathcal{D}}$), l'anneau des opérateurs différentiels linéaires en $\frac{\partial}{\partial x_i}$ $(1 \leq i \leq n)$ à coefficients dans $\mathcal{O}$ (resp. $\hat{\mathcal{O}}$). Si l'on filtre $\mathcal{D}$ par le degré en $\frac{\partial}{\partial x}$, on a $gr(\mathcal{D}) = \mathcal{O}[\frac{\partial}{\partial x_1},\ldots,\frac{\partial}{\partial x_n}]$; il en résulte immédiatement que $\mathcal{D}$ est noethérien à gauche et à droite ; de même pour $\hat{\mathcal{D}}$.

Par définition de $\mathcal{D}$, $\mathcal{O}$ est un $\mathcal{D}$-module à gauche ; pour $P \in \mathcal{D}$, $f \in \mathcal{O}$, on note $P(f)$ ou Pf l'action de $\mathcal{D}$ sur $\mathcal{O}$; d'autre part, $\mathcal{O}$ s'identifie à l'espace $\mathcal{D}^{\circ}$ des opérateurs différentiels de degré 0 ; nous noterons alors $P \circ f$ (ou quelquefois Pf si aucune confusion n'est possible) l'opérateur différentiel composé, i.e. $g \mapsto P(fg)$, $g \in \mathcal{O}$.

Soit Ω^n l'espace des n-formes différentielles à coefficients dans $\mathcal{O}$; on munit naturellement Ω^n d'une structure de $\mathcal{D}$-module à droite de la manière suivante :

1) Pour $\omega \in \Omega^n$, $f \in \mathcal{O}$, on pose $\omega f = f\omega$

2) Pour $\omega \in \Omega^n$, $\xi \in \mathcal{D}_0^1$, espace des opérateurs différentiels de degré 1 sans terme constant (= espace des champs de vecteurs), on pose $\omega\xi = -\theta(\xi)\omega$, θ désignant la dérivée de Lie. Nous laissons le soin au lecteur de vérifier que ces opérations se prolongent de manière unique en une structure de $\mathcal{D}$-module à droite sur Ω^n.

Une autre manière de définir cette structure est la suivante : posant $dx = dx_1 \wedge \ldots \wedge dx_n$, toute forme $\omega \in \Omega^n$ s'écrit de manière unique gdx , $g \in \mathcal{O}$; désignant par P^* l'adjoint, au sens usuel de la théorie des équations aux dérivées partielles de P , on a : $\omega P = (P^* g)dx$.

[Les vérifications sont laissées au lecteur ; rappelons que l'adjoint est défini par les propriétés suivantes :
1) si $f \in \mathcal{O}$, $f^* = f$; 2) $(\frac{\partial}{\partial x_i})^* = -\frac{\partial}{\partial x_i}$; 3) $(PQ)^* = QP$].

D'une façon générale, si M est un $\mathcal{D}$-module à gauche, $M \underset{\mathcal{O}}{\otimes} \Omega^n$ est muni naturellement d'une structure de $\mathcal{D}$-module à droite (cf. Bernstein [1]), qui peut être définie ainsi :

1) $(m \otimes \omega)f = m \otimes \omega f$ pour $f \in \mathcal{O}$.

2) $(m \otimes \omega)\xi = -\xi m \otimes \omega + m \otimes \omega \xi$, pour $\xi \in \mathcal{D}_0^1$.

On obtient ainsi une bijection entre $\mathcal{D}$-modules à gauche et $\mathcal{D}$-modules à droite : par exemple, soit $\mathcal{E}'^n$ le dual sur $\mathbb{C}$ de $\hat{\mathcal{O}}$ pour la topologie définie par la filtration de $\hat{\mathcal{O}}$, c'est-à-dire l'espace des courants de degré n au sens de de Rham à support l'origine ; $\mathcal{E}'^n$ est naturellement un $\mathcal{D}$-module à droite par la formule $\langle SP, f\rangle = \langle S, Pf\rangle$, $S \in \mathcal{E}'^n$, $P \in \mathcal{D}$, $f \in \hat{\mathcal{O}}$. Le $\mathcal{D}$-module à gauche qui lui correspond sera noté $\mathcal{E}'$ (c'est aussi le dual de $\hat{\Omega}^n$, et l'espace des courants de degré 0 à support l'origine). Un générateur de $\mathcal{E}'$ sur $\mathcal{D}$ est, par exemple, l'élément δ qui est défini par la formule suivante $\langle \delta \otimes dx , f\rangle = f(0)$, ou, si l'on préfère : si $\omega = fdx$, $\langle \delta, \omega\rangle = f(0)$.

Cohomologie de de Rham.

Soit Ω^p l'espace des p-formes différentielles à coefficients dans $\mathcal{O}$, et soit M un $\mathcal{D}$-module à gauche. On considère le complexe :

$$DR(M) \qquad 0 \to M \xrightarrow{d} \Omega^1 \underset{\mathcal{O}}{\otimes} M \xrightarrow{d} \ldots \xrightarrow{d} \Omega^n \underset{\mathcal{O}}{\otimes} M \to 0$$

dont la différentielle est définie ainsi :

$$d(dx_{i_1} \wedge \ldots \wedge dx_{i_p} \otimes m) = \sum_j dx_j \wedge dx_{i_1} \wedge \ldots \wedge dx_{i_p} \otimes \frac{\partial}{\partial x_j} m \quad .$$

On vérifie comme d'habitude que $d^2 = 0$. Les groupes de cohomologie de DR(M) seront notés $H^p(M)$.

Proposition (2.1) :

Il existe des isomorphismes $H^p(M) \simeq \mathrm{Ext}^p_{\mathcal{D}}(\mathcal{O}, M) \simeq \mathrm{Tor}^{\mathcal{D}}_{n-p}(\Omega^n, M)$.

Pour établir le premier isomorphisme, on part de l'isomorphisme évident $\mathcal{O} = \mathcal{D}/\Sigma \mathcal{D}\frac{\partial}{\partial x_i}$; on considère alors le complexe de Koszul "gauche" $K_g(\frac{\partial}{\partial x_i}, \mathcal{D})$, c'est-à-dire qu'on fait la construction suivante ; on considère $C = \mathbb{C}[\frac{\partial}{\partial x_1}, \ldots, \frac{\partial}{\partial x_n}]$, qui peut être plongé dans $\mathcal{D}$ comme l'anneau des opérateurs différentiels à coefficients constants. Comme tout élément de $\mathcal{D}$ s'écrit d'une manière et d'une seule $\Sigma f_\alpha D^\alpha$, $f_\alpha \in \mathcal{O}$, $D^\alpha = (\frac{\partial}{\partial x_1})^{d_1} \ldots (\frac{\partial}{\partial x_n})^{\alpha_n}$ et que les D^α forment une base de C , l'application $j : \mathcal{O} \otimes_{\mathbb{C}} C \to \mathcal{D}$ définie par $f \otimes P \mapsto fP$ est un isomorphisme de C modules à droite ; $\mathcal{D}$ est donc libre en tant que C module à droite ; de même on voit que $\mathcal{D}$ est un C-module à gauche libre. Le complexe de Koszul "gauche" $K_g(\frac{\partial}{\partial x_i}, \mathcal{D})$ est par définition l'image par j de $\mathcal{O} \otimes_{\mathbb{C}} K(\frac{\partial}{\partial x_i}, C)$; comme $K(\frac{\partial}{\partial x_i}, C)$ est une résolution de $\mathbb{C}$, $K_g(\frac{\partial}{\partial x_i}, \mathcal{D})$ est une résolution de $\mathcal{O}$; d'autre part, la définition du complexe de Koszul montre immédiatement que le complexe DR(M) n'est autre que $\mathrm{Hom}_{\mathcal{D}}(K_g(\frac{\partial}{\partial x_i}, \mathcal{D}), M)$, d'où le premier isomorphisme. L'isomorphisme $H^p(M) \simeq \mathrm{Tor}^{\mathcal{D}}_{n-p}(\Omega^n, M)$ s'obtient de manière analogue en remarquant que l'application $P \mapsto (dx)P$ donne un isomorphisme $\Omega^n \simeq \mathcal{D}/\Sigma \frac{\partial}{\partial x_i}\mathcal{D}$, et en raisonnant comme ci-dessus avec le complexe de Koszul "droit" $K_d(\frac{\partial}{\partial x_i}, \mathcal{D})$.

Remarque 1.2 :

Les isomorphismes précédents sont en fait invariants par un changement analytique de coordonnées. (Ceci n'est pas tout à fait évident ; comme nous ne nous servirons pas de ce résultat, nous laisserons le lecteur essayer de s'en convaincre par lui-même).

Corollaire (2.3) :

$$H^p(\mathcal{D}) = \begin{cases} 0 & \text{si } p \neq n \\ \Omega^n & \text{si } p = n \end{cases}$$

Il suffit d'appliquer le second isomorphisme à $M = \mathcal{D}$ (ou, directement, de remarquer que $K_d(\frac{\partial}{\partial x_i}, \mathcal{D})$ est une résolution de Ω^n).

Corollaire (2.4) :

$$\mathrm{Ext}^p_{\mathcal{D}}(\mathcal{O},\mathcal{O}) = \begin{cases} \mathbb{C} & \text{si } p = 0 \\ 0 & \text{si } p \neq 0 \end{cases}$$

Il suffit d'appliquer le premier isomorphisme à $M = \mathcal{O}$; on est alors ramené au lemme de Poincaré.

$\mathcal{D}$-modules de support l'origine.

Les résultats qui suivent sont essentiellement dûs à Kashiwara (non publié) ; ils joueront un rôle essentiel dans la suite.

Définition (2.5) :

Un $\mathcal{D}$-module M (fini ou non) est dit "de support l'origine" si, pour tout $m \in M$ et tout i , il existe p tel qu'on ait $x_i^p m = 0$.

Par exemple, $\mathcal{E}'$ est de support l'origine ; en effet tout élément de $\mathcal{E}'$ s'écrit d'une manière et d'une seule $P\delta$, avec $P \in \mathbb{C}$; si p est le degré de P , on a $x_i^{p+1} P\delta = 0$. D'une façon générale, pour que M soit de support l'origine, il suffit que la condition 1.5 soit vérifiée pour les m appartenant à un système de générateurs de M . Supposons en effet qu'on ait, pour un m : $x_i^p m = 0$, $i = 1,\dots,n$. Pour $f \in \mathcal{O}$, on a alors $x_i^p fm = 0$; pour $i \neq j$ on a $x_i^p \frac{\partial}{\partial x_j} m = \frac{\partial}{\partial x_j} x_i^p m = 0$; enfin, on a :

$$x_i^{p+1} \frac{\partial}{\partial x_i} m = \frac{\partial}{\partial x_i} x_i^{p+1} m - (p+1) x_i^p m = 0 .$$

La structure des $\mathcal{D}$-modules de support l'origine va résulter des deux lemmes suivants :

Lemme (2.6) :

On a $H^p(\mathcal{E}') = \begin{cases} 0 & \text{si } p \neq n \\ \mathbb{C} & \text{si } p = n \end{cases}$

D'après la proposition 2.1, il suffit de calculer les $\mathrm{Tor}_k^{\mathcal{D}}(\Omega^n, \mathcal{E}')$; pour le faire on remarque que l'application $P \mapsto P\delta$ établit un isomorphisme $\mathcal{E}' \simeq \mathcal{D}/\Sigma\,\mathcal{D}x_i$; on considère alors le complexe de Koszul "gauche" $K_g(x_i, \mathcal{D})$ dont on voit en raisonnant de manière analogue à (2.1) qu'il est une résolution de $\mathcal{E}'$; les groupes de cohomologie cherchés sont donc ceux de $\Omega^n \underset{\mathcal{D}}{\otimes} K_g(x_i, \mathcal{D})$, c'est-à-dire de $K(x_i, \Omega^n)$; d'où immédiatement le résultat.

Lemme (2.7) :

On a $\mathrm{Ext}^p_{\mathcal{D}}(\mathcal{E}', \mathcal{E}') = \begin{cases} \mathbb{C} & \text{si } p = 0 \\ 0 & \text{si } p \neq 0 . \end{cases}$

Il suffit de calculer les groupes de cohomologie du complexe $\mathrm{Hom}(K_g(x_i, \mathcal{D}), \mathcal{E}')$; mais ce complexe (au remplacement près de p par $n-p$) n'est autre que $K(x_i, \mathcal{E}')$; le résultat se déduit alors du fait connu que $\mathcal{E}'$ est un $\mathcal{O}$-module (ou un $\hat{\mathcal{O}}$-module) injectif ; il peut aussi se démontrer par un calcul direct : l'application $P \to P\delta$ établit un isomorphisme entre $\mathcal{C} = \mathbb{C}\left[\frac{\partial}{\partial x_1}, \dots, \frac{\partial}{\partial x_n}\right]$ et $\mathcal{E}'$; posons $\xi_j = \frac{\partial}{\partial x_j}$; dans cet isomorphisme, l'action de x_i sur $\mathcal{E}'$ est transformée dans l'action de $-\frac{\partial}{\partial \xi_j}$ sur $\mathcal{C}$ (à cause des "relations de commutation" $[x_i, \frac{\partial}{\partial x_j}] = -\delta_{ij}$ et des relations : $x_i\delta = 0$). On est alors ramené au lemme de Poincaré pour l'anneau $\mathcal{C}$ des polynômes.

Théorème (2.8) (Kashiwara).

Soit M un $\mathcal{D}$-module fini, de support l'origine. Alors M est isomorphe à $\mathcal{E}'^{\ell}$ pour un entier ℓ.

Soit $m \in M$, $m \neq 0$ et $\mathcal{O}m$ le $\mathcal{O}$-module qu'il engendre ; notons $\mathfrak{m}$ l'idéal maximal de $\mathcal{O}$; par hypothèse, on a $\mathfrak{m}^k m = 0$; donc il existe $n \in \mathcal{O}m$, $n \neq 0$ tel qu'on ait $x_i n = 0$ pour $i = 1, \dots, n$; donc $\mathcal{D}n$ est isomorphe à un quotient de $\mathcal{E}'$; or, on vérifie facilement que tout élément $\neq 0$ de $\mathcal{E}'$ engendre $\mathcal{E}'$ sur $\mathcal{D}$, donc que $\mathcal{E}'$ est un $\mathcal{D}$-module simple ; par suite $\mathcal{D}n$ est isomorphe à $\mathcal{E}'$; on applique le même résultat à $M/\mathcal{D}n$, et ainsi de suite ; comme $\mathcal{D}$ est noethérien à gauche, on s'arrête au bout d'un nombre fini d'étapes. Par récurrence, on peut alors supposer que $M/\mathcal{D}n$ est isomorphe à $\mathcal{E}^{\ell}$, pour un certain ℓ ; le lemme (2.7) nous montre alors que M est isomorphe à $\mathcal{E}^{\ell+1}$.

Corollaire (2.9) :

Avec les notations du théorème 1.8, on a $H^p(M) = \begin{cases} 0 & , p \neq n \\ \mathbb{C}^{\ell} & , p = n \end{cases}$

Cela résulte immédiatement du lemme 1.6.

Corollaire (2.10) :

Soit M un $\mathcal{D}$-module de support l'origine ; alors :

1) on a $H^p(M) = 0$ pour $p \neq n$;

2) pour que M soit fini sur $\mathcal{D}$, il faut et il suffit que $H^n(M)$ soit fini sur $\mathbb{C}$.

La première assertion est immédiate. Pour démontrer la seconde, notons que le théorème (2.8) et la simplicité de $\mathcal{E}'$ entraînent ceci : si M est un $\mathcal{D}$-module de type fini et de support l'origine alors "$H^n(M) = \mathbb{C}^{\ell}$" équivaut à " M est de longueur ℓ ". Soit alors M de support l'origine, avec $H^n(M)$ fini sur $\mathbb{C}$, disons égal à $\mathbb{C}^{\ell}$; si M' est un sous-module de type fini de M, la suite exacte de cohomologie

$$0 = H^{n-1}(M/M') \to H^n(M') \to H^n(M) \to H^n(M/M') \to 0$$

montre que M' est de longueur $\leq \ell$. Donc M est lui-même de longueur $\leq \ell$, donc il est fini sur $\mathcal{D}$. D'où le résultat.

3. LE POLYNOME DE BERNSTEIN LOCAL : GENERALITES

Soit $f \in \mathcal{O}$, fixé une fois pour toutes, avec $f(0) = 0$, $f \neq 0$. On considère deux indéterminées s et T et le module libre sur $\mathcal{O}[f^{-1}, s]$ de base T qu'on notera $\mathcal{O}[f^{-1}, s]T$; on fait agir $\mathcal{D}$ à gauche sur ce module en faisant agir trivialement $\mathcal{D}$ sur s et en posant :

$$\frac{\partial}{\partial x_i}(gT) = \frac{\partial g}{\partial x_i} T + sgf^{-1}\frac{\partial f}{\partial x_i} T \quad , \quad g \in \mathcal{O}[f^{-1},s] .$$

Dans la suite, on écrira f^s au lieu de T et f^{s+k} au lieu de $f^k T$ $(k \in \mathbb{Z})$. Il est clair que, si l'on donne à s un valeur entière, on retombe sur les formules usuelles. En faisant agir s de la manière évidente, l'action de $\mathcal{D}$ s'étend en une action de $\mathcal{D}[s]$ sur $\mathcal{O}[f^{-1},s]f^s$; nous désignerons une fois pour toutes par M le sous-module $\mathcal{D}[s]f^s$ engendré par f^s. Rappelons le résultat suivant, démontré par Bernstein [1] si f est un polynôme, et étendu au cas général par Björk [2].

<u>Théorème 3.1</u> :

> <u>Il existe</u> $B \in \mathbb{C}[s]$, $B \neq 0$, <u>et</u> $P \in \mathcal{D}[s]$ <u>tels qu'on ait</u> $Pf^s = Bf^{s-1}$.

Il est visible que l'ensemble des $B \in \mathbb{C}[s]$ tel qu'il existe $P \in \mathcal{D}[s]$ avec $Pf^s = Bf^{s-1}$ est un idéal de $\mathbb{C}[s]$; le générateur (unitaire) de cet idéal sera noté b et sera appelé "le polynôme de Bernstein de f". Il est immédiat aussi qu'on a $b(0) = 0$ (faire $s=0$ dans l'identité précédente). On posera dans la suite $b = s\tilde{b}$.

Soit d'autre part $\mathfrak{J}$ l'idéal à gauche de $\mathcal{D}[s]$ formé des P qui vérifient $Pf^s = 0$; on a alors un isomorphisme $M \simeq \mathcal{D}/\mathfrak{J}$.

Le but de cet article est d'examiner les conjectures suivantes.

<u>Conjecture (1)</u> :

> <u>Les zéros de</u> b <u>sont rationnels</u>.

<u>Conjecture (1bis)</u> :

> <u>Les zéros de</u> b <u>sont</u> < 1.

<u>Conjecture (2)</u> (Kashiwara)

> M <u>est fini sur</u> $\mathcal{D}$.

Notons que la conjecture (2) est équivalente à la suivante : <u>il existe un</u> $P \in \mathfrak{J}$ <u>de la forme</u> $P = s^k + \sum_1^k a_i s^{k-i}$, $a_i \in \mathcal{D}$. En fait, Kashiwara [5] conjecture un résultat plus fort :

<u>Conjecture (2 bis)</u> :

<u>Il existe un</u> $P \in \mathfrak{J}$ <u>de la forme</u> $P = s^k + \sum_1^k a_i s^{k-i}$, <u>avec</u> $a_i \in \mathcal{D}^i$ (i.e. le degré de a_i par rapport aux $\frac{\partial}{\partial x_j}$ est $\leq i$).

Dans la suite, nous démontrerons (1), (1bis) et (2) mais non (2bis), <u>lorsque</u> f <u>a une singularité isolée en</u> 0 . Pour l'instant, nous ne faisons pas cette dernière hypothèse. L'idée essentielle, qui avait été considérée dans un cas particulier dans Malgrange [6], consiste à étudier les relations entre la cohomologie de de Rham de M et la "monodromie locale" de f .

Pour simplifier les notations, remarquons d'abord que $\mathcal{O}[f^{-1},s]f^s$ est égal au localisé $M[f^{-1}]$ de M par rapport à f . On définit alors une structure de $\mathbb{C}\{t\}$-module sur $M[f^{-1}]$ en posant, pour $g(s) \in \mathcal{O}[f^{-1},s]$: $t[g(s)f^s] = g(s+1)f^{s+1}$; et plus généralement, si $\varphi = \sum_{k \geq 0} \lambda_k t^k \in \mathbb{C}\{t\}$, $\varphi[g(s)f^s] = \sum \lambda_k g(s+k)f^{s+k}$ (cette série converge à cause du théorème de dérivation des séries convergentes). Notons que cette action ne commute pas à celle de $\mathcal{D}[s]$ sur $M[f^{-1}]$, mais qu'on a les propriétés suivantes :

(3.2) <u>Si</u> $P \in \mathcal{D}$, $\varphi \in \mathbb{C}\{t\}$, $[P,\varphi] = 0$.

(3.3) <u>On a</u> $(s+1)t = ts$; <u>et, plus généralement, si</u> $\varphi \in \mathbb{C}\{t\}$, <u>on a</u> $\varphi s = s\varphi + t\varphi'$.

En remarquant que t est une bijection sur $M[f^{-1}]$, donc que t^{-1} est défini sur $M[f^{-1}]$ cette dernière propriété signifie ceci :

(3.3') $\nabla_{\frac{d}{dt}} = -t^{-1}(s+1)$ <u>est une connexion sur le</u> $\mathbb{C}\{t\}$<u>-module</u> $M[f^{-1}]$.

Posons maintenant, pour tout $k \in \mathbb{Z}$: $M_k = \mathcal{D}[s]f^{s+k}$; on a $M_o = M$, $M_k \subset M_{k-1}$, $\cup_k M_k = M[f^{-1}]$. D'autre part, les M_k sont stables par t (avec des notations évidentes, $t(P(s)f^s) = P(s+1)f^{s+1}$, $P(s) \in \mathcal{D}[s]$), et même par $\mathbb{C}\{t\}$, à cause du théorème de dérivation des séries convergentes. On a $tM_k = M_{k+1}$, $M[f^{-1}] = \cup M_k = M[t^{-1}]$, et $\nabla\frac{d}{dt} M_k \subset M_{k-1}$. En général, M_k n'est pas stable par $\nabla\frac{d}{dt}$ (voir au paragraphe 4).

Si b est le polynôme de Bernstein de f , on a $bf^{s-1} \in M$; on en déduit immédiatement qu'on a alors $bM_{-1} \subset M$; et réciproquement ; par conséquent en remplaçant s par $(s+1)$ et en tenant compte de l'égalité $M_1 = tM$, on a ceci :

(3.4) b <u>est le polynôme minimal de l'action de</u> $(s+1)$ <u>sur</u> M/tM .

<u>Proposition (3.5)</u> :

M/tM <u>est fini sur</u> $\mathcal{D}$.

En effet M , donc M/tM , est fini sur $\mathcal{D}[s]$; le théroème (3.1) montre qu'il existe un $b \in \mathbb{C}[s]$, $b \neq 0$ qui annule M/tM ; la proposition en résulte immédiatement.

<u>Remarque (3.6)</u> :

Il est probable que M/tM est de dimension minimale sur $\mathcal{D}$ (i.e. de dimension n ; voir [2] ou [8]), et qu'en outre il est "régulier" ou "fuchsien" en un sens qu'il faudrait préciser. La principale difficulté semble ici de trouver quelle devrait être la définition d'un $\mathcal{D}$-module fuchsien : l'hypothèse de dimension minimale, manifestement nécessaire, est tout aussi manifestement insuffisante. Il me semble que ce dernier problème présente un grand intérêt : on la retrouve dans des sujets très variés, qui dépassent de loin le problème particulier des polynômes de Bernstein.

Considérons $M[t^{-1}]$ comme un $\mathcal{D}$-module ; on peut alors considérer ses groupes de cohomologie de de Rham $H^p(M[t^{-1}])$; comme l'action de $\mathbb{C}\{t\}$ et celle de $\nabla\frac{d}{dt}$ commutent à l'action de $\mathcal{D}$ sur $M[t^{-1}]$, $H^p(M[t^{-1}])$ est muni d'une structure de $\mathbb{C}\{t\}$-module avec connexion. Pour interpréter géométriquement ces faits, considérons le complexe $\Omega^\bullet_{rel}$ des formes différentielles relatives de l'application $f : \mathbb{C}^n \to \mathbb{C}$ (voir par exemple Deligne [4]) ; rappelons que par définition $\Omega^p_{rel} = \Omega^p[f^{-1}]/df \wedge \Omega^{p-1}[f^{-1}]$, la différentielle relative d_{rel} étant obtenue par passage au quotient à partir de $d : \Omega^p[f^{-1}] \to \Omega^{p+1}[f^{-1}]$, et que les groupes $H^p(\Omega^\circ_{rel})$ sont munis naturellement d'une structure de $\mathbb{C}\{t\}$-module avec une connexion (la "connexion de Gauss-Manin").

Théorème (3.7) :

On a un isomorphisme de $\mathbb{C}\{t\}$-modules avec connexion $H^p(\Omega^\bullet_{rel}) \simeq H^{p+1}(M[t^{-1}])$.

Un élément de $\Omega^p \otimes M[t^{-1}]$ s'écrit $\omega(s)f^s$, avec $\omega(s) \in \Omega^p[f^{-1},s]$, et la différentielle de $DR(M[t^{-1}])$ s'écrit $d(\omega(s)f^s = [s\frac{df}{f} \wedge \omega(s) + d\omega(s)]f^s$, avec $d\omega(s)$ la différentielle usuelle sur les composantes en s^k de $\omega(s)$. Il revient donc au même d'étudier le complexe $\Omega^\circ[f^{-1},s]$, muni de la différentielle $\delta\omega(s) = s\frac{df}{f} \wedge \omega(s) + d\omega(s)$.

Filtrons alors ce dernier complexe par le degré en s.

Lemme 3.8 :

Le complexe $\Omega^\bullet[f^{-1},s]$, muni de la différentielle $\omega \mapsto \frac{df}{f} \wedge \omega$ est acyclique en tous degrés.

Ce complexe n'est autre que le complexe de Koszul $K(f^{-1}\frac{\partial f}{\partial x_i}, \mathcal{O}[f^{-1},s])$; pour démontrer son acyclicité, il suffit par un résultat connu de trouver des $b_i \in \mathcal{O}[f^{-1}]$ tels qu'on ait $\Sigma b_i f^{-1} \frac{\partial f}{\partial x_i} = 1$; or cela résulte immédiatement du fait, tout aussi connu, qu'il existe un $k \geq 0$ tel qu'on ait $f^{k+1} = \Sigma a_i \frac{\partial f}{\partial x_i}$, $a_i \in \mathcal{O}$.

Le lemme précédent montre que la cohomologie du complexe $\Omega^\bullet[f^{-1},s]$ muni de δ est isomorphe à celle de son sous-complexe $A^\bullet$ formé des ω indépendants de s, et vérifiant $\frac{df}{f} \wedge \omega = 0$, ou $df \wedge \omega = 0$; mais alors l'application $\omega \to df \wedge \omega$ établit un isomorphisme entre Ω^{p-1}_{rel} et A^p, puisque le lemme 3.8 est vrai aussi pour $\Omega^\bullet[f^{-1}]$. Cet isomorphisme est manifestement compatible avec les différentielles extérieures. D'où l'isomorphisme cherché.

Cet isomorphisme est compatible avec les structures de $\mathbb{C}\{t\}$-modules : en effet, si l'on prend un $\omega \in A^p$, son image dans $\Omega^p \otimes M[t^{-1}]$ s'écrit ωf^s ; comme ω est indépendant de s, on a $t(\omega f^s) = \omega f^{s+1} = (\omega f) f^s$; d'autre part, par définition, l'action de t dans $\Omega^\bullet_{rel}$ est la multiplication par f.

Enfin, cet isomorphisme est compatible avec les connexions : prenons $\omega \in A^p$; on a $\omega = df \wedge \alpha$, $\alpha \in \Omega^{p-1}[f^{-1}]$; d'où $(s+1)\omega = \delta(\alpha f) - fd\alpha$, on a $df \wedge d\alpha = d\omega = 0$ d'où immédiatement $fd\alpha \in A^p$, et $(s+1)\omega$ est homologue à $-fd\alpha$; par suite $-t^{-1}(s+1)\omega = \nabla \frac{d}{dt}\omega$ est homologue à $d\alpha$; or, si l'on pose $d\alpha = df \wedge \beta$ on sait que l'application $\alpha \mapsto \beta$ donne précisément par passage à la cohomologie la connexion de Gauss-Manin. D'où le théorème.

Nous terminerons ce paragraphe par la remarque suivante : puisque l'action de t (ou t^{-1}) commute à $\mathcal{D}$, on a $H^p(M[t^{-1}]) \simeq H^p(M)[t^{-1}]$; par suite $H^p(M)/\text{torsion}$ est un sous-module de $H^p(M[t^{-1}])$ dont le saturé par t^{-1} est $H^p(M[t^{-1}])$ tout entier. En admettant (ce qui paraît probable) que ce dernier espace est fini sur le corps $\mathbb{C}\{t\}[t^{-1}]$, il serait intéressant de savoir si $H^p(M)/\text{torsion}$ en est un <u>réseau</u> : nous verrons plus loin que c'est bien le cas lorsque f a une singularité isolée.

4. QUELQUES CALCULS AUXILIAIRES

Nous ne nous servirons pas directement du théorème (3.7) mais d'une variante ; notons $\mathcal{O}_{x,t}$ l'anneau $\mathbb{C}\{x_1,\dots,x_n,t\}$, $\mathcal{D}_{x,t}$ l'anneau des opérateurs différentiels en $\frac{\partial}{\partial x_i}$, $\frac{\partial}{\partial t}$ à coefficients dans $\mathcal{O}_{x,t}$, et faisons opérer $\mathcal{D}_{x,t}$ dans $M[t^{-1}]$ de la manière suivante : les $\frac{\partial}{\partial x_i}$ opèrent comme ci-dessus, et $\frac{\partial}{\partial t} = \nabla \frac{d}{dt}$; enfin, si $\varphi = \Sigma a_k(x) t^k \in \mathcal{O}_{x,t}$, on pose $\varphi[g(s)f^s] = \Sigma a_k(x)g(s+k)f^{s+k}$ (série qui converge encore puisque, pour tout $\ell \geq 0$, $\Sigma a_k(x)k^\ell t^k$ converge). Comme ces opérations étendent celles qu'on a précédemment définies, on a bien encore $(s+1) = -t\frac{\partial}{\partial t}$. Posons $N = \mathcal{D}_{x,t} f^s$; on a évidemment $M \subset N \subset M[t^{-1}]$; ces inclusions sont strictes en général. Pour que l'on ait $M = N$, il faut que l'on ait $-\frac{\partial}{\partial t}f^s = sf^{s-1} \in M$, autrement dit qu'il existe $P \in \mathcal{D}[s]$ tel qu'on ait $Pf^s = sf^{s-1}$, ou encore que le polynôme de Bernstein de f soit réduit à s ; ceci est vérifié (et c'est vraisemblablement le seul cas) si f est non singulière en 0 ; inversement, si f est non singulière en 0, on peut supposer $f = x_1$; alors on a $\frac{\partial}{\partial t}x_1^s = -\frac{\partial}{\partial x_1}x_1^s$, et $\frac{\partial}{\partial t}[P(s)x_1^s] = -P(s-1)\frac{\partial}{\partial x_1}x_1^s$ $(P(s) \in \mathcal{D}[s])$, donc on a $\frac{\partial}{\partial t}M \subset M$, et $N = M$.

Revenons au cas général ; on a $(f-t)f^s = 0$, $(\frac{\partial}{\partial x_i} + \frac{\partial f}{\partial x_i}\frac{\partial}{\partial t})f^s = 0$; montrons que les éléments précédents engendrent l'annulateur de f^s dans $\mathcal{D}_{x,t}$. Il suffit pour cela d'établir le lemme suivant.

Lemme (4.1) :

> Dans $\mathcal{D}_{x,t}$, l'idéal à gauche $\mathcal{J}$ engendré par $f-t$ et les $\frac{\partial}{\partial x_i} + \frac{\partial f}{\partial x_i}\frac{\partial}{\partial t}$ est maximal.

Le changement de coordonnées $x_i \mapsto x_i$, $t \mapsto f+t$ nous ramène au cas où $f = 0$. Alors, toute classe modulo $\mathcal{J}$ a un représentant unique de la forme $P = \sum_0^\ell a_k(x)\frac{\partial^k}{\partial t^k}$, $a_k \in \mathcal{O}\ (=\mathcal{O}_x)$; considérons maintenant un tel $P \neq 0$, et examinons l'idéal engendré par P et $\mathcal{J}$; supposons par exemple $a_\ell \neq 0$; en prenant ℓ fois le crochet $[t, \ldots, [t,P]]$, on voit qu'il doit contenir a_ℓ ; et comme $[\frac{\partial}{\partial x_i}, a_\ell] = \frac{\partial a_\ell}{\partial x_i}$, en prenant les crochets successivement avec les $\frac{\partial}{\partial x_i}$, on trouve qu'il contient toutes les dérivées partielles de a_ℓ, en particulier une fonction inversible. D'où le résultat.

Notons que $\mathcal{D}_{x,t}f^s$ est isomorphe à $\mathcal{D}_{x,t}\delta(f-t)$ en désignant par $\delta(f-t)$ la classe de $\frac{1}{f-t}$ dans $K_{x,t}/\mathcal{O}_{x,t}$, $K_{x,t}$ le corps des fonctions de $\mathcal{O}_{x,t}$ (voir par exemple [8]). Il est immédiat de voir que tout élément de $N = \mathcal{D}_{x,t}f^s$ s'écrit d'une manière unique sous la forme $\Sigma g_k(x)\frac{\partial^k}{\partial t^k}f^s$.

Considérons N comme $\mathcal{D}(=\mathcal{D}_X)$-module ; la différentielle est alors donnée par

$$d(g_k \frac{\partial^k}{\partial t^k} f^s) = dg_k \frac{\partial^k}{\partial t^k} f^s - g_k df \frac{\partial^{k+1}}{\partial t^{k+1}} f^s .$$

Il revient au même, en posant $\frac{\partial}{\partial t} = -\tau$ de considérer le $\mathcal{D}$-module $\mathcal{O}[\tau]$ muni de la différentielle $\delta(g\tau^k) = \tau^k dg + \tau^{k+1} gdf$, ou encore $\mathcal{O}[\tau]e^{\tau f}$ muni de la différentielle $d(g\tau^k e^{\tau f}) = \tau^k e^{\tau f} dg + \tau^{k+1} ge^{\tau f} df$. Dans la suite, nous passerons librement d'une forme à une autre.

<u>Supposons à partir de maintenant que</u> f <u>a une singularité isolée en</u> 0 <u>et qu'on a</u> $n \geq 2$; pour calculer $H^*(N)$, nous allons utiliser les résultats de [7].

<u>Proposition (4.2)</u> :

<u>Pour</u> $p \neq 1,n$ <u>on a</u> $H^p(N) = 0$, <u>et</u> $H^1(N) = \mathbb{C}\{t\}$.

Pour $p > n$, c'est évident ; pour $p < n$, filtrons le complexe DR(N) par le degré en τ ; en remarquant que le complexe de Koszul $K(\frac{\partial f}{\partial x_i}, \mathcal{O})$ (translaté de n) est une résolution de $\mathcal{O}/(\frac{\partial f}{\partial x_i})$, donc est acyclique en tous degrés $\neq n$, on voit que la cohomologie se réduit à celle du sous-complexe des formes de degré zéro en τ qui vérifient $df \wedge \omega = 0$, ou encore du complexe $df \wedge \Omega^\bullet$. Or on sait (cf. [3] ou [7]) que la cohomologie de ce complexe est nulle en degrés $\neq 1,n$, et égale à $\mathbb{C}\{t\}$ en degré 1 .

Reste le cas $p = n$. Filtrons encore N par le degré en τ ; on obtient une filtration sur $H^n(N)$; pour simplifier les notations, on écrira $H^n(N)^k = G^k$, et $G^\circ = G$.

1) Calculons G ; il faut pour cela déterminer les $a \in \Omega^n$ de la forme $\delta(b_o + b_1\tau + \ldots + b_\ell \tau^\ell) = db_o + (df \wedge b_o + db_1)\tau + \ldots + (df \wedge b_\ell)\tau^{\ell+1}$; on doit avoir $df \wedge b_\ell = 0$, d'où $b_\ell = df \wedge \tau_\ell$, $\tau_\ell \in \Omega^{n-2}$ (par le même argument de complexe de Koszul que plus haut) ; en retranchant $\delta(\tau_\ell \tau^{\ell-1})$ à $b_o + \ldots + b_\ell \tau^\ell$, on peut supposer $b_\ell = 0$, et ainsi de suite ; finalement, on

voit qu'on doit avoir $a = \delta b$, ou encore $a = db$, $df \wedge b = 0$, ou enfin, en posant $b = df \wedge c$: $a = df \wedge dc$; finalement, on trouve que l'on a $G = \Omega^n / df \wedge d\Omega^{n-2}$.

2) Calculons G^k , pour $k > 0$; prenons un $a = a_o + \ldots + a_k \tau^k$, avec $\delta a = 0$; en posant $a_o = db_o$, on voit que $a - \delta b_o$ est sans terme constant et ainsi de suite. On peut donc supposer $a_o = \ldots = a_{k-1} = 0$ sans changer la classe de a ; ensuite, si $a_k \tau^k = \delta(b_o + \ldots + b_\ell \tau^\ell)$, on voit d'abord comme en 1) qu'on peut supposer $b_{k+1} = \ldots = b_\ell = 0$; ensuite, comme $db_o = 0$, le lemme de Poincaré montre qu'on a $b_o = dc_o$; en retranchant δc_o à $b_o + \ldots + b_k \tau^k$, on peut supposer $b_o = 0$, et ainsi de suite ; finalement, il reste $a_k \tau^k = \delta(b_k \tau^k)$, c'est-à-dire $a_k = \delta(b_k)$; finalement, on trouve que τ^k (qui agit évidemment sur $H^n(N)$) est un isomorphisme $G \simeq G^k$.

3) Soit F l'ensemble des $\alpha \in G$ tels qu'on ait $\tau\alpha \in G$; on sait (voir [7], ou exercice) que l'on a $F = df \wedge \Omega^{n-1} / df \wedge d\Omega^{n-2}$, et que τ est un isomorphisme $F \to G$; en considérant G comme un $\mathbb{C}[\tau^{-1}]$-module , on voit qu'on a finalement le résultat suivant.

Proposition (4.3.) :

On a $H^n(N) = G \otimes_{\mathbb{C}[\tau^{-1}]} \mathbb{C}[\tau, \tau^{-1}]$.

D'autre part, $H^n(N)$ est un $\mathbb{C}\{t\}$-module ; G est l'image dans $H^n(N)$ de $\Omega^n f^s$; dont sur G , l'action de t coïncide avec la multiplication par f et redonne donc la structure usuelle de $\mathbb{C}\{t\}$-module de G . Rappelons les propriétés suivantes (voir [3], [9], [7]) :

1) Soit $\mu = \dim_{\mathbb{C}} \mathcal{O} / (\frac{\partial f}{\partial x_i})$ le "nombre de Milnor" de f . Alors G est libre de rang μ sur $\mathbb{C}\{t\}$.

2) Soit K le corps des fractions de G ; la connexion $\frac{\partial}{\partial t}$ (noté aussi τ) $F \to G$ s'étend en une K-connexion sur $G \otimes_{\mathbb{C}\{t\}} K$, connexion qui est régulière.

3) Soit $\tilde{F}$ (resp. $\tilde{G}$) le saturé de F (resp. G) pour $t\frac{\partial}{\partial t} = -(s+1)$ dans $G \underset{\mathbb{C}\{t\}}{\otimes} K$; alors $\tilde{F}$ et $\tilde{G}$ sont encore des réseaux (i.e. sont libres de rang μ sur $\mathbb{C}\{t\}$), et $t\frac{\partial}{\partial t}$ est une bijection $\tilde{F} \to \tilde{F}$.

Comme $\frac{\partial}{\partial t}$ est une bijection $F \to G$, il est immédiat qu'il envoie encore $\tilde{F}$ sur $\tilde{G}$; d'autre part, $t\frac{\partial}{\partial t}$ étant injectif sur $\tilde{F}$, $\frac{\partial}{\partial t}$ l'est aussi ; finalement $\frac{\partial}{\partial t}$ est une bijection $\tilde{F} \to \tilde{G}$, et t est une bijection $\tilde{G} \to \tilde{F}$. On sait (voir [7]) que sur $\tilde{G}$, les filtrations définies par les $t^k\tilde{G}$ et les $\tau^{-k}\tilde{G}$ coïncident ; par suite, on a $\tau^{-k}\tilde{G} \subset G$ pour k assez grand ; de là résulte facilement qu'il existe une application unique de $\tilde{G}$ dans $H^n(N)$ qui coïncide avec l'identité sur G et soit compatible avec les structures de $\mathbb{C}\{t\}$-module muni d'une connexion (ou, si l'on préfère, de $\mathcal{D}_t$-modules). L'image de cette application est évidemment le saturé de G dans $H^n(N)$; il résulte de ce qui précède qu'il est sans torsion sur $\mathbb{C}\{t\}$.

Remarquons que, par contre, $H^n(N)$ peut avoir de la torsion. Soit $\alpha \in G$; pour qu'on ait (dans $H^n(N)$) $t\frac{\partial^k}{\partial t^k}\alpha = 0$, il faut et il suffit qu'on ait $\frac{\partial^{k-1}}{\partial t^{k-1}}[t\frac{\partial \alpha}{\partial t} - (k-1)\alpha] = 0$, ou encore, puisque $\frac{\partial}{\partial t} = -\tau$ est bijectif dans $H^n(N)$: $t\frac{\partial \alpha}{\partial t} = (k-1)\alpha$ [ou encore $(s+k)\alpha = 0$], ce qui se produit lorsque la monodromie de f admet la valeur propre + 1 .

<u>Remarque (4.4)</u> :

On peut voir aisément, à partir du résultat qui précède que $H^n(N)$ est fini sur $\mathcal{D}_t$, l'anneau des opérateurs différentiels en $\frac{\partial}{\partial t}$ à coefficients dans $\mathbb{C}\{t\}$; le théorème de régularité rappelé plus haut montre alors que ce module est "fuchsien", en un sens qu'il est facile de définir (parce qu'on est en dimension 1 ; la remarque (3.6) concerne les dimensions supérieures). Il est probable que le même résultat subsiste pour les $H^p(M)$ dans le cas d'une singularité non isolée ; comme, d'autre part, toute définition raisonnable devrait considérer N comme fuchsien sur $\mathcal{D}_{X,t}$, il paraît même possible qu'il existe un résultat assez général sur la conservation du caractère fuchsien par image directe.

5. LE POLYNOME DE BERNSTEIN D'UNE SINGULARITE ISOLEE

Nous continuons à supposer que f a une singularité isolée, et qu'on a $n \geq 2$ (le cas $n = 1$ est trivial, et obligerait à modifier légèrement les énoncés).

Lemme (5.1) :

Le $\mathcal{D}$-module N/M est de support l'origine.

Comme les $a \frac{\partial^k}{\partial t^k} f^s$ engendrent N sur $\mathcal{D}$, $a \in \mathcal{O}_{x,t}$, il suffit de démontrer que, pour tout k, on peut trouver un ℓ tel que les $x_i^\ell a \frac{\partial^k}{\partial t^k} f^s$ appartiennent à M ; comme $\mathcal{O}_{x,t} M \subset M$, il suffit, pour un k fixé, d'établir le résultat pour $a = 1$. Procédons alors par récurrence, et supposons le résultat établi pour k ; par hypothèse, il existe un entier, par exemple μ tel que $x_i^\mu \in (\frac{\partial f}{\partial x_1}, \ldots, \frac{\partial f}{\partial x_n})$.

De la relation $\frac{\partial}{\partial x_i} f^s = -\frac{\partial f}{\partial x_i} \frac{\partial}{\partial t} f^s$, on tire $x_j^{\ell+1} \frac{\partial f}{\partial x_i} \frac{\partial^{k+1}}{\partial t^{k+1}} f^s = -x_j^{\ell+1} \frac{\partial}{\partial x_i} \frac{\partial^k}{\partial t^k} f^s$; si $j \neq i$, le second membre vaut $-\frac{\partial}{\partial x_i} x_j^{\ell+1} \frac{\partial^k}{\partial t^k} f^s$, qui appartient à M par hypothèse de récurrence ; si $j = i$, il vaut $-\frac{\partial}{\partial x_i} x_i^{\ell+1} \frac{\partial^k}{\partial t^k} f^s + (\ell+1) x_i^\ell \frac{\partial^k}{\partial t^k} f^s$ qui appartient encore à M. En appliquant ce résultat à j fixé avec $i = 1, \ldots, n$, on trouve donc qu'on a pour tout j : $x_j^{\ell+\mu+1} \frac{\partial^{k+1}}{\partial t^{k+1}} f^s \in M$; d'où le résultat.

Notons que ce lemme est intuitivement évident du fait que, en dehors de 0, f est non singulière et par conséquent $N = M$ (cet argument pourrait être facilement rendu correct si l'on avait considéré des faisceaux cohérents au lieu seulement de germes à l'origine comme on le fait ici).

Considérons maintenant la suite exacte $0 \to M \to N \to M/N \to 0$; en appliquant (2.10) et la suite exacte de cohomologie, on trouve que

$H^p(M) \to H^p(N)$ est bijectif pour $p < n$, et injectif pour $p = n$; en appliquant (4.2) on trouve en particulier ceci :

Proposition (5.2) :

On a $H^1(M) = \mathbb{C}\{t\}$, et $H^p(M) = 0$ pour $p \neq 1, n$.

Déterminons l'image de $H^n(M)$ dans $H^n(N)$: par (2.1), $H^n(M)$ est isomorphe à $\Omega^n \otimes_{\mathcal{D}} M$, donc à $\Omega^n[s]/\Omega^n[s]\mathfrak{J}$ (on rappelle que $\mathfrak{J}$ désigne l'idéal de $\mathcal{D}[s]$ qui annule f^s, et qu'on a donc $M = \mathcal{D}[s]/\mathfrak{J}$) ; l'injection $H^n(M) \to H^n(N)$ se relève donc en une application $\Omega^n[s] \to H^n(N) =$ $\Omega^n[s] \to H^n(N) = G \otimes_{\mathbb{C}\{\tau^{-1}\}} \mathbb{C}\{\tau,\tau^{-1}\}$; il est immédiat que l'image de Ω^n par cette application est G ; et par conséquent l'image de $\Omega^n[s]$ est le saturé $\tilde{G}$ de G pour s. D'où la proposition suivante.

Proposition (5.3) :

On a $H^n(M) \simeq \tilde{G}$ (et naturellement, les structures de $\mathcal{D}_t$-module des deux membres coïncident).

Désignons par a le polynôme minimal de l'action de $(s+1)$ sur $\tilde{G}/t\tilde{G} = \tilde{G}/\tilde{F}$. Le résultat principal est le théorème suivant.

Théorème (5.4) :

On a $\tilde{b} = a$ [ou encore $b(s) = sa(s)$].

De (3.4) et (5.3), on déduit seulement que b est un multiple de a, donc il faut regarder des choses d'un peu plus près.

1) Soit $P(s) \in \mathcal{D}[s]$ tel qu'on ait $P(s)f^s = b(s)f^{s-1}$; en faisant $s = 0$ on trouve $P(0)1 = b(0)f^{-1}$; ceci n'est possible que si $b(0) = 0$ (d'où $b = s\tilde{b}$, comme on l'a déjà vu), et $P(0)1 = 0$. Donc $P(0)$ est sans terme constant, donc P s'écrit $P(s) = sQ(s) + \sum R_i \frac{\partial}{\partial x_i}$, $R_i \in \mathcal{D}$, $Q(s) \in \mathcal{D}[s]$; on a alors $sQ(s)f^s + \sum sR_i(\frac{\partial f}{\partial x_i} f^{s-1}) = s\tilde{b}(s)f^{s-1}$, ou encore $Q(s+1)f^{s+1} + \sum R_i(\frac{\partial f}{\partial x_i} f^s) = \tilde{b}(s+1)f^s$, ou enfin $\tilde{b}(s+1) \in \mathfrak{J} + \mathcal{D}[s]f + \sum \mathcal{D}[s]\frac{\partial f}{\partial x_i}$.

Appelons L le $\mathcal{D}$-module $\mathcal{D}[s]/\mathfrak{J} + \mathcal{D}[s]f + \Sigma\mathcal{D}[s]\frac{\partial f}{\partial x_i}$. Alors, ce qui précède montre que $\tilde{b}$ est le polynôme minimal de l'action de $(s+1)$ sur L .

2) L est un quotient de $M/tM = \mathcal{D}[s]/\mathfrak{J} + \mathcal{D}[s]f$. Comme ce dernier module est fini sur $\mathcal{D}$ (proposition (3.5)), L est aussi fini sur $\mathcal{D}$. D'autre part, L est manifestement de support l'origine ; on déduit alors de (2.8) et (2.9) le résultat suivant (dû lui aussi à Kashiwara) $\tilde{b}$ <u>est le polynôme minimal de</u> $H^n(L)$.

3) Pour terminer la démonstration, il suffit de voir que l'application naturelle $\tilde{G}/t\tilde{G} = H^n(M/tM) \to H^n(L)$ est bijective. Par (2.1), on a $H^n(L) = \Omega^n[s]/\Omega^n[s]\mathfrak{J} + \Omega^n[s]f + \Sigma\Omega^n[s]\frac{\partial f}{\partial x_i}$, et aussi $\tilde{G}/t\tilde{G} = H^n(M/tM) = \Omega^n[s]/\Omega^n[s]\mathfrak{J} + \Omega^n[s]f$. Pour établir le résultat, il suffit de montrer ceci : <u>l'application</u> $u : \Omega^n[s] \to \tilde{G}$ <u>définie ci-dessus</u> (après 5.2)) <u>envoie</u> $\Omega^n[s]\frac{\partial f}{\partial x_i}$ <u>dans</u> $t\tilde{G} = \tilde{F}$.

Prenons d'abord $\omega \in \Omega^n$; on a $\frac{\partial f}{\partial x_i}\omega \in df \wedge \Omega^{n-1}$, donc $u(\omega\frac{\partial f}{\partial x_i}) \in F$ (par définition de F) et, a fortiori, $u(\omega\frac{\partial f}{\partial x_i}) \in \tilde{F}$. Dans le cas général, soit $\omega(s) \in \Omega^n[s]$; de l'égalité $\frac{\partial}{\partial x_i}f^{s+1} = (s+1)\frac{\partial f}{\partial x_i}f^s$, on déduit qu'on a $(s+1)\frac{\partial f}{\partial x_i} - \frac{\partial}{\partial x_i}\circ f \in \mathfrak{J}$, donc $(s+1)\frac{\partial f}{\partial x_i} \in \mathfrak{J} + \mathcal{D}[s]f$; on en déduit qu'on a $\omega(s)\frac{\partial f}{\partial x_i} = \omega(-1)\frac{\partial f}{\partial x_i} \bmod \Omega^n[s]\mathfrak{J} + \Omega^n[s]f$ ou encore $u[\omega(s)\frac{\partial f}{\partial x_i}] = u[\omega(-1)\frac{\partial f}{\partial x_i}] \bmod \tilde{F}$. Ceci achève la démonstration.

Le théorème (5.4) entraîne immédiatement la conjecture (1), dans le cas des singularités isolées, à cause du <u>théorème de monodromie</u> qui nous affirme précisément que les zéros de a sont rationnels. Pour démontrer (1bis), il faut établir que les valeurs propres de $(s+1) = -t\frac{\partial}{\partial t}$ dans $\tilde{G}/t\tilde{G}$ sont < 1 , ou encore que celles de $s = -\frac{\partial}{\partial t}\circ t$ sont < 0 ; après multiplication par t , il suffit de voir que les valeurs propres de $t\frac{\partial}{\partial t}$ dans $\tilde{F}/t\tilde{F}$ sont > 0 . Ceci résulte immédiatement de la "positivité des exposants caractéristiques dans $\tilde{F}$ " (voir [7], démonstration du lemme (5.6)) et du lemme élémentaire suivant :

Lemme (5.5) :

Soit M un $\mathbb{C}\{t\}[t^{-1}]$ module de type fini, muni d'une connexion (notée $\frac{d}{dt}$) régulière. Soit E un réseau stable par $t\frac{d}{dt}$, et A le résidu dans E de cette connexion, i.e. l'action de $t\frac{d}{dt}$ sur E/tE. Soit enfin λ une valeur propre de A telle que $\lambda-k$ (entier >0) ne soit jamais valeur propre de A. Dans ces conditions, il existe un vecteur $e \in E$ vérifiant $t\frac{d}{dt}e = \lambda e$.

Il suffit de chercher un vecteur $e \in \tilde{E} = E \otimes_{\mathbb{C}\{t\}} \mathbb{C}([t])$ vérifiant $t\frac{d}{dt}e = \lambda e$; en effet, la connexion étant régulière e sera nécessairement convergent. Par hypothèse, il existe $e_o \in E$ tel qu'on ait $t\frac{d}{dt}e_o = \lambda e_o \mod tE$; montrons par récurrence qu'il existe $e_1,\dots,e_k \in E$ tels qu'on ait $t\frac{d}{dt}(e_o+te_1+\dots+t^k e_k) = \lambda(e_o+te_1+\dots+t^k e_k) \mod t^{k+1}E$.

Supposons trouvés $e_1,\dots,e_k$ et cherchons e_{k+1} ; on a $t\frac{d}{dt}(e_o+te_1+\dots+t^k e_k) = \lambda(e_o+te_1+\dots+t^k e_k) + t^{k+1}f \mod t^{k+2}E$; donc e_{k+1} doit vérifier $t\frac{d}{dt}(t^{k+1}e_{k+1}) = \lambda t^{k+1}e_{k+1} + t^{k+1}f \mod t^{k+2}E$, ou encore $t\frac{d}{dt}e_{k+1} = (\lambda-k-1)e_{k+1} + f \mod tE$; l'hypothèse faite sur λ montre justement qu'il existe un tel e_{k+1} (d'ailleurs unique modulo tE) ; d'où le lemme en prenant $e = \Sigma t^k e_k$.

Démontrons pour terminer la conjecture (2).

Théorème (5.6) :

M est fini sur $\mathcal{B}$.

Filtrons $\mathcal{B}[s]$ par le degré en s, et filtrons $M = \mathcal{B}[s]/\mathfrak{J}$ par la filtration quotient. Evidemment $M° = \mathcal{B}/\mathfrak{J} \cap \mathcal{B}$ est fini sur $\mathcal{B}$ (notons qu'il est élémentaire de démontrer que $\mathfrak{J}\cap\mathcal{B}$ est engendré par les $\frac{\partial f}{\partial x_i}\frac{\partial}{\partial x_j} - \frac{\partial f}{\partial x_j}\frac{\partial}{\partial x_i}$, mais ce fait ne nous servira pas). Tout revient donc à montrer que $M/M°$ est fini sur $\mathcal{B}$.

Pour cela, montrons que $M/M°$ est de support l'origine ; il suffit d'établir que, pour tout k, M^{k+1}/M^k est de support l'origine. Or, ce module a par exemple pour générateur sur $\mathcal{D}$ la classe de l'élément $(s+1)^{k+1}$ de $\mathcal{D}[s]$; alors la formule $(s+1)\frac{\partial f}{\partial x_i} - \frac{\partial}{\partial x_i}\circ f \in \mathfrak{J}$ montre qu'on a $\frac{\partial f}{\partial x_i}(s+1)^{k+1} \in \mathfrak{J} + \mathcal{D}[s]^k$, donc que la classe de $\frac{\partial f}{\partial x_i}(s+1)^{k+1}$ modulo $\mathfrak{J}$ appartient à M^k ; d'où le résultat.

Pour démontrer que $M/M°$ est fini sur $\mathcal{D}$, il suffit donc, par (2.10), de démontrer que $H^n(M/M°)$ est fini sur $\mathbb{C}$. Appliquant (2.10.1) et la suite exacte de cohomologie, on trouve une suite exacte

$$0 \to H^n(M°) \to H^n(M) \to H^n(M/M°) \to 0$$

on a $H^n(M) = \tilde{G}$, et l'image de $H^n(M°)$ dans $\tilde{G}$ est manifestement égale à G on a donc $H^n(M/M°) = \tilde{G}/G$ (et accessoirement $H^n(M°) = G$). Or, le fait que $\tilde{G}/G$ soit fini sur $\mathbb{C}$ résulte de ce que $\tilde{G}$ est un réseau, ce qui d'ailleurs revient à dire que la connexion de Gauss-Manin est régulière. D'où le théorème.

BIBLIOGRAPHIE

[1] BERNSTEIN, I.N. - Prolongement analytique des fonctions généralisées avec paramètres (en russe).
Funkts. Analyz 6.4 (1972), p. 26-40

[2] BJÖRK, J.E. - Dimensions over algebras of differential operators.
A paraître.

[3] BRIESKORN, E. - Die Monodromie der isolierten singularitäten von hyperflächen.
Man. Math. 2, (1970) p. 103-161.

[4] DELIGNE, P. - Equations différentielles à points singuliers réguliers.
Lect. notes in Math. n° 163, Springer-Verlag (1970).

[5] KASHIWARA, M. - Papiers non publiés (en japonais).

[6] MALGRANGE, B. - Sur les polynômes de I.N. Bernstein.
Uspekhi Mat. Nauk 29-4 (1974), p. 81-88 (voir aussi Séminaire Goulaouic-Schwartz 1973-1974).

[7] MALGRANGE, B. - Intégrales asymptotiques et monodromie.
A paraître aux Ann. Ecole Norm. Sup.

[8] SATO, M., KAWAI, T., KASHIWARA, M. - Hyperfunctions and pseudodifferential equations.
Lect. notes in Math. n° 287, Springer-Verlag (1973), p. 264-529.

[9] SEBASTIANI, M. - Preuve d'une conjecture de Brieskorn.
Man. Math. 2 (1970), p. 301-308.

FOURIER INTEGRAL OPERATORS WITH COMPLEX-VALUED PHASE FUNCTIONS

Anders Melin and Johannes Sjöstrand

Contents

0. Introduction.

In this paper we shall present what we think is the natural extension of Hörmander's theory of Fourier integral operators to the case of complex valued phase functions. It is a well known phenomenon that complex phase functions appear in general, when one tries to construct parametrices or singular homogeneous solutions for operators of principal type with non-real principal symbol. It is therefore desirable to dispose a systematic theory for Fourier integral operators with complex valued phase functions.

Our paper follows very much the article [3] of Hörmander and we shall assume that the reader is acquainted with this paper. We shall also use the same notations as in [3] for function spaces, symbols and so on. Let us shortly describe one of the new difficulties one meets when trying to generalize the theory.

A Fourier integral distribution (or Fourier distribution for short) should be a distribution $A \in \mathcal{D}'(\mathbb{R}^n)$ which in a suitable sense is given (microlocally) by

$$A(x) = \int e^{i\varphi(x,\theta)} a(x,\theta)\, d\theta, \quad x \in \mathbb{R}^n,\ \theta \in \mathbb{R}^N. \qquad (0.1)$$

Here $\varphi \in C^\infty(V)$ and $V \subset \mathbb{R}^n \times (\mathbb{R}^N \setminus \{0\})$ is an open set, conic with respect to the variable θ. Moreover $a \in S^m_{1-\delta}(\mathbb{R}^n \times \mathbb{R}^N)$, $\delta < 1/2$, and has its support in a closed cone in $\mathbb{R}^n \times (\mathbb{R}^N \setminus \{0\})$, contained in V. Recall from [3], that φ is called a non-degenerate phase function if

(i) φ is real valued,

(ii) $d\varphi \neq 0$ everywhere in V,

(iii) $\varphi(x,\lambda\theta) = \lambda\varphi(x,\theta)$, $(x,\theta)\in V$, $\lambda \in \mathbb{R}_+$,

(iv) $d(\partial\varphi/\partial\theta_1),\ldots, d(\partial\varphi/\partial\theta_N)$ are linearly independent at $C_\varphi = \{(x,\theta) \in V \,;\, \varphi'_\theta(x,\theta) = 0\}$.

Clearly under these conditions C_φ is a manifold of dimension n and we know from [3] that the image $\Lambda_\varphi \subset \mathbb{R}^n \times (\mathbb{R}^n \setminus \{0\}) = T^*(\mathbb{R}^n)\setminus 0$ under the map

$$C_\varphi \ni (x,\theta) \longrightarrow (x,\varphi'_x(x,\theta)) \in T^*(\mathbb{R}^n)\setminus 0 \qquad (0.2)$$

is locally a Lagrangean manifold,which is an extremely important invariant attached to the Fourier integral distribution A in (0.1).

Now the natural generalization of the notion of non-degenerate phase function is the following: We say that $\varphi \in C^\infty(V)$ is a positive regular phase function (or "regular phase function of positive type ") if (ii) and (iii) are valid, and instead of (i) ,(iv) we have

(i)' $\mathrm{Im}\, \varphi \geq 0$,

(iv)' $d(\partial\varphi/\partial\theta_1),\ldots, d(\partial\varphi/\partial\theta_N)$ are linearly independent as complex vectors at $C_\varphi = C_{\varphi\mathbb{R}} = \{(x,\theta) \in V \,;\, \varphi'_\theta(x,\theta)=0\}$.

With such phase functions there is no new difficulty in defining distributions of the type (0.1).However there is a geometric difficulty : $C_{\varphi\mathbb{R}}$ is in general not a manifold.This difficulty is avoided by working with almost analytic extensions.These were first introduced by Hörmander [4] and later by Nirenberg [12] in a different context.If $\tilde{V} \subset \mathbb{C}^n \times (\mathbb{C}^N \setminus \{0\})$ is open with $\tilde{V} \cap (\mathbb{R}^n \times (\mathbb{R}^N \setminus \{0\})) = V$,we say that $\tilde{\varphi} \in C^\infty(\tilde{V})$ is an almost analytic extension of φ ,if $\bar{\partial}\tilde{\varphi}$ vanishes to infinite order at V and $\tilde{\varphi}|_V = \varphi$. Such extensions always exist, and if we put

$$C_{\tilde{\varphi}} = \left\{(x,\theta) \in \tilde{V} \;;\; \frac{\partial\tilde{\varphi}}{\partial\theta}(x,\theta) = 0\right\},$$

where $d_\theta\tilde{\varphi} = (\partial\tilde{\varphi}/\partial\theta)d\theta + (\partial\tilde{\varphi}/\partial\bar{\theta})\, d\bar{\theta}$, it is easy to see that $C_{\tilde{\varphi}}$ is a manifold near the real domain.We can now define a "Lagrangean manifold" $\Lambda_{\tilde{\varphi}} \subset \mathbb{C}^n \times (\mathbb{C}^n \setminus \{0\})$ as the image of the mapping

$$C_{\tilde{\varphi}} \ni (x,\theta) \longrightarrow (x, \frac{\partial\tilde{\varphi}}{\partial x}(x,\theta)) \in \mathbb{C}^n \times (\mathbb{C}^n \setminus \{0\}), \qquad (0.3)$$

restricted to a sufficiently small neighbourhood of $C_{\varphi\mathbb{R}}$.It turns out that for different choices of the extension $\tilde{\varphi}$, the corresponding manifolds $\Lambda_{\tilde{\varphi}}$ are equivalent with respect to a very natural equivalence relation.

The plan of the paper is the following:

In section 1 we introduce the almost analytic terminology we need,and prove some simple general results for almost analytic objects.This is necessary to do before we can even state the results of the calculus.

In section 2 we present an extension of the stationary phase method to the case of complex valued phase functions.The stationary phase method is essentially the only tool of analysis which is needed in order to build up the theory.In our case we have some new difficulties because the critical point of the phase function may disappear (out to the complex domain) after a small perturbation.Here it seems necessary to work with almost analytic extensions.At the end of this section we present an application which might be of independent interest.(Cf. [7,Lemma 1.3.1] and [13 ,Lemma 5.1].)

In section 3 we introduce so called positive(almost analytic) Lagrangean manifolds and we establish the connection between these and positive regular phase functions,given by (0.3).

In section 4 we prove the fundamental result of the theory:If two phase functions φ and ψ give equivalent Lagrangean manifolds $\Lambda_{\tilde{\varphi}}$ and $\Lambda_{\tilde{\psi}}$ near a point then microlocally they give rise to the same classes of distributions defined by formula (0.1).This result permits us to develop the global theory on paracompact manifolds.The converse of this result is also true if only classical symbols are considered ,but not in general.This is proved in section 5.

In section 6 we define the principal symbol as a section of a suitable "Maslov line bundle ", and in section 7 we prove the results about composition of Fourier integral operators ,completely along the lines of [3].

In section 8 finally we present two applications.The second one is perhaps the most interesting one.It treats in a particular case the following general problem: Given an operator P of principal type ,construct a distribution $u \notin C^{\infty}$ such that $Pu \in C^{\infty}$ (for instance with given Cauchy data on some hypersurface).This is actually the main application that we had in mind for our calculus, and in as general cases as possible we would like to solve the problem by a "geometrical optics " construction in two steps:

I . If $\tilde{p}$ is an almost analytic extension of the principal symbol,construct a positive Lagrangean manifold Λ contained in $(\tilde{p})^{-1}(0)$.

II. Try with u as a Fourier distribution corresponding to Λ . Then the principal symbol of u as well as the lower order terms in the symbol should be obtained by solving certain transport equations on Λ.

The details of the construction in section 8 as well as of another particular case of our general problem will bè given elsewhere.

In his lecture [11]at the international congress in Nice, Maslov has stated a generalization to the complex case of his theory of the Canonical operator.We have studied [11] without having quite understood everything.However we think it is a question of an approximate theory.Recently Kucherenko [8,9] has published two very interesting notes .The first one treats the Hamilton-Jacobi integration theory in the complex case.He works with approximate almost analytic extensions f for which the $\bar{\partial} f$ vanish to high but finite order.He also has a definition reminding of our definition of positive Lagrangean manifolds.In the second note he states a "Canonical Operator " theory and applies it to construct oscillating approximate solutions of a Cauchy problem.

A few weeks after this colloquium in Nice there appeared a more complete work of Kucherenko in Mat.Sbornik 94(136):1(5) which we have not yet got the time to translate and read . There is also a recent book of Maslov , treating the case of complex phase functions. We thank professor Bojarski who showed it to us during the colloquium and translated some parts of it .

Finally we would like to thank professor Hörmander who suggested us to this work and who also read an earlier version of this manuscript and gave us suggestions for improvements.

1. Almost analytic machinery.

Hörmander [4] and Nirenberg [12] have in different contexts introduced the notion of almost analytic functions. This notion will be of fundamental importance for us in the present paper, and so will the notion of almost analytic manifolds.

We introduce the following notations: If f is a smooth function in an open set in $\mathbb{C}^m$ then $df = \partial f + \bar{\partial} f$, where $\partial f = \frac{\partial f}{\partial z} dz = \sum \frac{\partial f}{\partial z_j} dz_j$ and $\bar{\partial} f = \frac{\partial f}{\partial \bar{z}} d\bar{z} = \sum \frac{\partial f}{\partial \bar{z}_j} d\bar{z}_j$, is the decomposition of df in its complex linear and complex anti-linear parts. We shall often write f'_z instead of $\frac{\partial f}{\partial z}$. If we have chosen $\tau_1, \ldots, \tau_m$ as notations for the standard coordinates for $\mathbb{R}^m$ then the corresponding coordinates for $\mathbb{C}^m$ will frequently be denoted $\tilde{\tau}_1, \ldots, \tilde{\tau}_m$.

Definition 1.1. Let $\omega \subset \mathbb{C}^n$ be an open set and let $\Gamma \subset \omega$ be closed. If $f \in C^\infty(\omega)$, we say that f is almost analytic at Γ if $\bar{\partial} f$ vanishes to infinite order there. When $\Gamma = \omega_{\mathbb{R}} = \omega \cap \mathbb{R}^n$ we simply say that f is almost analytic.

We point out that this definition will be generalized later.

Definition 1.2. Let $f_1, f_2 \in C^\infty(\omega)$ with ω, Γ as above. We say that f_1 and f_2 are equivalent at Γ if $f_1 - f_2$ vanishes to infinite order there. When $\Gamma = \omega_{\mathbb{R}}$ we just say that they are equivalent and write $f_1 \sim f_2$.

It is quite elementary to prove that every $f \in C^\infty(\omega_{\mathbb{R}})$ has an almost analytic extension, uniquely determined up to equivalence. We give in fact a slightly stronger result due to Hörmander [4].

Theorem 1.3 . Suppose that $\tilde{\Gamma} \subset \mathbb{C}^n \times (\mathbb{C}^M \setminus \{0\})$ is an open cone in the sense that $(z,\zeta) \in \tilde{\Gamma}$, $\lambda \in \mathbb{R}_+ \Rightarrow (z,\lambda\zeta) \in \tilde{\Gamma}$ and let Γ be the intersection with the real domain. If $a \in S^m_{1-\delta}(\Gamma)$, there exists an extension $\tilde{a}(z,\zeta) \in S^m_{1-\delta}(\tilde{\Gamma})$ vanishing when

$$|\zeta|^{\delta}|y| + |\zeta|^{\delta-1}|\eta| \geq 1 \text{ , } z = x+iy \text{ , } \zeta = \theta+i\eta \text{ ,}$$

and such that for $|\zeta| \geq 1$

$$|\bar{\partial}_z \tilde{a}(z,\zeta)| + |\zeta||\bar{\partial}_\zeta \tilde{a}(z,\zeta)| \leq C_{\Gamma',N} |\zeta|^{m+\delta+N\delta} |(y, \eta/|\theta|)|^N \qquad (1.1)$$

in Γ' for all open cones $\Gamma' \subset\subset \tilde{\Gamma}$ and $N \in \mathbb{Z}_+$. (We write $\Gamma' \subset\subset \tilde{\Gamma}$ if we have the corresponding relations between the intersections with $\mathbb{C}^n \times S^{2M-1}$.) Any other almost analytic extension $b(z,\zeta) \in S^m_{1-\delta}(\tilde{\Gamma})$ of a is equivalent to $\tilde{a}(z,\zeta)$ in the sense that

$$|b(z,\zeta) - \tilde{a}(z,\zeta)| \leq C'_{\Gamma',N} |\zeta|^{m+\delta N} |(y, \eta/|\theta|)|^N \text{ , } |\zeta| \geq 1, \qquad (1.2)$$

in Γ' for all $\Gamma' \subset\subset \tilde{\Gamma}$, $N \in \mathbb{Z}_+$.

Finally if $a \in S^m(\Gamma)$ is positively homogeneous of degree m , we can choose $\tilde{a}$ homogeneous in the sense that for all $\Gamma' \subset\subset \Gamma$:

$$\tilde{a}(z,\lambda\zeta) = \lambda^m \tilde{a}(z,\zeta) \text{ , } (z,\zeta) \in \Gamma' \text{ , } \operatorname{Re}\lambda > 0 \qquad (1.3)$$

$$|\operatorname{Im}\lambda|/\operatorname{Re}\lambda < \varepsilon_{\Gamma'} \text{ , } \quad |(y, \eta/|\theta|)| \leq \varepsilon_{\Gamma'}$$

where $\varepsilon_{\Gamma'}$ is a positive constant depending on Γ' .

Proof. (Cf [4]) It suffices to prove the theorem in the case when $\tilde{\Gamma} = \mathbb{C}^n \times (\mathbb{C}^M \setminus \{0\})$ and when $a(x,\theta)$ vanishes for x outside some compact subset of $\mathbb{R}^n$. This follows from a partition of unity argument (for by the construction below it follows that we have almost analytic partitions of unity). For simplicity we shall also assume that $a(x,\theta)$ vanishes for $|\theta| \leq 1$.

Let $\chi(y,\eta) \in C_0^\infty(\mathbb{R}^{n+M})$ be equal to one in a neighbourhood of the origin and vanish for $|(y,\eta)| \geq 1$. Put

$$\tilde{a}(z,\zeta) = a(x+iy,\theta+i\eta) = \tag{1.4}$$

$$\sum a_{(\alpha)}^{(\beta)}(x,\theta)(iy)^\alpha(i\eta)^\beta \chi(t_{\alpha,\beta}(|\theta|^\delta y, |\theta|^{\delta-1}\eta))/\alpha!\beta!$$

where the numbers $t_{\alpha,\beta} \geq 1$ will be chosen very large later so that the series converges. Put $(\tilde{y},\tilde{\eta}) = (|\theta|^\delta y, |\theta|^{\delta-1}\eta)$. Then

$$\tilde{a}(z,\zeta) = \sum b_{\alpha,\beta}(x,\theta)\, \chi_{\alpha,\beta}(t_{\alpha,\beta}(\tilde{y},\tilde{\eta}))\, t_{\alpha,\beta}^{-(|\alpha|+|\beta|)} \tag{1.5}$$

where $b_{\alpha,\beta}(x,\theta) = a_{(\alpha)}^{(\beta)}(x,\theta)\, |\theta|^{-\delta|\alpha|+(1-\delta)|\beta|} \in S^m_{1-\delta}(\mathbb{R}^n \times \mathbb{R}^M)$

and $\chi_{\alpha,\beta}(y,\eta) = (iy)^\alpha(i\eta)^\beta\chi(y,\eta)/\alpha!\beta!$. Now

$$D_{\tilde{y}}^\mu D_{\tilde{\eta}}^\nu \chi_{\alpha,\beta}(t_{\alpha,\beta}(\tilde{y},\tilde{\eta})) = t_{\alpha,\beta}^{|\mu|+|\nu|}\chi_{\alpha,\beta}^{\mu,\nu}(t_{\alpha,\beta}(\tilde{y},\tilde{\eta})),$$

where $\chi_{\alpha,\beta}^{\mu,\nu} = D_y^\mu D_\eta^\nu \chi_{\alpha,\beta}$, so it follows that we have the estimates

$$|D_\theta^\gamma D_{\tilde{y}}^\mu D_{\tilde{\eta}}^\nu \chi_{\alpha,\beta}(t_{\alpha,\beta}(\tilde{y},\tilde{\eta}))| \leq C_{\alpha,\beta,\gamma,\mu,\nu}\, t_{\alpha,\beta}^{|\gamma|+|\mu|+|\nu|}|\theta|^{-(1-\delta)|\gamma|} \tag{1.6}$$

Thus if the $t_{\alpha,\beta}$ are chosen sufficiently large, we have

$$|D_x^\rho D_\theta^\gamma D_{\tilde{y}}^\mu D_{\tilde{\eta}}^\nu (b_{\alpha,\beta}(x,\theta)\chi_{\alpha,\beta}(t_{\alpha,\beta}(\tilde{y},\tilde{\eta}))\, t_{\alpha,\beta}^{-(|\alpha|+|\beta|)}| \leq \tag{1.7}$$
$$2^{-(|\alpha|+|\beta|)}\, |\theta|^{m+\delta|\rho|-(1-\delta)|\gamma|}, \text{ if } |\rho|+|\gamma|+|\mu|+|\nu| < |\alpha|+|\beta|.$$

It now follows from (1.5)-(1.7) that (1.4) converges geometrically with all its derivatives and that

$$|D_x^\rho D_\theta^\gamma D_{\tilde{y}}^\mu D_{\tilde{\eta}}^\nu \tilde{a}(z,\zeta)| \leq C_{\rho,\gamma,\mu,\nu}\, |\theta|^{m+\delta|\rho|-(1-\delta)|\gamma|}. \tag{1.8}$$

Now $|\zeta| \leq 2|\theta|$ in the support of $\tilde{a}$ and since $D_{\tilde{y}_j} = |\theta|^{-\delta} D_{y_j}$ and $D_{\tilde{\eta}_j} = |\theta|^{1-\delta} D_{\eta_j}$, it follows from (1.8) that $\tilde{a} \in S^m_{1-\delta}(\mathbb{C}^n \times \mathbb{C}^M)$. It is trivial that $|\zeta|^{\delta}|y| + |\zeta|^{\delta-1}|\eta| \leq 1$ in the support of $\tilde{a}$.

For $1 \leq j \leq n$, let e_j be the j:th unit vector in $\mathbb{R}^n$ considered as a multi-index. Then by a simple calculation

$$\partial\tilde{a}(z,\zeta)/\partial\bar{z}_j = \sum a^{(\beta)}_{(\alpha+e_j)}(x,\theta)(iy)^{\alpha}(i\eta)^{\beta}(\chi(t_{\alpha,\beta}(\tilde{y},\tilde{\eta}))-\chi(t_{\alpha+e_j,\beta}(\tilde{y},\tilde{\eta})))/2\alpha!\beta!$$
$$+ \sum a^{(\beta)}_{(\alpha)}(x,\theta)(iy)^{\alpha}(i\eta)^{\beta}\frac{\partial}{\partial\bar{z}_j}\chi(t_{\alpha,\beta}(\tilde{y},\tilde{\eta}))/\alpha!\beta!$$

and using the analogous formula for $\partial\tilde{a}/\partial\bar{\zeta}_j$, we see that $\bar{\partial}\tilde{a}$ vanishes to infinite order on $\mathbb{R}^n \times \mathbb{R}^M$. Since $\tilde{a} \in S^m_{1-\delta}(\mathbb{C}^n \times \mathbb{C}^M)$ it follows by a Taylor expansion of $\partial\tilde{a}/\partial\bar{z}$ and $\partial\tilde{a}/\partial\bar{\zeta}$ at $\mathbb{R}^n \times \mathbb{R}^M$ that (1.1) holds.

As for the uniqueness, let $b(z,\zeta) \in S^m_{1-\delta}(\mathbb{C}^n \times (\mathbb{C}^M \setminus \{0\}))$ be another almost analytic extension and put $c(z,\zeta) = \tilde{a}(z,\zeta) - b(z,\zeta)$. Then c vanishes on $\mathbb{R}^n \times (\mathbb{R}^M \setminus \{0\})$ and $\bar{\partial}c$ vanishes to infinite order there. It follows easily by induction that all derivatives of c vanish on $\mathbb{R}^n \times (\mathbb{R}^M \setminus \{0\})$ and we get (1.2) from Taylor's formula.

Now suppose that $a \in S^m(\mathbb{R}^n \times (\mathbb{R}^M \setminus \{0\}))$ is positively homogeneous of degree m in θ. Without any restriction we can assume that a has its support in a small cone where $\theta_M > 0$. Let $b(z, \zeta') \in C^{\infty}(\mathbb{C}^n \times \mathbb{C}^{M-1})$ be an almost analytic extension of $a(x,\theta',1)$ with compact support in ζ' and set $\tilde{a}(z,\zeta) = \zeta_M^m\, b(z, \zeta'/\zeta_M)$ for small $|\mathrm{Im}\,\zeta_M|/\mathrm{Re}\,\zeta_M$ when $\mathrm{Re}\,\zeta_M > 0$ and define $\tilde{a}(z,\zeta)$ suitably cut down to zero outside a conic neighbourhood of this domain. By a partition of unity we also get $\tilde{a}$ in the case when supp a is arbitrary.

We next introduce the notion of almost analytic manifolds. <u>By definition an almost analytic manifold is a C^{∞} manifold Λ satisfying one of the equivalent conditions in the following theorem.</u>

Theorem 1.4. Let $\omega \subset \mathbb{C}^n$ be an open set and let $\Lambda \subset \omega$ be a closed submanifold of real dimension $2k$. Then the following conditions are equivalent:

1^0. For all $\omega' \subset\subset \omega$ and $N \in \mathbb{Z}_+$ there is a constant C_N such that

$$d(T_z(\Lambda), iT_z(\Lambda)) \leq C_N |\mathrm{Im}\, z|^N , \quad z \in \omega' \cap \Lambda .$$

Here $i : T_z(\mathbb{C}^n) \longrightarrow T_z(\mathbb{C}^n)$ is the map induced by multiplication with the imaginary unit and $d(,)$ denotes some distance in the Grassmannian.

2^0. For every $z_0 \in \Lambda$ one can find a neighbourhood $\omega \subset \Lambda$ of z_0 and a C^∞ map $\omega \ni z \longrightarrow \hat{I}_z \in \mathcal{L}(T_z(\Lambda), T_z(\Lambda))$ such that for all $N \in \mathbb{Z}_+$ there is a constant C_N with

$$\| \hat{I}_z - i \| \leq C_N |\mathrm{Im}\, z|^N , \quad z \in \omega .$$

3^0. For every real point z_0 of Λ we can find a neighbourhood ω of z_0 in which after regrouping the coordinates in $\mathbb{C}^n$ the manifold Λ is given by $z'' = h(z')$, $z' = (z_1, \ldots, z_k)$, $z'' = (z_{k+1}, \ldots, z_n)$, where h is a C^∞ map satisfying the inequalities

$$|\bar{\partial}_{z'} h(z')| \leq C_N |\mathrm{Im}(z', h(z'))|^N , (z', h(z')) \in \omega ,$$

for all $N \in \mathbb{Z}_+$.

4^0. For every real point z_0 of Λ there exists a neighbourhood ω of z_0 and C^∞ functions $f_{k+1}, \ldots, f_n$ such that Λ is given by $f_{k+1}(z) = \ldots = f_n(z) = 0$ in ω and

$$|\bar{\partial} f_j(z)| \leq C_N (|\mathrm{Im}\, z|^N + \max_\nu |f_\nu(z)|^N) , \quad z \in \omega ,$$

for all $N \in \mathbb{Z}_+$ and the complex linear differentials $\partial f_{k+1}(z), \ldots, \partial f_n(z)$ are linearly independent over $\mathbb{C}$.

Proof. It is trivial that 2^0 implies 1^0 and that 3^0 implies 4^0. To prove that 1^0 implies 2^0, let $P_z: T_z(\mathbb{C}^n) \longrightarrow T_z(\Lambda)$, $z \in \Lambda$, be the (real) ortogonal projection. Then the conditions in 2^0 are fulfilled if we set $I_z^{\Lambda} = P_z \circ i$.

We next prove that 1^0 implies 3^0 so we let $x_0 \in \Lambda \cap \mathbb{R}^n$. Then $T_{x_0}(\Lambda)$ is a complex linear space given by an equation $A\,z = 0$ in $\mathbb{C}^n$ where $A: \mathbb{C}^n \longrightarrow \mathbb{C}^{n-k}$ is a surjective complex linear map.After regrouping the coordinates in $\mathbb{C}^n$ we have $A\,z = A'z' + A''z''$, where $z = (z', z'') \in \mathbb{C}^k \times \mathbb{C}^{n-k}$ and A'' is bijective. It follows that Λ is given near x_0 by a C^∞ equation $z'' = h(z')$ .A vector $(t_{z'}, t_{z''})$ is in the tangent space $T_z(\Lambda)$ for $z = (z', h(z'))$ if $t_{z''} = dh(z')\,t_{z'}$ and the condition 1^0 therefore implies that h satisfies the following inequalities in a neighbourhood of x_0'

$$\|i \circ dh(z') - dh(z') \circ i\| \leq C_N |\mathrm{Im}(z', h(z'))|^N$$

when $N \in \mathbb{Z}_+$. Then 3^0 follows if we observe that $[i, dh] = [i, \bar{\partial}_{z'} h] = 2i\,\bar{\partial}_{z'} h$. We leave to the reader to show that 4^0 implies 1^0.

Next we shall introduce an equivalence relation for (almost analytic) manifolds.In order to do that we shall first establish a technical lemma which will be used several times in the following

Lemma 1.5. Let $\Omega \subset \mathbb{R}^n$ be an open set and suppose that $u_\lambda(x) \in S^0_{1-\delta}(\Omega \times \mathbb{R}_+)$, $x \in \Omega$, $\lambda \in \mathbb{R}_+$. Let $v(x)$ be a Lipschitz continuous function on Ω so that $|v(x) - v(y)| \leq C_{\Omega'}|x-y|$ when $x, y \in \Omega'$ and $\Omega' \subset\subset \Omega$. Suppose that for all $N \in \mathbb{Z}_+$, we have

$$|u_\lambda(x)| \leq C_{N,\Omega'}\,\lambda^{N\delta}\,|v(x)|^N, \quad x \in \Omega', \ \lambda \geq 1. \qquad (1.9)$$

Then for all $\Omega' \subset\subset \Omega$, $N \in \mathbb{Z}_+$ and multi-indices α there is a constant $C_{N,\Omega',\alpha}$ such that

$$|v(x)|^{|\alpha|} |D_x^\alpha u_\lambda(x)| \leq C_{N,\Omega',\alpha} \lambda^{N\delta} |v(x)|^N, \quad x \in \Omega'. \qquad (1.10)$$

Proof. By induction it is sufficient to prove (1.10) when $|\alpha| = 1$. It is also enough to consider points x with $|v(x)| < \lambda^{-\delta}$. We shall use the following well-known and elementary inequalities valid for $f \in C^\infty(\mathbb{R}^n)$

$$r \sum_{|\alpha|=1} \sup_{B(r/2)} |D^\alpha f| \leq$$

$$\leq C_N (\sup_{B(r)} |f|)^{1-1/N} (\sup_{B(r)} |f| + r^N \sum_{|\beta|=N} \sup_{B(r)} |D^\beta f|)^{1/N}, \qquad (1.11)$$

when $N \in \mathbb{Z}_+$, $r > 0$ and $B(r) = \{ x \in \mathbb{R}^n ; |x| \leq r \}$. Since v is Lipschitz continuous we have $|v(x)|/2 \leq |v(y)| \leq 2|v(x)|$ in the ball $B(r,x) = \{ y \in \mathbb{R}^n ; |x - y| \leq r = \varepsilon |v(x)| \}$, if $\varepsilon > 0$ is small enough. Then by (1.9) we have inequalities

$$|u_\lambda(y)| \leq C'_N |v(x)|^N \lambda^{N\delta} \quad , \quad y \in B(r,x),$$

and since $u_\lambda \in S^0_{1-\delta}(\Omega \times \mathbb{R}_+)$ the same type of inequalities hold for $r^N D_y^\beta u_\lambda(y)$ in $B(r,x)$ when $|\beta| = N$. Using (1.11) we then get

$$\varepsilon |v(x)| \cdot |D_x^\alpha u_\lambda(x)| \leq C''_N |v(x)|^N \lambda^{N\delta}, \ |\alpha| = 1, \ N \in \mathbb{Z}_+ .$$

This completes the proof of the lemma since we may choose ε constant on each compact subset of Ω .

In this lemma note the special case when $u \in C^\infty(\Omega)$ is independent of λ . We can now define equivalence of manifolds:

Definition 1.6. Let Λ_1 and Λ_2 be two closed submanifolds of an open set $\omega \subset \mathbb{C}^n$. We say that Λ_1 and Λ_2 are equivalent (and we write $\Lambda_1 \sim \Lambda_2$) if they have the same intersection with $\mathbb{R}^n$ and the same dimension and if for every $\omega' \subset\subset \omega$ and $N \in \mathbb{Z}_+$ we have

$$d(z, \Lambda_2) \leq C_{N,\omega'} |\mathrm{Im}\, z|^N, \quad z \in \omega' \cap \Lambda_1,$$

for some constant $C_{N,\omega'}$. Here $d(z, \Lambda_2)$ denotes the distance from z to Λ_2.

It is trivial that $\sim$ is an equivalence relation and that Λ_1 and Λ_2 are tangential to infinite order in the real points when $\Lambda_1 \sim \Lambda_2$. (However , two manifolds may have the same real parts and be tangential to infinite order there without being equivalent.) Note that our notion of equivalence can be easily extended to the case when $\Lambda_j \subset \omega_j$, $j = 1,2$, if $\omega_1 \cap \mathbb{R}^n = \omega_2 \cap \mathbb{R}^n$. Also note that we can define the notion of local equivalence near a real point.

Proposition 1.7. Let $\Lambda_j \subset \omega \subset \mathbb{C}^n$, $j = 1,2$, be almost analytic manifolds of the same dimension and with $\Lambda_{1\mathbb{R}} = \Lambda_{2\mathbb{R}}$ (= the intersections with the real domain) . Suppose that Λ_j is given by $z'' = h_j(z')$, $j = 1,2$, $h_j \in C^\infty$, $z' \in \mathbb{C}^k$, $z'' \in \mathbb{C}^{n-k}$. Then the following conditions are equivalent :

(i) $\Lambda_1 \sim \Lambda_2$,

(ii) For all $\omega' \subset\subset \omega$ and $N \in \mathbb{Z}_+$ there is a constant $C_{N,\omega'}$ so that $|h_1(x') - h_2(x')| \leq C_{N,\omega'} |\mathrm{Im}\, h_2(x')|^N$ when $(x', h_j(x')) \in \omega'$, $x' \in \mathbb{R}^k$.

(continued)

(iii) For all $\omega' \subset\subset \omega$, $N \in \mathbb{Z}_+$ and multi-indices α there is a constant $C_{N,\omega',\alpha}$ so that $|D^{\alpha}_{x'}(h_1(x')-h_2(x'))| \leq$ $\leq C_{N,\omega',\alpha}$ $|\mathrm{Im}\ h_2(x')|^N$ when $(x', h_j(x')) \in \omega'$, $x' \in \mathbb{R}^k$.

(iv) For all $\omega' \subset\subset \omega$ and $N \in \mathbb{Z}_+$ there is a constant $C_{N,\omega'}$ so that $|h_1(z') - h_2(z')| \leq C_{N,\omega'}|\mathrm{Im}(z',h_2(z'))|^N$ when $(z',h_j(z')) \in \omega'$.

Proof. It is clear that (i) $\Longleftrightarrow$ (iv) $\Longrightarrow$ (ii) . Moreover (ii) $\Longrightarrow$ (iii) by the preceding lemma. It remains to prove that (iii) $\Longrightarrow$ (iv).Since Λ_j are almost analytic we have

$$|\bar{\partial}_{z'} \ h_j(z')| \leq C_N \ |\mathrm{Im}(z',h_j(z'))|^N \ ,(z',h_j(z')) \in \omega' ,$$

when $\omega' \subset\subset \omega$ and $N \in \mathbb{Z}_+$. By Lemma 1.6 we have the same estimate for all the derivatives of $\bar{\partial}_{z'} h_j$. Using this and (iii) we get recursively the local estimates

$$|D^{\alpha}_{y'}(h_1 - h_2)(x'+iy')| \leq C_{N,\alpha} \ |\mathrm{Im}\ h_2(x')|^N \ , \ x' \in \mathbb{R}^k, y' = 0,$$

for all multi-indices α . By Taylor's formula we then get locally for all $N \in \mathbb{Z}_+$:

$$|(h_1 - h_2)(z')| \leq C_N \ (\ |\mathrm{Im}\ h_2(x')|^N + |\mathrm{Im}\ z'|^N \ )$$

$$\leq C'_N \ (\ |\mathrm{Im}\ h_2(z')|^N + |\mathrm{Im}\ z'|^N \ ) \ .$$

This proves that (iii) $\Longrightarrow$ (iv) and the proof is complete.

Note that every almost analytic manifold Λ is locally equivalent to one given by almost analytic equations $f_{k+1}(z) =$ $= \dots = f_n(z) = 0$ with linearly independent differentials (over $\mathbb{C}$) at the real points. In fact, let Λ be given locally by the

C^∞ equation $z'' = h(z')$, $z' \in \mathbb{C}^k$, and let Λ' be the manifold $z'' = h'(z')$ where h' is an almost analytic extension of $h|_{\mathbb{R}^k}$. Then $\Lambda' \sim \Lambda$ locally by the proposition above. Also note that any manifold which is locally equivalent to an almost analytic manifold is also almost analytic.

Next we consider almost analytic functions on almost analytic manifolds.

Definition 1.8. Let $\Lambda \subset \omega \subset \mathbb{C}^n$ be almost analytic and let $f \in C^\infty(\Lambda)$. We say that f is almost analytic if for every $N \in \mathbb{Z}_+$ and every $\omega' \subset\subset \omega$ we have

$$\| i \circ df(z) - df(z) \circ I_z^\Lambda \| \leq C_{N,\omega'} |\operatorname{Im} z|^N , \quad z \in \Lambda \cap \omega'.$$

Here i denotes multiplication with the imaginary unit in $T_{f(z)}(\mathbb{C})$ and I_z^Λ is any operator as in condition 2^0 of Theorem 1.4.

Definition 1.9. Let Λ be almost analytic and let f_1 and f_2 be C^∞ functions on Λ. We say that f_1 and f_2 are equivalent and we write $f_1 \sim f_2$ if for every $\omega' \subset\subset \Lambda$ and $N \in \mathbb{Z}_+$ there is a constant $C_{N,\omega'}$ so that $|(f_1 - f_2)(z)| \leq$ $\leq C_{N,\omega'} |\operatorname{Im} z|^N$ when $z \in \omega'$.

It is clear by Lemma 1.5 that if $f_1 \sim f_2$ and f_2 is almost analytic then so is f_1. Also note that if $f_j \in C^\infty(\Lambda_j)$, $j=1,2$, and $\Lambda_1 \sim \Lambda_2$ then there is a natural way to extend the notion of equivalence so that we can define " $f_1 \sim f_2$" .

Now suppose that $f \in C^\infty(\Lambda)$ is almost analytic and that Λ is given by the C^∞ equation $z'' = h(z')$, so that for all $N \in \mathbb{Z}_+$ we have locally

$$|\bar{\partial}_{z'} h(z')| \leq C_N \, |\operatorname{Im}(z',h(z'))|^N , \quad z' \in \mathbb{C}^k .$$

Then we can regard f as a function of z' and the almost analyticity on Λ means that we have the local estimates

$$|\bar{\partial}_{z'} f(z')| \leq C_N \, |\operatorname{Im}(z',h(z'))|^N , \quad z' \in \mathbb{C}^k .$$

Let $g(z')$ be an almost analytic extension of $f|_{\mathbb{R}^k}$. It is easy to prove using Lemma 1.5 that

$$|f(z') - g(z')| \leq C_N \, |\operatorname{Im}(z',h(z'))|^N$$

locally for all $N \in \mathbb{Z}_+$. Now put $G(z) = g(z')$. Then G is almost analytic on $\mathbb{C}^n$ and $G|_\Lambda \sim f$. More generally , by a partition of unity we get

Lemma 1.10. If $\Lambda \subset \omega \subset \mathbb{C}^n$ is almost analytic and $f \in C^\infty(\Lambda)$ is almost analytic, then there exists $g \in C^\infty(\omega)$, almost analytic and such that $g|_\Lambda \sim f$.

Remark. Assume that the almost analytic manifold Λ above is also given by almost analytic equations $f_1(z) = \ldots = f_{n-k}(z) = 0$, where k is the number of z'-variables and the differentials df_i are linearly independent over $\mathbb{C}$. Assume that $g \in C^\infty(\omega)$ is almost analytic and that $g|_\Lambda \sim 0$. Then there are functions $t_j \in C^\infty(\omega_{\mathbb{R}})$ such that

$$g(x) = \sum t_j(x) \, f_j(x) \quad , \quad x \in \omega_{\mathbb{R}} . \tag{1.12}$$

In fact , with h as in the proof of Lemma 1.10 we get

$$g(z) = g(z',h(z')) + \int_0^1 \frac{\partial}{\partial s} g(z',h(z')+s(z''-h(z'))) \, ds$$

$$= \langle z''-h(z'), B(z)\rangle + \rho(z) \quad ,$$

where B is some almost analytic matrix-valued function and where $\rho(z) = \mathcal{O}(|\mathrm{Im}\, z|^N + |\mathrm{Im}\, h(z')|^N)$ for all $N \in \mathbb{Z}_+$. Since $f|_\Lambda \sim 0$, where $f = (f_1, \dots, f_{n-k})$ we also get

$$f(z) = C(z)(z''-h(z')) + \rho_f(z)$$

with the same type of remainder ρ_f as above and with an invertible $(n-k) \times (n-k)$ matrix C. Therefore we may replace $z'' - h(z')$ by f in (1.13). Now it is clear that $|\mathrm{Im}\, h(x')| = |\mathrm{Im}(x'' - h(x'))| = \mathcal{O}(|f(x)|)$, uniformly on every compact subset of $\omega_{\mathbb{R}}$, so using Lemma 1.5 we see that $\rho(x)$ contains the factor $|f(x)|^2 = \langle f(x), \overline{f(x)}\rangle$ in $C^\infty(\omega_{\mathbb{R}})$. This shows that $g(x)$ has the form (1.12)

Next we consider almost analytic mappings.

Lemma 1.11. Let $\Lambda \subset \omega \subset \mathbb{C}^n$ be almost analytic and let $\mathcal{H} : \Lambda \longrightarrow \mathbb{C}^m$ be an almost analytic map with an injective differential. If for every $\omega' \subset\subset \omega$ we can find positive constants C and k so that

$$|\mathrm{Im}\, \mathcal{H}(z)| \geq C\, |\mathrm{Im}\, z|^k \, , \qquad z \in \omega' \cap \Lambda \quad ,$$

then $\mathcal{H}(\Lambda)$ is locally an almost analytic manifold whose equivalence class does not change if we replace $(\Lambda, \mathcal{H})$ by $(\Lambda', \mathcal{H}')$ where $\Lambda \sim \Lambda'$, $\mathcal{H} \sim \mathcal{H}'$.

Proof. We adopt the notations of Theorem 1.4. Let $t \in i\, T_{\mathcal{H}(z)}(\mathcal{H}(\Lambda))$, $\|t\| = 1$, $z \in \Lambda$. Then $t = i \circ d\mathcal{H}(z)s$, $s \in T_z(\Lambda)$, and we can write $t = (\, i \cdot d\mathcal{H} - d\mathcal{H} \circ \hat{I}_z)s + d\mathcal{H} \circ \hat{I}_z\, s$.

Since the last term is in $T_{\mathcal{H}(z)}(\mathcal{H}(\Lambda))$ it follows that for all $N \in \mathbb{Z}_+$ we have the estimate

$$d(\, iT_{\mathcal{H}(z)}(\mathcal{H}(\Lambda)), T_{\mathcal{H}(z)}(\mathcal{H}(\Lambda)) \leq C_N |\mathrm{Im}\, z|^N$$

$$\leq C'_N |\mathrm{Im}\,\mathcal{H}(z)|^{N/k}, \quad z \in \omega' \cap \Lambda.$$

where $d(\ ,\ )$ is some distance in the Grassmannian .This shows that $\mathcal{H}(\Lambda)$ is locally almost analytic . The last assertion in the lemma follows easily from the definition of equivalence.

Now let $\mathcal{H}: \Lambda_1 \longrightarrow \Lambda_2$ be a C^∞map between two almost analytic manifolds. We say that $\mathcal{H}$ is almost analytic if the graph of $\mathcal{H}$ is almost analytic. We have the following easy lemma.

<u>Lemma 1.12</u>. Let $\mathcal{H}$, Λ_1, Λ_2 be as above and let f be an almost analytic function on Λ_2.Then $f \circ \mathcal{H}$ is almost analytic if for every $\omega \subset\subset \Lambda$ we can find positive constants C and k so that

$$|\mathrm{Im}\, \mathcal{H}(z)| \leq C\ |\mathrm{Im}\, z|^k, \quad z \in \omega.$$

We shall now generalize the notion of almost analytic manifolds. Let M be a n-dimensional real paracompact C^∞manifold. By definition an almost analytic manifold Λ associated to M (or formally : $\Lambda \subset \widetilde{M}$) is given by

1^0 A locally closed set $\Lambda_{\mathbb{R}} \subset M$. ("Locally closed" means that $\Lambda_{\mathbb{R}}$ is the intersection of a closed and an open set or equivalently that every point of $\Lambda_{\mathbb{R}}$ has an open neighbourhood ω in M such that $\Lambda_{\mathbb{R}} \cap \omega$ is closed in ω .)

2^0 A covering of $\Lambda_{\mathbb{R}}$ by open coordinate neighbourhoods

$$M \supset X_\alpha \xrightarrow{\mathcal{H}_\alpha} \Omega_\alpha \subset \mathbb{R}^n , \alpha \in J$$

and almost analytic manifolds $\Lambda_\alpha \subset \tilde{\Omega}_\alpha \subset \mathbb{C}^n$ with $\Lambda_{\alpha\mathbb{R}} = \mathcal{H}_\alpha(X_\alpha \cap \Lambda_{\mathbb{R}})$. Here $\tilde{\Omega}_\alpha \subset \mathbb{C}^n$ is some open set with $\tilde{\Omega}_\alpha \cap \mathbb{R}^n = \Omega_\alpha$ and the Λ_α shall satisfy the following compatibility conditions : If $\mathcal{H}_{\beta\alpha} = \mathcal{H}_\beta \circ \mathcal{H}_\alpha^{-1} : \mathcal{H}_\alpha(X_\alpha \cap X_\beta) \longrightarrow \mathcal{H}_\beta(X_\alpha \cap X_\beta)$ and if $\tilde{\mathcal{H}}_{\beta\alpha}$ is an almost analytic extension , then $\tilde{\mathcal{H}}_{\beta\alpha}(\Lambda_\alpha)$ and Λ_β are equivalent near all points of $\mathcal{H}_\beta(X_\alpha \cap X_\beta \cap \Lambda_{\mathbb{R}})$.

The Λ_α are called local representatives of Λ and we shall say that two almost analytic manifolds Λ , $\Lambda' \subset \tilde{M}$ are equivalent if $\Lambda_{\mathbb{R}} = \Lambda'_{\mathbb{R}}$ and if the corresponding local representatives are equivalent as in 2^0. Similarly we extend the notion of almost analytic functions and equivalence of almost analytic functions .

Remark 1.13 . Our terminology will be slightly abused in the following sections for we are not going to distinguish all the time between almost analytic objects and their equivalence classes. Thus we shall use the terms almost analytic function and almost analytic manifold also for equivalence classes of such objects.We do not think that there will be any serious confusion, and we point out that in the final results of the calculus , only the equivalence classes play any role and not the particular local representatives.

Occasionally we shall also need the notion of a manifold $\Lambda \subset \tilde{M}$. Such an object is also given by 1^0 and 2^0 but without any condition about almost analyticity for the local representatives.

In the case when $M = \mathbb{R}^n$ our (new) definition does not give anything essentially new. For if Λ is an almost analytic manifold over $\mathbb{R}^n$ in the new sense with the corresponding local representatives Λ_α as in 2^0, we can consider the almost analytic manifolds $\Lambda'_\alpha = \tilde{\mathcal{H}}_\alpha^{-1}(\Lambda_\alpha) \subset \mathbb{C}^n$ where $\tilde{\mathcal{H}}_\alpha^{-1}$ is an almost analytic extension of $\mathcal{H}_\alpha^{-1}$. The manifolds Λ'_α are then equivalent at all real points and by a rather simple partition of unity we can "glue together" the Λ'_α and obtain a concrete manifold $\Lambda' \subset \mathbb{C}^n$ which is locally equivalent to all the Λ'_α. (Here we use the following argument: If the equations $z'' = h_j(z')$, $z' \in \mathbb{C}^k$, define equivalent/almost analytic manifolds $\Lambda_j \subset \omega_j \subset \mathbb{C}^n$ and $\varphi_j \in C_0^\infty(\omega_j)$, $j = 1, 2$, then the equation $z'' = \varphi_1(z)h_1(z') + \varphi_2(z)h_2(z')$ defines an almost analytic manifold $\Lambda \subset \omega \subset \mathbb{C}^n$ for any open set $\omega \subset \mathbb{C}^n$ such that $\varphi_1 + \varphi_2 = 1$ on ω, and we have $\Lambda_j \sim \Lambda$ in $\omega_j \cap \omega$.)

We leave to the reader to define concepts such as "$\Lambda_1 \subset \Lambda_2$" and the notion of almost analytic vectorbundles on almost analytic manifolds.

To end this section we shall do some technical preparations for the proof of the equivalence of phase functions in section 4.

Definition 1.14. Let $\omega \subset \mathbb{C}^n$ be an open set and let $\Gamma \subset \omega$ be a closed submanifold. We say that Γ is totally real (cf [14]) if $T_z(\Gamma) \cap iT_z(\Gamma) = 0$ for all $z \in \Gamma$.

Clearly the real dimension of a totally real submanifold of $\mathbb{C}^n$ is $\leq n$. If Γ is totally real of maximal dimension and f is a holomorphic function vanishing on Γ , then f vanishes also in a neighbourhood.

Lemma 1.15. Let $\omega \subset \mathbb{C}^n \times (\mathbb{C}^N \setminus \{0\})$ be an open conic set and let $\Gamma \subset \omega$ be a closed conic totally real submanifold. Then every $p \in S^m_{1-\delta}(\Gamma)$ has an extension $\tilde{p} \in S^m_{1-\delta}(\omega)$ such that

$$|\bar{\partial} p_\lambda(x,\theta)| \leq C_{r,K}\, \lambda^{m+(r+1)\delta}\, d((x,\theta),\Gamma)^r , \tag{1.14}$$

$$(x,\theta) \in K \subset\subset \omega ,\ \lambda \geq 1,\ r \in \mathbb{Z}_+ .$$

Here $p_\lambda(x,\theta) = p(x,\lambda\theta)$ by definition and $d(\cdot,\Gamma)$ is the distance to Γ .

Proof. We note that in the case when Γ is an open cone in $\mathbb{R}^n \times (\mathbb{R}^N \setminus \{0\})$ this lemma is only a reformulation of Theorem 1.3 .

By a partition of unity in Γ we see that it is sufficient to construct $\tilde{p}$ in the case when p has its support in a small conic neighbourhood of a point (x_0,θ_0) , provided that we can find $\tilde{p}$ with support in an arbitrary small conic neighbourhood of supp(p) . We first suppose that Γ is of maximal dimension $= n+N$. Then we can find a homogeneous diffeomorphism $\mathcal{H}$ from some open cone Γ_0 in $\mathbb{R}^{n+N-1} \times \mathbb{R}_+$ onto an open conic neighbourhood of (x_0,θ_0) in Γ . Let $\tilde{\Gamma}_0 \subset \mathbb{C}^{n+N-1} \times (\mathbb{C} \setminus \{0\})$ be an open cone with $\tilde{\Gamma}_{0\mathbb{R}} = \Gamma_0$ and let $\tilde{\mathcal{H}} : \tilde{\Gamma}_0 \longrightarrow \omega$ be an almost analytic extension , homogeneous as in Theorem 1.3.The fact that Γ is totally real and of maximal dimension implies that

$\tilde{\mathcal{H}}$ is a diffeomorphism at least after Γ_0 and $\tilde{\Gamma}_0$ have been shrunk a little . Put $q = p \circ \mathcal{H}$ and let $\tilde{q}$ be an almost analytic extension as in Theorem 1.3 with small support . It is easy to verify that $\tilde{p} = \tilde{q} \circ \tilde{\mathcal{H}}^{-1}$ satisfies the requirements of the lemma .

Now in the case when $\dim \Gamma < n+N$, it is easy to construct locally a conic totally real manifold Γ' of maximal dimension which contains Γ . To construct $\tilde{p}$ we first extend p to Γ' and then apply the preceding construction . This completes the proof of the lemma .

Using Lemma 1.15 we shall now prove

Proposition 1.16. Let Γ be a closed conic totally real submanifold of ω , where ω is an open conic subset of $\mathbb{C}^n \times (\mathbb{C}^N \setminus \{0\})$. If $p \in S^m_{1-\delta}(\Gamma)$, there exists $q \in S^m_{1-\delta}(\omega)$, almost analytic (at $\omega_{\mathbb{R}}$) such that

$$|\bar{\partial} q_\lambda(x,\theta)| \leq C_{r,K} \lambda^{m+(r+1)\delta} |\operatorname{Im}(x,\theta)|^r , \tag{1.15}$$

$$(x,\theta) \in K \subset\subset \omega , \quad r \in \mathbb{Z}_+ , \quad \lambda \geq 1 ,$$

$$|(p_\lambda - q_\lambda)(x,\theta)| \leq C_{r,K} \lambda^{m+r\delta} |\operatorname{Im}(x,\theta)|^r , \tag{1.16}$$

$$(x,\theta) \in K \subset\subset \Gamma , \quad r \in \mathbb{Z}_+ , \quad \lambda \geq 1 .$$

Here $p_\lambda(x,\theta) = p(x, \lambda\theta)$, $q_\lambda(x,\theta) = q(x, \lambda\theta)$.

Proof. By a partition of unity in Γ it is easy to reduce the proof to the case when $\omega = \mathbb{C}^n \times (\mathbb{C}^N \setminus \{0\})$. Moreover , to simplify the notations we assume that $m = 0$. Put $g(x,\theta) = d((x,\theta),\Gamma)$ and let $\tilde{p}(x,\theta)$ be an extension of p as in the preceding lemma . Combination of (1.14) and Lemma 1.5 gives the inequalities

$$|D_x^\alpha D_\theta^\beta \bar{\partial} \tilde{p}_\lambda(x,\theta)| \le C_{r,K,\alpha,\beta}\, \lambda^{(r+1+|\alpha|+|\beta|)\delta}\, g(x,\theta)^r , \tag{1.17}$$

$$(x,\theta) \in K \subset\subset \mathbb{C}^n \times (\mathbb{C}^N \setminus \{0\}) ,\ \lambda \ge 1,$$

for all multi-indices α and β . By Lemma 1.15 and 1.5 we have an extension $q \in S^0_{1-\delta}(\mathbb{C}^n \times (\mathbb{C}^N \setminus \{0\}))$ of $\tilde{p}|_{\mathbb{R}^n \times (\mathbb{R}^N \setminus \{0\})}$ such that

$$|D_x^\alpha D_\theta^\beta \bar{\partial} q_\lambda(x,\theta)| \le C_{r,K,\alpha,\beta}\, \lambda^{(r+1+|\alpha|+|\beta|)\delta}\, |\mathrm{Im}\,(x,\theta)|^r \tag{1.18}$$

under the same conditions as in (1.17). Now put

$$u(x,\theta) = q(x,\theta) - \tilde{p}(x,\theta) \in S^0_{1-\delta}(\mathbb{C}^n \times (\mathbb{C}^N \setminus \{0\}) .$$

Then

$$u = 0 \quad \text{on} \quad \mathbb{R}^n \times (\mathbb{R}^N \setminus \{0\}) \tag{1.19}$$

and

$$|D_x^\alpha D_\theta^\beta \bar{\partial} u_\lambda(x,\theta)| \le C_{r,K,\alpha,\beta}\, \lambda^{(r+1+|\alpha|+|\beta|)\delta}\, (g(x,\theta)+|\mathrm{Im}(x,\theta)|)^r \tag{1.20}$$

for all multi-indices α and β . If we write $(x,\theta) = X + iY$ with real X and Y we get by induction from (1.19) and (1.20) that

$$|D_Y^\beta u_\lambda(x,\theta)| \le C_{r,\beta,K}\, \lambda^{(r+|\beta|)\delta}\, g(x,\theta)^r \tag{1.21}$$

$$r \in \mathbb{Z}_+ \cup \{0\},\ \lambda \ge 1,\ (x,\theta) \in K \subset\subset \mathbb{R}^n \times (\mathbb{R}^N \setminus \{0\})$$

If we expand $u(X+iY)$ in a Taylor series to order r

and use (1.21) we get

$$|u_\lambda(X+iY)| \leq \sum_{k=0}^{r-1} C_{r(k)} \; |Y|^k \; \lambda^{(r(k)+k)\delta} \; g(X)^{r(k)} \; + \qquad (1.22)$$
$$+ \; C_r \; |Y|^r \; \lambda^{r\delta},$$

where we are free to choose the $r(k)$ in $\mathbb{Z}_+ \cup \{0\}$. Putting $r(k) = r - k$ we get

$$|u_\lambda(X + iY)| \leq C_r \sum_0^r |Y|^k \; g(X)^{r-k} \, \lambda^{r\delta} \quad .$$

Now if $X+iY \in \Gamma$ we have $g(X) \leq |Y|$ and we get

$$u_\lambda(X+iY) \leq C'_r \lambda^{r\delta} \; |Y|^r \; , \quad X+iY \in \Gamma \quad ,$$

which proves the proposition .

2. The method of stationary phase.

In this section we shall present an extension of the method of stationary phase to the case when the phase function is complex-valued.

Let $a(x,w)$, $x \in \mathbb{R}^n$, $w \in \mathbb{R}^k$, be a C^∞ function defined in a neighbourhood of $(0,0)$. We suppose that $a'_x(0,0) = 0$, $\det(a''_{xx}(0,0)) \neq 0$ and that $\operatorname{Im} a \geq 0$ with equality at $(0,0)$, so in particular we have that $\operatorname{Im} a''_{xx}(0,0) \geq 0$. We shall examine the asymptotic behaviour of the integral

$$I_t(w) = \int e^{ita(x,w)} u_t(x)\, dx , \quad t \longrightarrow +\infty, \tag{2.1}$$

when $u_t(x) \in S^0_{1-\delta}(\mathbb{R}^n \times \mathbb{R}_+)$ has its support with respect to x in a small neighbourhood of 0 in $\mathbb{R}^n$.

In doing this we shall replace the integration along $\mathbb{R}^n$ by the integration along a suitable chain in the complex domain passing through "the critical point" of an almost analytic extension of a . The following lemma will be of fundamental importance later.

Lemma 2.1. Let a be as above and denote by $a(z,w)$, $z = x+iy$, $w \in \mathbb{C}^k$, also an almost analytic extension of a to a complex neighbourhood of $(0,0)$. Then the equations

$$\partial_z a(z,w) = 0 , \tag{2.2}$$

$$\operatorname{grad}_{(x,y)} \operatorname{Re} a(z,w) = 0 , \tag{2.3}$$

$$\operatorname{grad}_{(x,y)} \operatorname{Im} a(z,w) = 0 , \tag{2.4}$$

define three equivalent almost analytic manifolds in a neighbourhood of $(0,0)$ which have the form $z = Z(w)$. For any such manifold M there is a positive constant C such that near the origin we have

$$\operatorname{Im} a(z,w) \geq C|\operatorname{Im} z|^2 , \quad (z,w)\in M , \quad w\in\mathbb{R}^k. \tag{2.5}$$

Proof. That (2.2) defines an almost analytic manifold of the form $z = Z(w)$ is a consequence of the implicit function theorem (note that $a''_{z\bar{z}}(0,0) = 0$) and the fact that $\det(a''_{zz}) \neq 0$. That (2.3) , (2.4) define equivalent manifolds follows from the Cauchy-Riemann equations , which are satisfied to infinite order on $\mathbb{R}^n \times \mathbb{R}^k$.

To prove the last statement we need a lemma.

Lemma 2.2. Let A be a complex invertible and symmetric $n \times n$ matrix and assume that $\operatorname{Im} A \geq 0$. Then there is a positive constant $c(A)$ so that

$$\sup_{y\in\mathbb{R}^n,\ |y|<\varepsilon|x|} \|A^{-1}\| \operatorname{Im} \langle A(x+iy),x+iy\rangle \geq \varepsilon|x|^2 \tag{2.6}$$

when $0 < \varepsilon < c(A)$ and $x\in\mathbb{R}^n$.

Proof. Consider the analytic map $Q \longrightarrow S(Q) = (I + {}^tQ)A(I + Q)$ from the space of complex $n \times n$ matrices Q to the space of complex, symmetric $n \times n$ matrices S . The differential at the origin is the map $Q \longrightarrow {}^tQ\,A + A\,Q$ which has the right inverse $S \longrightarrow A^{-1} S/2$. Thus the map $Q \longrightarrow S(Q)$ has a right inverse Φ in a neighbourhood of A with $\Phi(A) = 0$, $d\,\Phi(A)\cdot S = A^{-1}S/2$. In particular, the equation

$$(I + {}^tQ)\,A\,(I + Q) = A + i\tilde{\varepsilon}I , \quad \tilde{\varepsilon} = \varepsilon\|A^{-1}\|^{-1}\, 6/5 \tag{2.7}$$

has a solution $Q = Q(\varepsilon)$ with $\|Q\| \leq 3\varepsilon/4$ when $\varepsilon > 0$ is small enough. Multiplying (2.7) from the left by $(I + \operatorname{Re}{}^tQ)^{-1}$ and from the right by $(I + \operatorname{Re} Q)^{-1}$ we get

$$(I + i\,{}^tT)\, A\, (I + iT) = (I + \mathrm{Re}\,{}^tQ)^{-1}(A + i\tilde{\varepsilon}\, I)\,(I + \mathrm{Re}Q)^{-1}$$

where $T = T(\varepsilon)$ is the real $n \times n$ matrix $(\mathrm{Im}\, Q)(I + \mathrm{Re}\, Q)^{-1}$. Using this equation and the fact that $Q(\varepsilon) \to 0$ when $\varepsilon \to 0$ we see that $\mathrm{Im} \langle A(x+iy), x+iy \rangle \geq \|A^{-1}\|^{-1}\cdot\varepsilon$ when $x \in \mathbb{R}^n$ and $y = Tx$, if ε is small enough. This completes the proof of the lemma if we observe that $\|T(\varepsilon)\| < \varepsilon$ for ε small.

An inspection of the proof shows that the conclusion in the lemma remains true for small perturbations of A and that the best constant $c(A)$ is a continuous function of A.

It follows readily from Proposition 1.7 that the inequality (2.5) is independent of whether M is given by (2.2), (2.3) or (2.4) . We can therefore assume that M is given by (2.2) and we let $(z_0, w_0) \in M$ be a point close to the origin and with w_0 real. Using the almost analyticity , we get by Taylor's formula:

$$|a(z,w_0) - (a(z_0,w_0) + \langle z-z_0\, ,\, a''_{zz}(z_0,w_0)\,(z-z_0)\rangle /2)|$$

$$\leq C(|\mathrm{Im}\, z_0|^3 + |z-z_0|^3)$$

for z , z_0 and w_0 small . (C denotes always a new constant .) Now if $z_0 = x_0 + iy_0$, we let z vary in the real ball $B(|y_0|, x_0)$ with radius $|y_0|$ and center at x_0. Then by the preceding lemma there is a positive constant C so that when z_0 and w_0 are close to the origin

$$\inf_{z \in B(|y_0|, x_0)} \mathrm{Im} \langle z-z_0, a''_{zz}(z_0,w_0)(z-z_0)\rangle/2 =$$

$$= \inf_{|x| \leq |y_0|} (- \mathrm{Im} \langle y_0+ix, a''_{xx}(z_0,w_0)(y_0+ix)\rangle/2) \leq -C\,|y_0|^2 .$$

Moreover $\operatorname{Im} a(z,w_0) \geq 0$ for z real so we deduce that

$$\operatorname{Im} a(z_0,w_0) \geq C\ |y_0|^2$$

for some positive constant C, when z_0 and w_0 are small. The proof is complete.

In the following, we let $z = Z(w)$ be the point defined by (2.2). The main result in this section is now

<u>Theorem 2.3</u>. Let a be as above . Then there are neighbourhoods U and V of the origin in $\mathbb{R}^n$ and $\mathbb{R}^k$ respectively and differential operators $C_{\nu,w}(D)$ of order $\leq 2\nu$ which are C^∞ functions of $w \in V$ such that we have the asymptotic expansion

$$\int e^{ita(x,w)}\ u_t(x)\ dx \sim \tag{2.8}$$

$$\sim \sum_{\nu=0}^{\infty} t^{-\nu-n/2}\ e^{ita(Z(w),w)}\ (C_{\nu,w}(D)\ u_t)(Z(w)) \quad \text{in } S^{-n/2}_{0,1}(V \times \mathbb{R}_+).$$

Here $u_t(x) \in S^0_{1-\delta}(\mathbb{R}^n \times \mathbb{R}_+)$ is an arbitrary symbol with support in $U \times \mathbb{R}_+$ and $\delta < 1/2$. In the right hand side u_t also denotes an almost analytic extension as in Theorem 1.3 .The function $(2\pi)^{-n/2}C_{0,w}$ is the branch of the square root of $\det\,(i^{-1}\ a''_{zz}(Z(w),w))^{-1}$ which is continuously deformed into 1 under the homotopy

$$[0,1] \ni s \longrightarrow i^{-1}\ (1-s)\ a''_{zz} + s\ I \in GL(n,\mathbb{C}) \quad .$$

<u>Proof</u>. In the first part of the proof we shall use the arguments of [3, p.151] to find new coordinates $\tilde{z}$ in $\mathbb{C}^n$ for which $a(z,w) - a(Z(w),w)$ differs from a quadratic form in $\tilde{z}$ with a very small error .

Thus let us consider the function $h(z,w) = a(z+Z(w),w)-a(Z(w),w)$,

$z \in \mathbb{C}^n$, $w \in \mathbb{R}^k$. Then h is defined in a neighbourhood of $(0,0)$ and using Taylor's formula

$$h(z,w) = \frac{d}{dt} h(tz,w)\Big|_{t=0} + \int_0^1 (1-t) \frac{d^2}{dt^2} h(tz,w)\, dt$$

we can write h in the form

$$h(z,w) = \langle z, R(z,w)\cdot z\rangle/2 + \varrho(z+Z(w),w) \ . \tag{2.9}$$

Here $R(z,w) = 2 \int_0^1 (1-t)\, h''_{zz}(tz,w)\, dt$ and there are constants C_N such that

$$|\bar\partial_z R(z,w)| + |\varrho(z,w)| \leq C_N (|\mathrm{Im}\, z|^N + |\mathrm{Im}\, Z(w)|^N), \ N \in \mathbb{Z}_+ \tag{2.10}$$

when z and w are close to the origin.

Since all non-degenerate quadratic forms on $\mathbb{C}^n$ are equivalent there is a matrix A such that ${}^tA\, R(0,0)\, A = iI$. We want to find matrices $Q(z,w)$ such that

$$i\,{}^tQ(z,w)\, Q(z,w) = R(z,w) \ , \quad Q(0,0) = A^{-1}. \tag{2.11}$$

Since the map $GL(n,\mathbb{C}) \ni Q \to i\,{}^tQQ$ into the set of symmetric matrices is analytic with surjective differential at $Q = A^{-1}$, the equation (2.11) has a C^∞ solution $Q(z,w)$ defined near $(0,0)$ which is an analytic function of $R = R(z,w)$ and therefore satisfies the inequalities

$$|\bar\partial_z Q(z,w)| \leq C_N (|\mathrm{Im}\, z|^N + |\mathrm{Im}\, Z(w)|^N) \ , \quad N \in \mathbb{Z}_+ \tag{2.12}$$

with some new constants C_N. The map $z \longrightarrow \tilde z = \tilde z(z) = Q(z-Z(w),w)\cdot(z-Z(w))$ defines new coordinates for $\mathbb{C}^n$ in a neighbourhood of the origin when w is small for $Q(0,0) \in GL(n,\mathbb{C})$. In these coordinates we have

$$a(z,w) = a(Z(w),w) + i\langle \tilde z, \tilde z\rangle/2 + \varrho(z,w) \ . \tag{2.13}$$

(We have to keep in mind in the following that the new coordinates also depend on w.)

Note that $\langle\tilde{z},\tilde{z}\rangle = |\tilde{x}|^2 - |\tilde{y}|^2 + 2i\langle\tilde{x},\tilde{y}\rangle$ if $\tilde{z} = \tilde{x}+i\tilde{y}$. Since $\operatorname{Im} a \geq 0$ on $\mathbb{R}^n \times \mathbb{R}^k$ with equality at $(0,0)$, it follows that $|\tilde{x}|^2 \geq |\tilde{y}|^2$ on the tangent space of $\mathbb{R}^n$ at 0 for $w = 0$. Thus in the new coordinates $\mathbb{R}^n$ is given by an equation $\tilde{y} = g(\tilde{x},w)$, with a C^∞ function g defined in a neighbourhood of $(0,0)$.

Let $z(\tilde{z})$ be the inverse of the map $z \to \tilde{z}(z)$. In the next step we shall examine the behaviour of a along the chains $\Gamma_{w,s}$:

$$\mathbb{R}^n \ni \tilde{x} \longrightarrow z(\tilde{z}_s) \quad , \; \tilde{z}_s = \tilde{z}_s(\tilde{x}) = \tilde{x} + isg(\tilde{x},w) \; , \; 0 \leq s \leq 1.$$

<u>Lemma 2.4</u>. There are positive constants C and C' such that for $0 \leq s \leq 1$ and w and $\tilde{x}$ sufficiently small

$$\operatorname{Im} a(z(\tilde{z}_s),w) \geq C(1-s)\,(|\operatorname{Im} Z(w)|^2 + |\tilde{x}|^2) \geq C'|\operatorname{Im} z(\tilde{z}_s)|^2 \quad (2.14)$$

<u>Proof</u>. We first note that the following inequalities are valid for small w and x with a positive constant C:

$$C^{-1}|g(0,w)| \leq |\operatorname{Im} Z(w)| \leq C|g(0,w)| \quad , \qquad (2.15)$$

$$|\operatorname{Im} z(\tilde{z}_s)| \leq C\,(1-s)(|\operatorname{Im} Z(w)| + |\tilde{x}|). \qquad (2.16)$$

In fact, (2.15) says only that $g(0,w)$ measures the distance from $Z(w)$ to $\mathbb{R}^n$, which is evident, and (2.16) follows from (2.15) because we have with some (new) constants C and C'

$$|\operatorname{Im} z(\tilde{z}_s)| = |\operatorname{Im} z(\tilde{x}+ig(\tilde{x},w)) - \operatorname{Im} z(\tilde{x}+isg(\tilde{x},w))| \leq$$

$$\leq C\,(1-s)|g(\tilde{x},w)| \leq C\,C'\,(1-s)(|g(0,w)| + |\tilde{x}|)\,.$$

From (2.13) we get the identity

$$\operatorname{Im} a(z(\tilde{z}_s),w) = s^2 \operatorname{Im} a(z(\tilde{z}_1),w) + (1-s^2)\,(\operatorname{Im} a(Z(w),w) + |\tilde{x}|^2/2)$$

$$+ (1-s^2)\operatorname{Im}\varrho(z(\tilde{z}_1),w) + \operatorname{Im}(\varrho(z(\tilde{z}_s),w) - \varrho(z(\tilde{z}_1),w)) \qquad (2.17)$$

Using (2.16) and the fact that the estimate (2.9) also holds for any derivative of ϱ (Lemma 1.5) we also get an inequality

$$|\varrho(z(\tilde{z}_s),w) - \varrho(z(\tilde{z}_1),w)| \leq C(1-s)\,(|\operatorname{Im} Z(w)|^3 + |\tilde{x}|^3) .$$

Combining this with (2.17) , (2.10), (2.5) and (2.16) we get (2.14) and the lemma is proved.

We shall now replace the integration over $\mathbb{R}^n$ in (2.1) by an integration along $\Gamma_{w,0}$. By multiplying Q from the left by a real orthogonal matrix (so that the first equation in (2.11) is unaffected) we may change the sign of $\det(\frac{\partial z(\tilde{x}+ig(\tilde{x},0))}{\partial \tilde{x}}(0))$. Therefore we may assume that the maps $\Gamma_{w,1}$ are orientation preserving from $\mathbb{R}^n$ to $\mathbb{R}^n$.

Now let $u_t(z) \in S^0_{1-\delta}(\mathbb{C}^n \times \mathbb{R}_+)$, $\delta < 1/2$, be an extension of u_t which is almost analytic with respect to z in the sense of Theorem 1.3 and suppose that the support of u_t with respect to z is contained in a small fixed neighbourhood of the origin. Then we can write

$$\int_{\mathbb{R}^n} e^{ita(x,w)}\, u_t(x)\, dx = \int_{\Gamma_{w,1}} e^{ita(z,w)}\, u_t(z)\, dz_1\wedge\ldots\wedge dz_n . \qquad (2.18)$$

Lemma 2.5. Let u_t be as above . Then for every $N \in \mathbb{Z}_+$ there exists a constant C_N such that

$$\left| \int_{\mathbb{R}^n} e^{ita(x,w)} u_t(x)\, dx - \int_{\Gamma_{w,0}} e^{ita(z,w)} u_t(z)\, dz_1\wedge\ldots\wedge dz_n \right|$$

$$\leq C_N\, t^{-N} \ , \quad \text{when } w\in W \ , \ t\in \mathbb{R}_+ . \qquad (2.19)$$

Here $W \subset \mathbb{R}^k$ is a fixed neighbourhood of the origin (independent of u_t).

Proof. Let $U \subset \mathbb{R}^n$ be a small neighbourhood of the origin so that the maps

$$\gamma_w :\ U \times I \ni (\tilde{x},s) \longrightarrow \Gamma_{w,s}(\tilde{x}) = z(\tilde{z}_s) \quad , \ I = [0,1] \ ,$$

are defined for small w. Then if $f(z)$ is a C^∞ function with support near the origin in $\mathbb{C}^n$ and if we write $\omega = f(z)dz_1\wedge\ldots\wedge dz_n$, it follows from Stokes' formula that

$$\int_{\Gamma_{w,1}} \omega - \int_{\Gamma_{w,0}} \omega = \int_{\gamma_w} d\omega = \int_{\gamma_w} \bar{\partial} f \wedge dz_1\wedge\ldots\wedge dz_n .$$

Hence there is a constant C such that

$$\left| \int_{\Gamma_{w,1}} \omega - \int_{\Gamma_{w,0}} \omega \right| \leq C \sup_{\substack{\tilde{x}\in U \\ 0\leq s\leq 1}} |\bar{\partial} f(z(\tilde{z}_s))| .$$

With $f(z) = e^{ita(z,w)} u_t(z)$ we have

$$\bar{\partial}_z f = e^{ita(z,w)} (u_t(z)\, it\bar{\partial}_z a(z,w) + \bar{\partial}_z u_t(z)).$$

Using (2.14) and the estimates for a and u_t corresponding to (1.1) we get the following inequalities where c is a positive constant:

$$\bar{\partial} f(z(\tilde{z}_s)) \leq C_N\, e^{-ct|\mathrm{Im}\, z(\tilde{z}_s)|^2}\, t^{N\delta+1}\, |\mathrm{Im}\, z(\tilde{z}_s)|^N$$

$$\leq C_N'\, t^{1-N(1/2-\delta)} \left[(t\,|\mathrm{Im}\, z(\tilde{z}_s)|^2)^{N/2}\, e^{-ct\,|\mathrm{Im}\, z(\tilde{z}_s)|^2} \right].$$

This implies (2.19) since $\delta < 1/2$.

In order to complete the proof of Theorem 2.3 , we now have to compute

$$\int_{\Gamma_{w,0}} e^{ita(z,w)}\, u_t(z)\, dz_1 \wedge \dots \wedge dz_n \quad = \qquad (2.20)$$

$$\int_U e^{it(a(Z(w),w) + i|\tilde{x}|^2/2 + \rho(z(\tilde{x}),w))}\, u_t(z(\tilde{x}))\, J(\tilde{x}) d\tilde{x} \ ,$$

where $J(\tilde{x}) = \det\left(\frac{dz}{d\tilde{x}}\right)$.(Note that J also depends on w .)If we replace ρ by $s\rho$, $0 \leq s \leq 1$, then the derivative with respect to s of the integrand above can be estimated by a constant times

$$t\,(|\mathrm{Im}\, Z(w)|^N + |\tilde{x}|^N)\ e^{-t\,c(|\tilde{x}|^2 + |Z(w)|^2)} \ , \ N \in \mathbb{Z}_+ \ ,$$

where c is a positive constant . This follows from (2.10), (2.16) and (2.14) . The corresponding integral is then bounded by any negative power of t and the presence of ρ is therefore irrelevant for the asymptotic behaviour of the integral (2.20) ,so from now on we assume that $\rho = 0$.

Using the inequalities

$$\left| e^{-|\eta|^2/2t} - \sum_0^{k-1} (-1)^\nu (|\eta|^2/2t)^\nu/\nu! \right| \leq (|\eta|^2/2t)^k/k!$$

and the formula

$$\int e^{-t|x|^2/2} f(x)\, dx = (2\pi t)^{-n/2} \int e^{-|\eta|^2/2t} \hat{f}(\eta)\, d\eta \ ,$$

we get the following inequalities for $f \in C_0^\infty(\mathbb{R}^n)$

$$\left| \int e^{-t|x|^2/2} f(x)\, dx - \sum_0^{k-1} (2\pi/t)^{n/2} (\Delta/2t)^\nu f(0)/\nu! \right|$$
$$\leq C_k' \ t^{-k-n/2} \sum_{|\alpha| \leq 2k+n+1} \int |D^\alpha f|\, dx.$$

Thus replacing f by $u_t J = u_t(z(x,w))\, J(x)$ we see that there are differential operators $C_{\nu,w}(D)$ of order $\leq 2\nu$ such that

$$\left|\int e^{-t|\tilde{x}|^2/2}\, u_t(z(\tilde{x}))\, J(\tilde{x})\, d\tilde{x} - \sum_0^{k-1} t^{-\nu-n/2}(C_{\nu,w}(D)\, u_t)(Z(w))\right|$$
$$\leq C_k\, t^{-k-n/2}\, t^{(2k+n+1)\delta} = C_k\, t^{-(k+n/2)(1-2\delta)+\delta} \qquad (2.21)$$

In view of Lemma 2.5 and (2.20) we see that the differences between the left hand side of (2.8) and sufficiently high partial sums of the right hand side are bounded by arbitrarily large negative powers of t. An application of Theorem 2.9 in Hörmander [6] then shows that the asymptotic formula (2.8) is valid.

To complete the proof of the theorem, it remains to calculate $C_{0,w}(D) = (2\pi)^{n/2} J(0) = (2\pi)^{n/2} \det\left(\frac{dz}{d\tilde{x}}\right)$. Calculating the Hessians of the two sides of the equation

$$a(z(\tilde{z}),w) = a(Z(w),w) + i\langle\tilde{z},\tilde{z}\rangle/2 + \rho(z,w)$$

at the point $\tilde{z} = 0$, we get

$$^t\left(\frac{\partial z(\tilde{z})}{\partial \tilde{z}}\right) \circ i^{-1}\, a''_{zz}(Z(w),w) \circ \frac{\partial z(\tilde{z})}{\partial \tilde{z}} \approx I\,. \qquad (2.22)$$

Here we write in general $f(w) \approx g(w)$ if $f(w) - g(w) = \mathcal{O}(|\operatorname{Im} Z(w)|^N)$ for all $N > 0$. Since the asymptotic expansion (2.8) does not change if we replace $C_{0,w}$ by some new $C'_{0,w} \approx C_{0,w}$, we can assume by (2.22) that

$$(2\pi)^{-n/2}\, C_{0,w} = \pm\,(\det i^{-1} a''_{zz}(Z(w),w))^{-1/2}$$

The problem is to choose the right branch. Since our constructions above always give continuous functions of w, it suffices to choose the good branch for $w = 0$.

Let $\Omega^+ = \{Q \in GL(n,\mathbb{C})\ ;\ {}^tQ = Q\,,\ \operatorname{Re} Q \geq 0\}$.

Then Ω^+ is a closed contractible subset of $GL(n,\mathbb{C})$ and for $Q \in \Omega^+$ we can define $(\det Q)^{1/2}$ in a unique continuous way so that $(\det Q)^{1/2} = 1$ for $Q = I$. It is easy to see that this definition can be continuously extended to a small neighbourhood of Ω^+ in $GL(n,\mathbb{C})$.(As such a neighbourhood take for instance the set of $Q \in GL(n,\mathbb{C})$, having no real negative eigenvalues. This set is contractible by the homotopy $[0,1] \ni s \longrightarrow (1-s)Q + sI$.)

Now we consider the case $w = 0$ and we let $K_s : \mathbb{C}^n \longrightarrow \mathbb{C}^n$ be the complexification of the differential at the origin of the chain $\Gamma_{0,s}$ constructed above .Moreover we let $A = a''_{zz}(0,0)$. Our problem is to calculate $J(0) = \det(\frac{dz}{d\tilde{x}}) = \det K_0$.Since $\operatorname{Im} a \geqslant 0$ on the image of $\Gamma_{0,s}$ with equality at the origin, we know that

$$\operatorname{Re} {}^tK_s\, i^{-1}A\, K_s \geqslant 0 , \quad 0 \leq s \leq 1 .$$

By the construction above we know that K_1 is real with a positive determinant. Let $[1,2] \ni s \longrightarrow K_s$ be a curve in the real invertible matrices, joining K_1 to I . Then we have

$$\operatorname{Re} {}^tK_s\, i^{-1}A\, K_s \geqslant 0 , \; 0 \leq s \leq 2 , \tag{2.23}$$

and (2.22) implies

$${}^tK_0\, i^{-1}A\, K_0 = I \tag{2.24}$$

Put $\mathcal{A}_s = {}^tK_s\, i^{-1}A\, K_s \in \Omega^+$. Then

$$\mathcal{A}_s = {}^tB_s\, B_s \quad , \quad 0 \leq s \leq 2 , \tag{2.25}$$

if $B_s = K_0^{-1}K_s$ .Moreover $B_0 = I, \mathcal{A}_0 = I$ and $B_2 = K_0^{-1}$, $\mathcal{A}_2 = i^{-1}A$, so (2.25) implies that $\det K_0$ is the branch of $(\det i^{-1}A)^{-1/2}$ which is continuously deformed into 1 when $i^{-1}A$ is deformed into I in Ω^+. This completes the proof of Theorem 2.3 .

To conclude this section, we shall apply Theorem 2.3 to prove a result which we think is of independent interest and has not been proved in general before as far as we know. Let $P \in L^m_{1-\delta}(\mathbb{R}^n)$, $\delta < 1/2$, and let $\varphi \in C^\infty(\mathbb{R}^n)$ satisfy $\operatorname{Im}\varphi \geq 0$ and $d\varphi \neq 0$ where $\operatorname{Im}\varphi = 0$. We shall calculate the asymptotic behaviour of $P(u(x)e^{it\varphi(x)})$ as t tends to $+\infty$ when $u \in C_0^\infty(\mathbb{R}^n)$. This asymptotic behaviour is wellknown when φ is real-valued and we are now going to prove essentially the same asymptotic formula as in the real case. (Cf. [7, Lemma 1.3.1] and [13, Lemma 5.1])

If $p(x,\eta)$ is the symbol of P, we get after a substitution of variables

$$P(u\, e^{it\varphi}) = t^n \iint p(x,t\eta)\, e^{it(\langle x-y,\eta\rangle + \varphi(y))} u(y)\,dy\,d\eta/(2\pi)^n. \tag{2.26}$$

Partial integrations with respect to η show that the behaviour of u outside a neighbourhood of a point $x = x_0$ does not influence the asymptotic behaviour of $P(u\,e^{it\varphi})$ near that point. It is therefore no restriction to assume that supp u is contained in a small neighbourhood of x_0 and that $\operatorname{Im}\varphi(x_0) = 0$. We consider $\langle x-y,\eta\rangle + \varphi(y)$ as a function $a_x(y,\eta)$ of (y,η) with x as a parameter. Then it is clear that $a_x(y,\eta)$ has a unique critical point $(y,\eta) = (x, \varphi'_x(x))$ and that $\det(\operatorname{Hess}_{(y,\eta)} a_x) = \pm 1$ at that point. By suitable partial integrations we see that the contribution to the integral (2.26) for η far away from φ'_x does not influence the asymptotic behaviour. We can therefore apply Theorem 2.3 and get

$$P(u\,e^{it\ }) \sim \sum_0^\infty e^{it\varphi(x)}\, t^{-\nu} (C_{\nu,x}(D_y,D_\eta)u(y)\,\tilde{p}(x,t\eta))\Big|_{(y,\eta)=(x,\varphi'_x(x))} \tag{2.27}$$

where $\tilde{p}$ is an almost analytic extension of p as in Theorem 1.3, $\tilde{p} \in S^m_{1-\delta}(\mathbb{R}^n \times \mathbb{C}^n)$, and $C_{\nu,x}$ is a differential operator of order $\leq 2\nu$ whose coefficients are C^∞ functions of x. (Actually we need the slight (and evident!) extension of Theorem 2.3 to the case where the symbol $u_t(x)$ may vary in a bounded subset of $S^m_{1-\delta}(\mathbb{R}^n \times \mathbb{R}_+)$ with the parameter w.)

To explicit the formula (2.27) we note that in the case when P is a differential operator we have the usual formula:

$$P(u\, e^{it\varphi}) = \sum p^{(\alpha)}(x, t\varphi'_x) D_y^\alpha (u(y)\, e^{it\rho(x,y)})\Big|_{y=x} e^{it\varphi(x)} /\alpha! =$$

$$\sum ((t^{-1}\partial/\partial\eta)^\alpha\, p(x, t\eta))\Big|_{\eta=\varphi'_x} D_y^\alpha (u(y)\, e^{it\rho(x,y)})\Big|_{y=x} e^{it\varphi(x)} /\alpha! =$$

$$\sum_0^\infty t^{-\nu} (E_{\nu,x}(D_y, D_\eta) u(y)\, p(x, t\eta))\Big|_{\substack{(y,\eta)= \\ (x,\varphi'_x)}} e^{it\varphi(x)} .$$

Here $\rho(x,y) = \varphi(y) - \varphi(x) - \langle y-x, \varphi'_x(x)\rangle$ so we see that $E_{\nu,x}$ is of the degree $\leq 2\nu$. If we compare this formula with (2.27) in the case when p is a homogeneous polynomial in η and use Lemma 5.1 in section 5, we conclude that

$$\left((C_{\nu,x}(D_y, D_\eta) - E_{\nu,x}(D_y, D_\eta))\, (u(y)\tilde{p}(x,\eta))\right)\Big|_{\substack{(y,\eta)= \\ (x,\varphi'_x(x))}} = \mathcal{O}((\operatorname{Im}\varphi)^M)$$

for all $u \in C_0^\infty(\mathbb{R}^n)$, $M \in \mathbb{Z}_+$. It follows that all the coefficients of $C_{\nu,x} - E_{\nu,x}$ are $\mathcal{O}((\operatorname{Im}\varphi)^M)$ when $M \in \mathbb{Z}_+$ and we can therefore replace $C_{\nu,x}$ by $E_{\nu,x}$ in (2.27) also when p is an arbitrary symbol. From (2.27) we therefore get the "usual" asymptotic formula

$$P(u\, e^{it\varphi(x)}) \sim \sum e^{it\varphi(x)} \tilde{p}^{(\alpha)}(x, t\varphi'_x)\, D_y^\alpha (u(y) e^{it\rho(x,y)})/\alpha! \Big|_{y=x} \tag{2.28}$$

with asymptotic convergence in $S^m_{0,1}(\mathbb{R}^n \times \mathbb{R}_+)$

3. Lagrangean manifolds and phase functions.

Let us first recall some notions in the real case. Let M be a real symplectic manifold of dimension 2n , that is, there is a distinguished closed non-degenerate 2-form σ on M . We say that the local coordinates $(x, \xi) = (x_1, \dots, x_n, \xi_1, \dots, \xi_n)$ on M are symplectic if σ takes the form $\sigma = \sum d\xi_j \wedge dx_j$ in these coordinates. We say that a submanifold $\Lambda \subset M$ is Lagrangean if $\dim \Lambda = n$ and $\sigma|_\Lambda = 0$. It is rather easy to prove that if Λ is Lagrangean and if (x, ξ) are symplectic coordinates such that the projection $\Lambda \ni (x, \xi) \longrightarrow x \in \mathbb{R}^n$ is regular , then Λ is locally of the form $\xi = \partial g(x)/\partial x$ for some real valued C^∞ function g . Conversely every such manifold is Lagrangean.

We now want to extend this notion to almost analytic manifolds so we assume that $\Lambda \subset \tilde{M}$ is an almost analytic manifold, containing the real point $\rho_0 \in M$. Let W be a coordinate neighbourhood of ρ_0 which we identify with an open set in $\mathbb{R}^{2n}$. Let $\tilde{W} \subset \mathbb{C}^{2n}$ be an open set with $\tilde{W} \cap \mathbb{R}^{2n} = W$. Since we are going to make a local discussion we identify Λ with a local representative in $\tilde{W}$. Let (x, ξ) be real symplectic coordinates near ρ_0 and denote by $(\tilde{x}, \tilde{\xi})$ some almost analytic extensions to $\tilde{W}$. Then the functions $(\tilde{x}, \tilde{\xi})$ give a diffeomorphism from $\tilde{W}$ onto an open set in $\mathbb{C}^{2n}$ (if W and $\tilde{W}$ are suitably restricted). Suppose that Λ is given by

$$\tilde{\xi} = \partial g(\tilde{x})/\partial \tilde{x} \quad , \tilde{x} \in \mathbb{C}^n, \tag{3.1}$$

in a neighbourhood of ρ_0 , where $g \in C^\infty$ is almost analytic and satisfies $\operatorname{Im} g \geq 0$ on $\mathbb{R}^n$ with equality at $x_0 = x(\rho_0)$.

(Here $\partial g/\partial \tilde{x}$ is defined by the equation $\partial_x g = \partial g/\partial \tilde{x} \; d\tilde{x}$ as in section 1 .)

Let (y,η) be some other real symplectic coordinates near ρ_0 and denote by $(\tilde{y},\tilde{\eta})$ some almost analytic extensions to $\tilde{W}$. We suppose that the projection $\Lambda \ni (\tilde{y},\tilde{\eta}) \longrightarrow \tilde{y} \in \mathbb{C}^n$ has a surjective differential at ρ_0, so that Λ is given by

$$\tilde{\eta} = H(\tilde{y}) \ , \quad \tilde{y} \in \mathbb{C}^n \ , \tag{3.2}$$

near ρ_0, where $H: \mathbb{C}^n \longrightarrow \mathbb{C}^n$ is a C^∞ map.

Theorem 3.1. Under the assumptions above Λ is locally equivalent to a manifold $\tilde{\eta} = \partial h(\tilde{y})/\partial\tilde{y}$, $\tilde{y} \in \mathbb{C}^n$, where h is an almost analytic function and $\operatorname{Im} h \geqslant 0$ on $\mathbb{R}^n$ with equality at $y_0 = y(\rho_0)$.

Proof. We first make a reduction to the case when (x,y) are local coordinates for M near ρ_0. Thus suppose that this is not true .If we look at $T_{\rho_0}(M)$, this means that the Lagrangean planes $t_x = 0$ and $t_y = 0$ do not intersect transversally. Now it is wellknown that there are Lagrangean planes arbitrarily close to $t_y = 0$ which are transversal to both $t_x = 0$ and $t_y = 0$.(Cf. the proof of Lemma 3.4 below.) Thus we can find new real symplectic coordinates (z,ζ) as small perturbations of (y,η) such that $t_z = 0$ intersects both $t_x = 0$ and $t_y = 0$ transversally. We also see that the projection $\Lambda \ni (\tilde{z},\tilde{\zeta}) \rightarrow \tilde{z} \in \mathbb{C}^n$ has a surjective differential at ρ_0 , if $(\tilde{z},\tilde{\zeta})$ denotes some almost analytic extensions.This completes our reduction .

Since $d(\sum \xi_j\, dx_j - \sum \eta_j dy_j) = \sum d\xi_j \wedge dx_j - \sum d\eta_j \wedge dy_j = \sigma - \sigma = 0$, there is locally a C^∞ function $G = G(x,y)$ on M such that

$$\sum \xi_j dx_j \quad - \sum \eta_j\, dy_j \;=\; dG \quad . \tag{3.3}$$

The equation (3.3) means that $\xi = \partial G(x,y)/\partial x$, $\eta = -\partial G(x,y)/\partial y$ and G is called a generating function for the canonical transformation $(x,\xi) \longrightarrow (y,\eta)$. Note that

$$\det(G''_{xy})(x_0,y_0) \neq 0 \tag{3.4}$$

for $\det G''_{xy} = \det \frac{\partial \xi}{\partial y} = \det \frac{\partial(x,\xi)}{\partial(x,y)}$.

We then let G also denote some almost analytic extension of G to $\tilde{W}$. Then

$$\tilde{\xi} \sim \partial G(\tilde{x},\tilde{y})/\partial \tilde{x} \;,\; \tilde{\eta} \sim -\partial G(\tilde{x},\tilde{y})/\partial \tilde{y} \quad . \tag{3.5}$$

The equation (3.3) implies that

$$\sum \tilde{\eta}_j d\tilde{y}_j \sim \sum \tilde{\xi}_j d\tilde{x}_j - dG \quad \text{in } \tilde{W} \tag{3.6}$$

and if we take the restrictions to Λ and use (3.1) we get

$$\sum \tilde{\eta}_j \, d\tilde{y}_j \sim \sum \frac{\partial g}{\partial \tilde{x}_j} d\tilde{x}_j - dG \sim d(g - G) \quad \text{on } \Lambda \tag{3.7}$$

Put $h' = (g-G)|_\Lambda$ and consider h' also as a function of $\tilde{y}$ in $\mathbb{C}^n_y$ .Since h' is almost analytic as a function on Λ , we have $h \sim h'$ on Λ if $h(\tilde{y}) \in C^\infty(\mathbb{C}^n_y)$ is an almost analytic extension of $h'|_{\mathbb{R}^n_y}$. Thus (3.7) implies

$$\sum \tilde{\eta}_j d\tilde{y}_j \sim dh \quad \text{on } \Lambda . \tag{3.8}$$

Now $dh \sim \sum \frac{\partial h}{\partial \tilde{y}_j} d\tilde{y}_j$, so it follows from (3.8) that the manifold $\tilde{\eta} = \partial h/\partial \tilde{y}$ is equivalent to Λ . It remains only to prove that $\operatorname{Im} h \geq 0$ on $\mathbb{R}^n$.(It is trivial that $\operatorname{Im} h(y_0) = 0$.)

We observe that

$$\frac{\partial}{\partial \tilde{x}}(g(\tilde{x}) - G(\tilde{x},\tilde{y})) \sim 0 \quad \text{on } \Lambda. \tag{3.9}$$

in view of (3.1) and (3.5) . Moreover we state that

$$\det\left(\frac{\partial^2}{\partial x^2} (g(x) - G(x,y)) \right) \neq 0 \quad \text{at } (x_0, y_0). \qquad (3.10)$$

In fact , suppose that $t_x \neq 0$ is a complex tangent vector such that $(g''_{xx} - G''_{xx})\, t_x = 0$ and let (t_x, t_ξ) $(= (t_y, t_\eta))$ be the corresponding tangent vector to Λ so that $t_\xi = g''_{xx}\, t_x$ by (3.1) and $t_\xi = G''_{xx}\, t_x + G''_{xy}\, t_y$ by (3.5) .These three equations give a contradiction since we know that $t_y \neq 0$ and that G''_{xy} is non-singular .

The fundamental Lemma 2.1 and (3.9),(3.10) now show that there is a positive constant C such that in a neighbourhood of ρ_0 in Λ we have $\operatorname{Im}(g(\tilde{x}) - G(\tilde{x},y)) \geq C\, |\operatorname{Im} \tilde{x}|^2$ when y is real . It follows in particular that $\operatorname{Im} h(y) \geq 0$ for real y and Theorem 3.1 is proved.

Definition 3.2. An almost analytic manifold $\Lambda \subset \tilde{M}$ satisfying the conditions of Theorem 3.1 at every real point for some real symplectic coordinates (x,ξ) , is called a positive Lagrangean manifold.

There is an important special case:

Definition 3.3. An almost analytic manifold $\Lambda \subset \tilde{M}$ is called a strictly positive Lagrangean manifold if

(i) $\dim_{\mathbb{R}} \Lambda = 2n$,

(ii) $\Lambda_{\mathbb{R}}$ is a submanifold of M ,

(iii) $\sigma_\alpha|_{\Lambda_\alpha} \sim 0$ for all local representatives Λ_α of Λ and all local almost analytic extensions σ_α of σ .

(iv) $i^{-1}\sigma(v,\bar{v}) > 0$ for all $v \in T_\rho(\Lambda) \setminus \widetilde{T_\rho(\Lambda_{\mathbb{R}})}$, $\rho \in \Lambda_{\mathbb{R}}$.
Here σ is regarded as a bilinear form on $T_\rho(\tilde{M})$ and by $\widetilde{T_\rho(\Lambda_{\mathbb{R}})}$ we denote the complexification of $T_\rho(\Lambda_{\mathbb{R}})$.

This definition has essentially been proposed in Hörmander [5]. To verify that this is a special case we assume that $\Lambda \subset \tilde{M}$ is strictly positive and Lagrangean . By elementary symplectic geometry we can choose real symplectic coordinates (x,ξ) so that $\Lambda_{\mathbb{R}}$ is given by $\xi = 0$, $x'' = 0$, where $x = (x',x'')$, $x' \in \mathbb{R}^{n-k}$, $x'' \in \mathbb{R}^k$. From (iv) it follows that the projection

$$T_\rho(\Lambda) \ni (t_x, t_\xi) \longrightarrow t_x \in \mathbb{C}^n \tag{3.11}$$

is surjective. In fact, if this map is not surjective then one can find $0 \neq t \in \mathbb{C}^n$ ortogonal to the image.Then $(0,t) \in T_\rho(\Lambda)$ since this space is Lagrangean and condition (iv) then implies that $(0,t) \in T_\rho(\Lambda_{\mathbb{R}})$ which is impossible by the choice of coordinates above. Since (3.11) is surjective we see using (iii) and (iv) that $T_\rho(\Lambda)$ is given by the equations

$$t_{\xi'} = 0 \; , \quad t_{\xi''} = A \, t_{x''} \; , \tag{3.12}$$

where A is symmetric and $\operatorname{Im} A > 0$.

Since we are reasoning locally, we can identify Λ with a local representative in the coordinates $(\tilde{x},\tilde{\xi})$.Clearly Λ is of the form $\tilde{\xi} = H(\tilde{x})$, $\tilde{x} \in \mathbb{C}^n$, for some almost analytic map $H\colon \mathbb{C}^n \rightarrow \mathbb{C}^n$, and the condition (iii) means that

$$\sum dH_j(\tilde{x}) \wedge d\tilde{x}_j = \mathcal{O}((|\operatorname{Im} \tilde{x}| + |\operatorname{Im} H |)^N)$$

for all $N > 0$. Restricted to real x this implies that

$$\partial H_j/\partial x_k - \partial H_k/\partial x_j = \mathcal{O}(|\operatorname{Im} H|^N) \; , \; N \in \mathbb{Z}_+, x \in \mathbb{R}^n. \tag{3.13}$$

From (3.12) it follows that

$$C^{-1} |x''| \leq \operatorname{Im} H(x) \leq C \, |x''| \, , \; x \in \mathbb{R}^n, \tag{3.14}$$

locally for some positive constant C, so if we put

$$h(\tilde{x}) = \int_0^1 \langle \tilde{x}, H(t\tilde{x}) \rangle \, dt$$

it follows by an elementary calculus using (3. 3) that $\partial h(x)/\partial x - H(x) = \mathcal{O}(|x''|^N)$, $x \in \mathbb{R}^n$, for all N. Thus by Proposition 1.7 the manifold Λ is equivalent to $\tilde{\xi} = \partial h(\tilde{x})/\partial \tilde{x}$, $\tilde{x} \in \mathbb{C}^n$, if h also denotes some almost analytic extension. By (3.12) we know that $\operatorname{Im} h''_{xx} \geq 0$ and this proves that Λ is a positive Lagrangean manifold since $\partial h(x)/\partial x$ and $h(x)$ vanish when $x'' = 0$, $x \in \mathbb{R}^n$.

Now assume that $M = T^*(X) \setminus 0$ for some n-dimensional paracompact manifold X (with the usual symplectic form). We leave to the reader to define the notion of conic almost analytic manifolds $\Lambda \subset (T^*(X) \setminus 0)^{\sim}$.

Lemma 3.4. Let $\Lambda \subset (T^*(X) \setminus 0)^{\sim}$ be a conic n-dimensional almost analytic manifold such that $\sigma_\alpha|_{\Lambda_\alpha} \sim 0$ for all local representatives Λ_α and all local almost analytic extensions σ_α of σ. Then for every $\rho_0 \in \Lambda_{\mathbb{R}}$ there are local coordinates x in X such that Λ has a local representative near ρ_0 of the form

$$\tilde{x} = \partial g(\tilde{\xi})/\partial \tilde{\xi} .$$

Here ξ are the dual coordinates to x and $(\tilde{x}, \tilde{\xi})$ denotes some almost analytic extensions. The function g is almost analytic and positively homogeneous of degree 1.

Proof. We want to find local coordinates x in X such that the corresponding projection

$$\Lambda \ni (\tilde{x}, \tilde{\xi}) \longrightarrow \tilde{\xi} \in \mathbb{C}^n$$

is regular near ρ_0. (Here we identify Λ with some local representative .)Using the arguments of Hörmander [3, p. 136] we see that it suffices to find a real Lagrangean plane L in $T_{\rho_0}(T^*(X))$ whose complexification $\tilde{L}$ is transversal to the tangent spaces of Λ and the fiber .The existence of such an L follows if we can prove that the set of complex symmetric matrices M such that the Lagrangean plane $\Lambda_M = \{(z, Mz);\ z \in \mathbb{C}^n\}$ is not transversal to a given Lagrangean plane μ in $\mathbb{C}^n \times \mathbb{C}^n$, is a proper algebraic subset of the space of symmetric $n \times n$ matrices.

Suppose that μ is given by $Ax + B\xi = 0$, $(x, \xi) \in \mathbb{C}^n \times \mathbb{C}^n$, where $A, B : \mathbb{C}^n \to \mathbb{C}^n$ are linear maps and (A, B) is of maximal rank . After permuting the x-coordinates and the corresponding permutation of the ξ-coordinates , we can assume that $\det(A_k, B_{n-k}) \neq 0$ where A_k is the $n \times k$ matrix of the first k columns of A and B_{n-k} is the $n \times (n-k)$ matrix of the last n-k columns of B . In fact, let n-k be the rank of the projection $\mu \ni (x, \xi) \longrightarrow x \in \mathbb{C}^n$ and rearrange the coordinates so that $\mu \ni (x, \xi) \longrightarrow x'' \in \mathbb{C}^{n-k}$ is surjective . Then we have $x' = 0$ if $(x, \xi) \in \mu$ and $x'' = 0$. This implies that the projection $\mu \ni (x, \xi) \longrightarrow (x'', \xi')$ is bijective, for if $(x', 0, 0, \xi'') \in \mu$ we have seen that $x' = 0$ and since μ is Lagrangean it follows that ξ'' is orthogonal to the range of the map $(x, \xi) \to x''$, so $\xi'' = 0$. Thus Λ is given by an equation

$$(x', \xi'') = \mathcal{A}(x'', \xi') \quad \text{(equivalent to } Ax + B\xi = 0)$$

and this shows that $\det(A_k, B_{n-k}) \neq 0$.

Now the transversality of Λ_M and μ is equivalent to

$$\det\begin{pmatrix} -M & I \\ A & B \end{pmatrix} \neq 0$$

which is a non trivial polynomial in the coefficients of M. In fact, let

$$M = \begin{pmatrix} \lambda_1 & & \\ & \ddots & \\ & & \lambda_m \end{pmatrix}.$$

Then the coefficient of $\lambda_{k+1} \cdots \lambda_n$ is $\pm \det(A_k, B_{n-k}) \neq 0$. This proves the existence of L and we can therefore find real local coordinates in X such that the map $\Lambda \ni (\tilde{x}, \tilde{\xi}) \longrightarrow \tilde{\xi} \in \mathbb{C}^n$ is regular near ρ_0.

Now it is clear that Λ is represented in a conic neighbourhood of ρ_0 by an equation

$$\tilde{x} = H(\tilde{\xi})$$

where H is positively homogeneous of degree 0 and almost analytic. The condition that $\sigma_\alpha|_{\Lambda_\alpha} \sim 0$ for all σ_α and Λ_α as in the lemma now implies that we have the inequalities

$$|\partial H_k(\xi)/\partial \xi_j - \partial H_j(\xi)/\partial \xi_k| \leq C_N \,|\operatorname{Im} H(\xi)|^N, \quad \xi \in \mathbb{R}^n, \quad (3.15)$$

in a conic neighbourhood of $\xi(\rho_0)$ for all $N \in \mathbb{Z}_+$. If we put $g(\xi) = \sum \xi_j H_j(\xi)$ it follows from (3.15) and the Euler homogeneity relations that we have the inequalities

$$|\partial g(\xi)/\partial \xi_j - H_j(\xi)| \leq C'_N \,|\operatorname{Im} H(\xi)|^N$$

in the same set as above when $N \in \mathbb{Z}_+$. Proposition 1.7 then implies that the equation $\tilde{x} = \partial g(\tilde{\xi})/\partial \tilde{\xi}$ also gives a representation for Λ in a conic neighbourhood of ρ_0. The proof of Lemma 3.4 is complete.

Note that Λ above is a positive Lagrangean manifold if and only if $\operatorname{Im} g(\xi) \leq 0$ for real ξ .

We now introduce the phase functions .

Definition 3.5. The C^∞function $\varphi = \varphi(x, \theta)$ defined in an open conic set $V \subset \mathbb{R}^n \times (\mathbb{R}^N \setminus \{0\})$ is called a regular phase function of positive type if

1^0 φ has no critical points,

2^0 $\varphi(x,\theta)$ is positively homogeneous of degree 1 in the θ variable ,

3^0 the differentials $d(\partial\varphi/\partial\theta_1),\dots, d(\partial\varphi/\partial\theta_N)$ are linearly independent over the complex numbers on

$$C_{\varphi\mathbb{R}} = \{(x,\theta) \in V \ ; \ \varphi'_\theta = 0\} ,$$

4^0 $\operatorname{Im} \varphi(x,\theta) \geq 0$ on V .

Let $\tilde{\varphi}(\tilde{x},\tilde{\theta})$ be an almost analytic homogeneous extension of φ , defined in a conic neighbourhood in $\mathbb{C}^n \times (\mathbb{C}^N \setminus \{0\})$ of the point $(x_0,\theta_0) \in V$. Then the set

$$C_{\tilde{\varphi}} = \{(\tilde{x},\tilde{\theta}) \in \mathbb{C}^n \times (\mathbb{C}^N \setminus \{0\}) \ ; \ \tilde{\varphi}'_{\tilde{\theta}} \ (\tilde{x},\tilde{\theta}) = 0\}$$

is a conic almost analytic manifold of dimension 2n in a conic neighbourhood of (x_0,θ_0). The map

$$C_{\tilde{\varphi}} \ni (\tilde{x},\tilde{\theta}) \longrightarrow (\tilde{x}, \ \tilde{\varphi}'_{\tilde{x}} \ (\tilde{x},\tilde{\theta})) \in \mathbb{C}^n \times (\mathbb{C}^n \setminus \{0\}) \qquad (3.16)$$

is almost analytic and has an injective differential at (x_0,θ_0). Hence the image $\Lambda_{\tilde{\varphi}}$ is locally a manifold of dimension 2n near the point $\rho_0 = (x_0, \varphi'_x(x_0,\theta_0))$. Since $\operatorname{Im} \varphi$ is homogeneous and non-negative for (x,θ) real , the image of $C_{\varphi\mathbb{R}}$ =

$= C_{\tilde{\varphi}} \cap (\mathbb{R}^n \times \mathbb{R}^N)$ under (3.16) will be contained in $\Lambda_{\tilde{\varphi}\mathbb{R}} =$
$= \Lambda_{\tilde{\varphi}} \cap (\mathbb{R}^n \times \mathbb{R}^n)$.

Let us make a change of variables $x = \varkappa(y)$ in $\mathbb{R}^n$ and denote by $\tilde{x} = \tilde{\varkappa}(\tilde{y})$ an almost analytic extension . Clearly $\psi(y,\theta) = \varphi(x,\theta)$ is a regular phase function and if we define $\tilde{\psi}(\tilde{y},\tilde{\theta}) = \tilde{\varphi}(\tilde{x},\tilde{\theta})$ it is easily verified that the image of $\Lambda_{\tilde{\varphi}}$ under the almost analytic map $(\tilde{\varkappa}(\tilde{y}),\tilde{\xi}) \longrightarrow (\tilde{y}, {}^t\tilde{\varkappa}'_y(\tilde{y})\tilde{\xi})$ is equivalent to $\Lambda_{\tilde{\psi}}$. This means that $\Lambda_{\tilde{\varphi}}$ is a well defined manifold in the sense of section 1 in the complexification $(T^*(\mathbb{R}^n)\setminus 0)^{\sim}$, whose definition is independent of the choice of real coordinates in the base.

Theorem 3.6. Let φ be a regular phase function of positive type, defined in a conic neighbourhood of $(x_0,\theta_0) \in \mathbb{R}^n \times (\mathbb{R}^N\setminus\{0\})$. Then the image $\Lambda_{\tilde{\varphi}}$ under the map (3.16) is locally a conic positive Lagrangean manifold, whose equivalence class does not depend on the choice of almost analytic extension $\tilde{\varphi}$ (under the condition that the map (3.16) is restricted to a sufficiently small neighbourhood of the real domain) . Moreover $\Lambda_{\tilde{\varphi}\mathbb{R}}$ is precisely the image of $C_{\varphi\mathbb{R}}$.

Proof. We can of course assume that $(x_0,\theta_0) \in C_{\varphi\mathbb{R}}$ and we let $\rho_0 = (x_0, \xi_0)$ be the image of (x_0,θ_0) under (3.16) . It is easily verified that $T_{\rho_0}(\Lambda_{\tilde{\varphi}})$ is a Lagrangean plane so by the proof of Lemma 3.4 we can choose real coordinates x in $\mathbb{R}^n$ such that the projection

$$\Lambda_{\tilde{\varphi}} \ni (\tilde{x}, \tilde{\xi}) \longrightarrow \tilde{\xi} \in \mathbb{C}^n \setminus \{0\} \tag{3.17}$$

is regular at ρ_0 . (Here ξ are the corresponding dual coordinates .) Now composing (3.17) with the map (3.16) and using

3^0 in Definition 3.5, we deduce that the map

$$T_{(x_0,\theta_0)}(\mathbb{C}^n \times (\mathbb{C} \setminus \{0\})) \ni (t_x, t_\theta) \longrightarrow \begin{pmatrix} \varphi''_{xx} & \varphi''_{x\theta} \\ \varphi''_{\theta x} & \varphi''_{\theta\theta} \end{pmatrix} \begin{pmatrix} t_x \\ t_\theta \end{pmatrix} \in \mathbb{C}^n \times \mathbb{C}^N$$

is bijective, or equivalently that the Hessian of φ at (x_0, θ_0) is non-singular. (Note that $T_{(x_0,\theta_0)}(C_{\tilde{\varphi}})$ is given by the equation $\varphi''_{\theta x} t_x + \varphi''_{\theta\theta} t_\theta = 0$.)

Now let $\tilde{x} = \tilde{x}(\tilde{\xi})$ be the equation for $\Lambda_{\tilde{\varphi}}$ in a conic neighbourhood of ρ_0 , with the function $\tilde{x}(\tilde{\xi})$ positively homogeneous of degree 0 , and let $(\tilde{x}(\tilde{\xi}), \tilde{\theta}(\tilde{\xi}))$ be the point in $C_{\tilde{\varphi}}$ that corresponds to $(\tilde{x}(\tilde{\xi}), \tilde{\xi})$ under the map (3.16) . Thus we have $\tilde{\varphi}'_{\tilde{x}}(\tilde{x}(\tilde{\xi}), \tilde{\theta}(\tilde{\xi})) = \tilde{\xi}$ and $\tilde{\varphi}'_{\tilde{\theta}}(\tilde{x}(\tilde{\xi}), \tilde{\theta}(\tilde{\xi})) = 0$, so that $(\tilde{x}(\tilde{\xi}), \tilde{\theta}(\tilde{\xi}))$ is a non-degenerate critical point of the function

$$(\tilde{x}, \tilde{\theta}) \longrightarrow \tilde{\varphi}(\tilde{x}, \tilde{\theta}) - \langle \tilde{x}, \tilde{\xi} \rangle$$

as described in Lemma 2.1 (with $\tilde{\xi}$ as the parameter). Thus we have the inequality

$$\operatorname{Im}(\tilde{\varphi}(\tilde{x}(\xi), \tilde{\theta}(\xi)) - \langle \tilde{x}(\xi), \xi \rangle) \geq C \, | \operatorname{Im}(\tilde{x}(\xi), \tilde{\theta}(\xi)) |^2 \qquad (3.18)$$

for real ξ close to ξ_0 and with some positive constant C .
(C will always denote a new constant each time in the following.)
Since $\tilde{\varphi}(\tilde{x}, \tilde{\theta})$ is positively homogeneous of degree 1 in $\tilde{\theta}$ and $\tilde{\varphi}'_{\tilde{\theta}}(\tilde{x}(\tilde{\xi}), \tilde{\theta}(\tilde{\xi})) = 0$, we also have inequalities

$$|\tilde{\varphi}(\tilde{x}(\tilde{\xi}), \tilde{\theta}(\tilde{\xi}))| \leq C_M(|\operatorname{Im} \tilde{x}(\tilde{\xi})|^M + |\operatorname{Im} \tilde{\theta}(\tilde{\xi})|^M) \quad , M \in \mathbb{Z}_+, \quad (3.19)$$

for $\tilde{\xi}$ in a small neighbourhood of ξ_0. Combining this with (3.18) we get

$$\operatorname{Im} \langle \tilde{x}(\xi), \xi \rangle \leq -C(|\operatorname{Im} \tilde{x}(\xi)|^2 |\xi| + |\operatorname{Im} \tilde{\theta}(\xi)|^2 |\xi|^{-1}) \qquad (3.20)$$

with a positive constant C for real ξ in a conic neighbourhood of ξ_0. This implies that

$$|\mathrm{Im}\, \tilde{\theta}(\tilde{\xi})|^2 \leq C\, (|\tilde{\xi}|^2\, |\mathrm{Im}\, \tilde{x}(\tilde{\xi})| + |\tilde{\xi}|\, |\mathrm{Im}\, \tilde{\xi}|) \tag{3.21}$$

for complex $\tilde{\xi}$ close to ξ_0 (and with some new constant C). It follows that $\Lambda_{\tilde{\varphi}\,\mathbb{R}}$ is precisely the image of $C_{\varphi\mathbb{R}}$ under (3.16) (if this map is suitably restricted) and Lemma 1.11 and (3.21) show that $\Lambda_{\tilde{\varphi}}$ is locally an almost analytic manifold whose equivalence class does not depend on the choice of $\tilde{\varphi}$.

In general if f is a non-negative C^∞ function , we have locally an inequality

$$|df|^2 \leq C\, f\, .$$

Applying this to $\mathrm{Im}\,\varphi$ we get

$$|\mathrm{Im}\, \varphi'_x(x,\theta)|^2 \leq C\; \mathrm{Im}\,\varphi(x,\theta)$$

for real (x,θ) close to (x_0,θ_0) . In the complex domain we get locally

$$|\mathrm{Im}\; \tilde{\varphi}'_{\tilde{x}}(\tilde{x},\tilde{\theta})|^2 \leq C(\,|\mathrm{Im}\, \tilde{\varphi}(\tilde{x},\tilde{\theta})| + |\mathrm{Im}(\tilde{x},\tilde{\theta})|\,)\, .$$

Restricting this to $C_{\tilde{\varphi}}$ and using (3.19) we get

$$|\mathrm{Im}\, \tilde{\xi}|^2 \leq C\; |\mathrm{Im}(\tilde{x}(\tilde{\xi}),\tilde{\theta}(\tilde{\xi}))|\, . \tag{3.22}$$

The inequalities (3.21) and (3.22) can be rewritten as

$$C^{-1}|\mathrm{Im}(\tilde{x},\tilde{\xi})|^2 \leq |\mathrm{Im}(\tilde{x},\tilde{\theta})| \leq C\, |\mathrm{Im}(\tilde{x},\tilde{\xi})|^{1/2}\, , \tag{3.23}$$

where $(\tilde{x},\tilde{\xi})$ and $(\tilde{x},\tilde{\theta})$ are corresponding points on $\Lambda_{\tilde{\varphi}}$ and $C_{\tilde{\varphi}}$ respectively (and close to ξ_0 and (x_0,θ_0)).Lemma 1.12 and (3.23) permit us to identify almost analytic functions on

$C_{\tilde{\varphi}}$ and $\Lambda_{\tilde{\varphi}}$ by means of (3.16) .

We note that $\sum \tilde{\xi}_j \, d\tilde{x}_j |_{\Lambda_{\tilde{\varphi}}}$ can be identified with

$$\sum \partial\tilde{\varphi}/\partial\tilde{x}_j \, d\tilde{x}_j |_{C_{\tilde{\varphi}}} \sim d_{\tilde{x}} \tilde{\varphi} |_{C_{\tilde{\varphi}}} \sim 0$$

since the restriction of $\tilde{\varphi}$ to $C_{\tilde{\varphi}}$ is ~ 0 . Differentiating we get

$$\sum d\tilde{\xi}_j \wedge d\tilde{x}_j |_{\Lambda_{\tilde{\varphi}}} \sim 0$$

and Lemma 3.4 (and its proof) shows that $\Lambda_{\tilde{\varphi}}$ is represented by $\tilde{x} = \partial g/\partial \tilde{\xi}$ where $g(\tilde{\xi}) = \langle \tilde{x}(\tilde{\xi}), \tilde{\xi} \rangle$. Now $\operatorname{Im} g(\xi) \leq 0$ for real ξ close to ξ_0 by (3.20) , so it follows that $\Lambda_{\tilde{\varphi}}$ is locally a positive Lagrangean manifold .The proof of Theorem 3.6 is complete.

In the following sections we shall write Λ_{φ} instead of $\Lambda_{\tilde{\varphi}}$ for the image in (3.16) . The almost analytic manifold Λ_{φ} is then well defined up to equivalence.

4. Equivalence of phase functions and global definition of Fourier distributions .

Let $\varphi = \varphi(x,\theta) \in C^{\infty}(V)$ be a positive regular phase function where $V \subset \mathbb{R}^n \times (\mathbb{R}^N \setminus \{0\})$ is a conic open set . If $a = a(x,\theta) \in S^m_{1-\delta}(\mathbb{R}^n \times \mathbb{R}^N)$, $\delta < 1/2$, and supp a is contained in a closed conic subset of V, then we can define the distribution $A = I(a,\varphi) \in \mathcal{D}'(\mathbb{R}^n)$ by

$$\langle I(a,\varphi),u\rangle = \iint e^{i\varphi(x,\theta)} a(x,\theta)\, u(x)\, dx\, d\theta \ , \ u \in C_0^{\infty}(\mathbb{R}^n). \qquad (4.1)$$

Here the integral is an oscillatory integral as defined in Hörmander [3] . Sometimes we shall write formally:

$$I(a,\varphi) = \int e^{i\varphi(x,\theta)} a(x,\theta)\, d\theta \ .$$

The proof of Proposition 2.5.7 in [3] extends to the case of positive regular phase functions and gives the following result about the wave front set of A:

$$WF(A) \subset \{(x, \varphi'_x(x,\theta)) ; \ (x,\theta) \in \text{cone supp } a \cap C_{\varphi\mathbb{R}}\} . \qquad (4.2)$$

Now let $\varphi(x,\theta)$ and $\psi(x,w)$ be positive regular phase functions defined in small conic neighbourhoods V of $(x_0,\theta_0) \in \mathbb{R}^n \times (\mathbb{R}^N \setminus \{0\})$ and V' of $(x_0,w_0) \in \mathbb{R}^n \times (\mathbb{R}^M \setminus \{0\})$ respectively. We assume that $\varphi'_\theta(x_0,\theta_0)=0$, $\psi'_w(x_0,w_0) = 0$ and that $\varphi'_x(x_0,\theta_0) = \psi'_x(x_0,w_0) = \xi_0$ where the last equation is a definition . Put $\lambda_0 = (x_0,\xi_0)$. We shall say that two distributions $A,B \in \mathcal{D}'(\mathbb{R}^n)$ are microlocally equal near λ_0 if $\lambda_0 \notin WF(A - B)$. This is clearly an equivalence relation and we let $\mathcal{D}'_{\lambda_0}$ be the quotient of $\mathcal{D}'(\mathbb{R}^n)$ with respect to this equivalence relation. Now φ defines a map from $S^m_{1-\delta}(\mathbb{R}^n \times \mathbb{R}^N)$ to $\mathcal{D}'_{\lambda_0}$ as follows : For $a \in S^m_{1-\delta}(\mathbb{R}^n \times \mathbb{R}^N)$, let $a_0 \in S^m_{1-\delta}(\mathbb{R}^n \times \mathbb{R}^N)$ be equal to $a(x,\theta)$ for large θ in a small conic neighbourhood of $\{(x_0,t\theta_0)\ ;\ t>0\}$ and have its support close to this half-ray . Then $I(a_0,\varphi)$ defines an element of $\mathcal{D}'_{\lambda_0}$ which is independent of the choice of a_0.

Definition 4.1. We say that φ and ψ are equivalent at λ_0 for symbols of type $1-\delta$ if the corresponding images in $\mathcal{D}'_{\lambda_0}$ of $S^\infty_{1-\delta}(\mathbb{R}^n \times \mathbb{R}^N)$ and $S^\infty_{1-\delta}(\mathbb{R}^n \times \mathbb{R}^M)$ are the same. Similarly we define equivalence for classical symbols (that is symbols of type $1,0$ of the form $a \sim \sum_0^\infty a_{m-j}(x,\theta)$ where the a_{m-j} are positively homogeneous in θ of degree $m-j$.)

Theorem 4.2. Let φ and ψ be positive regular phase functions defined in small conic neighbourhoods of $(x_0,\theta_0) \in \mathbb{R}^n \times (\mathbb{R}^N \setminus \{0\})$ and $(x_0,w_0) \in \mathbb{R}^n \times (\mathbb{R}^M \setminus \{0\})$ respectively. Assume that Λ_φ and Λ_ψ are equivalent in a neighbourhood of (x_0, ξ_0) where $\xi_0 = \varphi'_x(x_0,\theta_0) = \psi'_x(x_0,w_0)$. Then φ and ψ are equivalent at (x_0,ξ_0) for symbols of type $1-\delta$, $\delta < 1/2$, as well as for classical symbols.

Proof. After a real change of coordinates in $\mathbb{R}^n$ we can assume that Λ_φ has the representative $\tilde{x} = \tilde{x}(\tilde{\xi})$ near (x_0,ξ_0) , where $\tilde{x}(\tilde{\xi})$ is almost analytic and positively homogeneous of degree 0 for real $\tilde{\xi}$. As in the proof of Theorem 3.6 , we introduce $\tilde{\theta}(\xi)$ for real ξ in a conic neighbourhood of ξ_0 so that $(\tilde{x}(\xi),\tilde{\theta}(\xi)) \in \mathbb{C}^n \times (\mathbb{C}^N \setminus \{0\})$ is a critical point (as in Lemma 2.1) of the function

$$(\tilde{x},\tilde{\theta}) \longrightarrow \tilde{\varphi}(\tilde{x},\tilde{\theta}) \quad - \langle \tilde{x},\xi\rangle \ .$$

Here $\tilde{\varphi}$ denotes some almost analytic extension of φ ,positively homogeneous of degree 1 in θ in the sense of Theorem 1.3.

Now suppose that $a(x,\theta) \in S^{m+(n-2N)/4}_{1-\delta}(\mathbb{R}^n \times \mathbb{R}^N)$ has its support in a small conic neighbourhood of (x_0,θ_0) . We shall study the Fourier transform of $A = I(a,\varphi)$ at $t\xi$ for large $t \in \mathbb{R}_+$ and for ξ in a small real neighbourhood of ξ_0 .This Fourier transform is given by

$$\hat{A}(t\xi) = t^N \iint e^{it(\varphi(x,\theta) - \langle x,\xi\rangle)} a_t(x,\theta)\, dx\, d\theta \ , \qquad (4.3)$$

where $a_t(x,\theta) = a(x,t\theta)$. Let $\chi \in C_0^\infty(\mathbb{R}^n \times \mathbb{R}^N)$ be equal to 1 in a neighbourhood of (x_0,θ_0) . Then the exponent in the integral

$$I_2(t\xi) = t^N \iint e^{it(\varphi(x,\theta) - \langle x,\xi\rangle)} a_t(x,\theta)(1 - \chi(x,\theta))dxd\theta \ (4.4)$$

has no critical points in the support of the integrand when ξ is in a small real neighbourhood ω of ξ_0. By suitable partial integrations (as in the proof of Theorem 3.2.4 in [3]) we see that $I_2(t\xi)$ is an element in $S^{-\infty}(\omega \times \mathbb{R}_+)$. Now $\hat{A}(t\xi) = I_1(t\xi) + I_2(t\xi)$ where

$$I_1(t\xi) = t^N \iint e^{it(\varphi(x,\theta) - \langle x,\xi\rangle)} a_t(x,\theta)\, \chi(x,\theta)\, dx\, d\theta, \qquad (4.5)$$

so if we apply Theorem 2.3 to this integral we get with asymptotic convergence in $S^{m-n/4}_{0,1}(\omega \times \mathbb{R}_+)$:

$$\hat{A}(t\xi) \sim \sum_0^\infty e^{it(\tilde{\varphi}(\tilde{x}(\xi),\tilde{\theta}(\xi)) - \langle \tilde{x}(\xi),\xi\rangle)} t^{(N-n)/2 - \nu}(C_{\nu,\xi}(D)\tilde{a}_t)(\tilde{x}(\xi),\tilde{\theta}(\xi)) \qquad (4.6)$$

Here $\tilde{a} \in S^{m+(n-2N)/4}_{1-\delta}(\mathbb{C}^n \times \mathbb{C}^N)$ denotes some almost analytic extension of a and $\tilde{a}_t(\tilde{x},\tilde{\theta}) = \tilde{a}(\tilde{x},t\tilde{\theta})$ also for complex $(\tilde{x},\tilde{\theta})$. Moreover the $C_{\nu,\xi}$ are as in Theorem 2.3 so if we define

$$\text{Hess}\ \tilde{\varphi}_{(\tilde{x},\tilde{\theta})} = \begin{pmatrix} \tilde{\varphi}''_{\tilde{x}\tilde{x}} & \tilde{\varphi}''_{\tilde{x}\tilde{\theta}} \\ \tilde{\varphi}''_{\tilde{\theta}\tilde{x}} & \tilde{\varphi}''_{\tilde{\theta}\tilde{\theta}} \end{pmatrix},$$

we have in particular

$$c_{0,\xi} = (\det((2\pi i)^{-1}\text{Hess}\ \tilde{\varphi}_{(\tilde{x}(\xi),\tilde{\theta}(\xi))}))^{-1/2}$$

with the branch chosen as in that theorem . It follows from (3.19),(3.20) that

$$e^{it(\tilde{\varphi}(\tilde{x}(\xi),\tilde{\theta}(\xi))-\langle\tilde{x}(\xi),\xi\rangle)} \equiv e^{-it\langle\tilde{x}(\xi),\xi\rangle} \mod S^{-\infty}(\omega\times\mathbb{R}_+)$$

for we have the following easy lemma:

Lemma 4.3. Let $\omega\subset\mathbb{R}^n$ be open and let $b_j\in C^\infty(\omega)$, $j=0,1$, where $\operatorname{Im} b_j \geq 0$ and

$$|b_0(\xi) - b_1(\xi)| \leq C_{N,K} \min_{j=0,1} (\operatorname{Im} b_j(\xi))^N , \ \xi\in K ,$$

for all $K\subset\subset\omega$ and all $N\in\mathbb{Z}_+$. Then $e^{itb_1(\xi)} - e^{itb_0(\xi)} \in S^{-\infty}(\omega\times\mathbb{R}_+)$.

(The converse is also true and follows from Lemma 5.2 below.)

Proof of the lemma. Put $b_s = sb_1 + (1-s)b_0$. Then

$$e^{itb_1(\xi)} - e^{itb_0(\xi)} = \int_0^1 \frac{\partial}{\partial s}(e^{itb_s(\xi)})\,ds$$

$$= \int_0^1 it(b_1(\xi) - b_0(\xi))\,e^{itb_s(\xi)}\,ds$$

and it is easy to see that the integrand is uniformly rapidly decreasing when $t\longrightarrow+\infty$ for ξ in any compact subset of ω .

Replacing $t\xi$ by ξ we now get from (4.6) with asymptotic convergence in $S^{m-n/4}_{0,1}(\mathbb{R}^n)$:

$$\hat{A}(\xi)\sim\sum_0^\infty e^{-i\langle\tilde{x}(\xi),\xi\rangle}\,|\xi|^{(N-n)/2-\nu}(c_{\nu,\xi/|\xi|}(D_{\tilde{x}},|\xi|D_{\tilde{\theta}})\tilde{a})(\tilde{x}(\xi),\tilde{\theta}(\xi))$$

for ξ in a small conic neighbourhood V of ξ_0, independent of a. (Here $D_{\tilde{x}} = (\partial/\partial\tilde{x}, \partial/\partial\bar{\tilde{x}})$ and $D_{\tilde{\theta}} = (\partial/\partial\tilde{\theta}, \partial/\partial\bar{\tilde{\theta}})$). The homogeneity of φ implies that

$$c_{0,\xi/|\xi|} = |\xi|^{(n-N)/2} \left(\det(2\pi i)^{-1} \operatorname{Hess} \tilde{\varphi}_{(\tilde{x}(\xi),\tilde{\theta}(\xi))}\right)^{-1/2} .$$

Now we can choose elements $Q_{-1}(\tilde{a}) \in S^{m-(1-2\delta)}_{1-\delta}(V)$ such that

$$Q_{-1}(\tilde{a}) \sim \sum_{1}^{\infty} (c_{0,\xi/|\xi|})^{-1} |\xi|^{-\nu} c_{\nu,\xi/|\xi|}(D_{\tilde{x}}, |\xi| D_{\tilde{\theta}})\, \tilde{a})(\tilde{x}(\xi),\tilde{\theta}(\xi)) .$$

This is clear since the ν:th term in the sum above belongs to $S^{m'-(1-2\delta)\nu}_{1-\delta}(V)$., $m' = m+(n-2N)/4$. Then we may write

$$\widehat{I(a,\varphi)}(\xi) \sim$$

$$e^{-i\langle\tilde{x}(\xi),\xi\rangle} \left(\det(2\pi i)^{-1} \operatorname{Hess}\tilde{\varphi}_{(\tilde{x}(\xi),\tilde{\theta}(\xi))}\right)^{-1/2} \left(\tilde{a}(\tilde{x}(\xi),\tilde{\theta}(\xi)) + Q_{-1}\tilde{a}(\xi)\right) \tag{4.7}$$

Also note that if a belongs to the space $S^k_c(\mathbb{R}^n \times \mathbb{R}^N)$ of classical symbols of order k then we can choose $Q_{-1}\tilde{a}$ in $S^{k-1}_c(V)$.

By Lemma 4.3 we know that the factor $e^{-i\langle\tilde{x}(\xi),\xi\rangle}$ is modulo $S^{-\infty}$ independent of the choice of local representatives of Λ_φ of the form $\tilde{x} = \tilde{x}(\tilde{\xi})$ with $\tilde{x}(\tilde{\xi})$ positively homogeneous of degree 0 for real $\tilde{\xi}$. Thus, since Λ_ψ is equivalent to Λ_φ near (x_0,ξ_0) , we have a formula for $\widehat{I(b,\psi)}(\xi)$ completely analogous to (4.7) with <u>the same factor</u> $e^{-i\langle\tilde{x}(\xi),\xi\rangle}$ if b $\in S^{m+(n-2M)/4}_{1-\delta}(\mathbb{R}^n \times \mathbb{R}^M)$ has its support in a small conic neighbourhood of (x_0,w_0). The theorem now follows in the case of symbols of type $1-\delta$, if we can prove that for each such symbol b there exists $a \in S^{m+(n-2N)/4}_{1-\delta}(\mathbb{R}^n \times \mathbb{R}^N)$ with support close to $\{(x_0,t\theta_0) ; t>0\}$ such that $\widehat{I(a,\varphi)}(\xi) \sim \widehat{I(b,\psi)}(\xi)$ in a conic neighbourhood of ξ_0 (and conversely that for such an a there exists b ...).

By formula (4.7) it is sufficient to prove that for any b

$\in S^m_{1-\delta}(\mathbb{R}^n)$ there exists an almost analytic symbol $\tilde{a} \in S^m_{1-\delta}(\mathbb{C}^n \times \mathbb{C}^N)$ such that

$$e^{-i\langle\tilde{x}(\xi),\xi\rangle}(\tilde{a}(\tilde{x}(\xi),\tilde{\theta}(\xi)) + Q_{-1}(\tilde{a})(\xi)) \sim e^{-i\langle\tilde{x}(\xi),\xi\rangle}b(\xi) \qquad (4.8)$$

in a conic neighbourhood of ξ_0. This follows if for every such b we can find $\tilde{a}$ such that

$$e^{-i\langle\tilde{x}(\xi),\xi\rangle}\, \tilde{a}(\tilde{x}(\xi),\tilde{\theta}(\xi)) \sim e^{-i\langle\tilde{x}(\xi),\xi\rangle}\, b(\xi) \quad . \qquad (4.9)$$

For then by successive approximations we can find a sequence $\tilde{a}_j \in S^{m-j(1-2\delta)}_{1-\delta}(\mathbb{C}^n \times \mathbb{C}^N)$ of almost analytic functions such that

$$\sum_0^\infty e^{-i\langle\tilde{x}(\xi),\xi\rangle}(\tilde{a}_j(\tilde{x}(\xi),\tilde{\theta}(\xi)) + Q_{-1}(\tilde{a}_j)(\xi)) \sim e^{-i\langle\tilde{x}(\xi),\xi\rangle}\, b(\xi)$$

in $S^m_{0,1}(\mathbb{R}^n)$ for ξ in a conic neighbourhood of ξ_0, and we are ready because we can take $\tilde{a} \in S^m_{1-\delta}(\mathbb{C}^n \times \mathbb{C}^N)$ almost analytic and such that $\tilde{a} \sim \sum_0^\infty \tilde{a}_j$. Then (4.8) holds .

In order to finish the proof we have to find an almost analytic symbol $\tilde{a}$ so that (4.9) is valid in a conic neighbourhood of ξ_0 (independent of b) . Now the set

$$\{(\tilde{x}(\xi),\tilde{\theta}(\xi)) \in \mathbb{C}^n \times \mathbb{C}^N \;\; ; \; \xi \in \mathbb{R}^n\}$$

is totally real for it is the inverse image of the totally real set $\{(x(\xi),\xi)\ ;\ \xi\in\mathbb{R}^n\}$ under the almost analytic diffeomorphism $C_\varphi \longrightarrow \Lambda_\varphi$, defined in the preceding section. Thus we can apply Proposition 1.16 and find $\tilde{a}\in S^m_{1-\delta}(\mathbb{C}^n \times \mathbb{C}^N)$, almost analytic and such that

$$|\tilde{a}(\tilde{x}(\xi),\tilde{\theta}(\xi)) - b(\xi)| \leq C_{r,b}\,|\xi|^{m+r\delta}\,|\mathrm{Im}(x(\xi),\theta(\xi)/|\xi|)|^r$$

in a fixed conic real neighbourhood of ξ_0 for all $r\in\mathbb{Z}_+$. Then it follows from (3.20) that (4.8) is valid and this completes the proof in the case of symbols of type $1-\delta$. The proof in the case of classical symbols is practically the same and we omit the details.

We can now define Fourier distributions globally. Let X be a paracompact C^∞ manifold of dimension n and let $\mathcal{D}'(X;\Omega_{1/2})$ be the space of distribution densities of order 1/2 on X (see [3 , p.117]).

In general we shall say that an almost analytic manifold $\Lambda \subset \tilde{M}$ is closed if $\Lambda_{\mathbb{R}}\subset M$ is a closed set. Now let $\Lambda\subset (T^*(X)\setminus 0)$ be a closed conic positive Lagrangean manifold.

<u>Definition 4.4</u>. By $I^m_{1-\delta}(X,\Lambda)$, $\delta< 1/2$, we denote the subspace of all $A\in\mathcal{D}'(X;\Omega_{1/2})$ which satisfy :

1^0 $WF(A)\subset\Lambda_{\mathbb{R}}$,

2^0 for every $\lambda_0\in\Lambda_{\mathbb{R}}$ and every choice of local coordinates $x_1,\dots,x_n$ near $\pi(\lambda_0)\in X$, A is microlocally of the form $I(a,\varphi)$ near λ_0, where $\varphi\in C^\infty(\mathbb{R}^n \times (\mathbb{R}^N\setminus\{0\}))$ is an arbitrary positive regular phase function generating Λ near λ_0 (in the chosen coordinates) and where

$a \in S^{m+(n-2N)/4}_{1-\delta}(\mathbb{R}^n \times \mathbb{R}^N)$ has its support in a small conic neighbourhood of $(x_0,\theta_0) \in C_{\varphi\mathbb{R}}$ - the point corresponding to λ_0.

Similarly we define $I^m_c(X,\Lambda)$ using classical symbols $a \in S^{m+(n-2N)/4}_c(\mathbb{R}^n \times \mathbb{R}^N)$. ($S^m_c$ is defined on the next page.)

We note that $I^m_{1-\delta}(X,\Lambda)$ is a module over the ring of properly supported pseudo-differential operators $P \in L^0_{1-\delta}(X)$; $P: \mathcal{D}'(X,\Omega_{1/2}) \to \mathcal{D}'(X,\Omega_{1/2})$. For if $A \in I^m_{1-\delta}(X,\Lambda)$ and $P \in L^0_{1-\delta}(X)$ then clearly $WF(PA) \subset \Lambda_{\mathbb{R}}$ and near a point $\lambda_0 \in \Lambda_{\mathbb{R}}$ we know that PA is microlocally of the form

$$P(x,D)\left(\int e^{i\varphi(x,\theta)}\, a(x,\theta)\, d\theta\right) = \int P(x,D)\left(a(x,\theta)e^{i\varphi(x,\theta)}\right)d\theta.$$

Now it follows from formula (2.28) that

$$P(x,D)\left(a(x,\theta)\, e^{i\varphi(x,\theta)}\right) \sim b(x,\theta)\, e^{i\varphi(x,\theta)},$$

where b is a symbol of type $1-\delta$ and of order less than or equal to the order of a. We omit the details. Similarly $I^m_c(X,\Lambda)$ is a module over the ring of properly supported classical pseudo-differential operators of order 0.

In particular, by a pseudo-differential partition of unity in $T^*(X) \setminus 0$, we are able to write any given $A \in I^m_{1-\delta}(X,\Lambda)$ as a locally finite sum of elements of the type $I(a,\varphi)$.

5. Necessary conditions for equivalence of phase functions .

We have seen in the preceding section that two positive regular phase functions φ and ψ are equivalent near a point ρ_0 in the cotangent space for symbols of type $1-\delta$ as well as for classical symbols when $\Lambda_\varphi \sim \Lambda_\psi$ near ρ_0. In the present section we shall prove that the converse is also true in the case of classical symbols , but not in general (not even for " nice"symbols of type 1,0 .The intuitive explanation of this phenomenon is that bad symbols may contain oscillations which can perturbe the corresponding Lagrangean manifolds Λ_φ and Λ_ψ. The essential part of the proof is given by the two following lemmas .

If $V \subset \mathbb{R}^n \times (\mathbb{R}^N \setminus \{0\})$ is an open conic set we write $S^m_c(V)$ for the set of elements in $S^m_1(V)$ which are asymptotic sums

$$p(x,\theta) \sim p_m(x,\theta) + p_{m-1}(x,\theta) + p_{m-2}(x,\theta) + \dots$$

where p_{m-j} are positively homogeneous of degree $m-j$ in θ . Moreover we write $S^m_{ac}(V)$ for the $p \in S^m_1(V)$ of the form

$$p(x,\theta) \sim \sum_{\nu \in D_p} p_\nu(x,\theta) \quad ,$$

where $D_p \subset (-\infty, m]$ is some discrete set and the p_ν are positively homogeneous of degree ν . (Here "c" stands for classical and "ac" for"almost classical " .) We write

$$S^\infty_c = \bigcup_{m \in \mathbb{R}} S^m_c \quad , \quad S^\infty_{ac} = \bigcup_{m \in \mathbb{R}} S^m_{ac} \quad .$$

Lemma 5.1. Let $a(x,t) \sim \sum_{\nu\in D_a} a_\nu(x)\, t^\nu \in S^\infty_{ac}(\mathbb{R}^n \times \mathbb{R}_+)$.
Suppose that $b \in C^\infty(\mathbb{R}^n)$, $\mathrm{Re}\, b \geq 0$ and that

$$e^{-tb(x)}\, a(x,t) \sim 0 . \tag{5.1}$$

Then for every $N \in \mathbb{Z}_+$, $\nu \in D_a$, we have $a_\nu(x) = \mathcal{O}((\mathrm{Re}\, b(x))^N)$ uniformly on every compact set.

Proof. We can assume that b is real-valued. Multiplying $a(x,t)$ if necessary by a negative power of t, we may also assume that $D_a \subset \mathbb{R}_-$. Put

$$D_N = \{ \nu\in D_a \; ; \; \nu \geq -N \} .$$

Then by (5.1) we have

$$\Big| e^{-tb(x)} \sum_{\nu\in D_N} a_\nu(x)\, t^\nu \Big| \leq C_{K,N}\, t^{-N} , \quad t > 1 ,$$

when $x \in K \subset\subset \mathbb{R}^n$, $N \in \mathbb{Z}_+$. We may assume that $b(x) < 1$, so putting $t = s/b(x)$ we get the following inequality in the domain where $b(x) \neq 0$:

$$\Big| \sum_{\nu\in D_N} a_\nu(x)\, b(x)^{-\nu} s^\nu \Big| \leq C'_{K,N}\, b(x)^N , \tag{5.2}$$

$$x \in K \subset\subset \mathbb{R}^n , \quad 1 \leq s \leq 2 .$$

Of course, (5.2) is trivial where $b(x)$ vanishes.

The functions s^ν, $\nu \in D_N$, form a linearly independent set on the interval $[1,2]$, so there are C^∞ functions $g_\nu(s) = g_{\nu,N}(s)$ satisfying

$$\int_1^2 g_\nu(s)\, s^\mu\, ds = \delta_{\nu,\mu} , \quad \nu,\mu \in D_N ,$$

where $\delta_{\nu,\mu}$ is the Kronecker delta. Integrating

$$\sum_{\nu \in D_N} a_\nu(x)\ b(x)^{-\nu} s^\nu\ g_\mu(s)$$

from 1 to 2 , we get an inequality

$$|a_\mu(x)\ b(x)^{-\mu}| \leq C_{K,N,\mu}\ b(x)^{\ N}\ ,\ \mu \in D_N\ ,\ x \in K \subset\subset \mathbb{R}^N\ ,$$

and the lemma follows .

Lemma 5.2. Let $b_j \in C^\infty(\mathbb{R}^n)$,$j = 1,2$, with $\operatorname{Im} b_j \geq 0$ and suppose that

$$S^\infty_{ac}(\mathbb{R}^n \times \mathbb{R}_+)\ e^{itb_1(x)} \sim S^\infty_{ac}(\mathbb{R}^n \times \mathbb{R}_+)\ e^{itb_2(x)}$$

in the sense that for every $a(x,t) \in S^\infty_{ac}(\mathbb{R}^n \times \mathbb{R}_+)$, there is a $c(x,t) \in S^\infty_{ac}(\mathbb{R}^n \times \mathbb{R}_+)$ such that

$$a(x,t)\ e^{itb_1(x)} \sim c(x,t)\ e^{itb_2(x)}$$

and conversely . Then for every $K \subset\subset \mathbb{R}^n$ and $N \in \mathbb{Z}_+$, there is a constant C_N such that

$$|b_1(x) - b_2(x)| \leq C_N(\operatorname{Im} b_j(x)\ )^N\ ,\ j = 1,2\ ,\ x \in K\ .$$

Proof. It is clear that $\operatorname{Im} b_1$ and $\operatorname{Im} b_2$ must have the same zeros. The condition in the lemma means precisely that there are symbols $a^{(12)}(x,t)$ and $a^{(21)}(x,t)$ in $S^\infty_{ac}(\mathbb{R}^n \times \mathbb{R}_+)$ such that

$$e^{itb_1(x)} \sim a^{(12)}(x,t)e^{itb_2(x)}\ ,\ e^{itb_2(x)} \sim a^{(21)}(x,t)e^{itb_1(x)} \tag{5.3}$$

and if we write

$$a^{(12)}(x,t) \sim \sum_{\nu\in D_{12}} a_\nu^{(12)}(x)\, t^\nu \quad , \quad a^{(21)}(x) \sim \sum_{\mu\in D_{21}} a_\mu^{(21)}(x,t)\, t^\mu$$

we see that $a_0^{(12)}(x) = a_0^{(21)}(x) = 1$ where $\operatorname{Im} b_1(x)$ (and $\operatorname{Im} b_2(x)$) vanishes . We shall prove below that $a^{(12)}$ and $a^{(21)}$ can be taken of order 0. Assuming this for the moment , we apply $\partial/\partial t - i\, b_1(x)$ to the first equation in (5.3) and get

$$0 \sim (\partial/\partial t - ib_1(x))\, (a^{(12)}(x,t)\, e^{itb_2(x)})$$

or equivalently

$$0 \sim (i(b_2(x) - b_1(x))\, a^{(12)}(x,t) + \partial a^{(12)}(x,t)/\partial t) e^{itb_2(x)}$$

and Lemma 5.1 shows that $(b_2(x) - b_1(x)) = \mathcal{O}((\operatorname{Im} b_2(x))^N)$ for all N , since $a_0^{(12)}(x) = 1$ where $\operatorname{Im} b_2$ vanishes. By symmetry, this gives the lemma when we have proved that $a^{(12)}$ and $a^{(21)}$ can be taken of order 0 .

Composing the two equations in (5.3) we get

$$(1 - a^{(12)}(x,t)a^{(21)}(x,t))e^{itb_1(x)} \sim 0 \quad . \tag{5.4}$$

Let us write in general $f(x) \equiv g(x)$ if f, g are C^∞ functions satisfying $f(x) - g(x) = \mathcal{O}((\operatorname{Im} b_1(x))^N)$ for all $N \in \mathbb{Z}_+$. Then (5.4) and Lemma 5.1 give

$$\sum_{\nu+\mu=k} a_\nu^{(12)}(x)\, a_\mu^{(21)}(x) \equiv \begin{cases} 0 & \text{if } k \in (D_{12}+D_{21}) \setminus \{0\} \\ 1 & \text{if } k = 0 \ . \end{cases} \tag{5.5}$$

We shall prove by induction that

$$a_\nu^{(12)}\, a_\mu^{(21)} \equiv 0 \quad \text{if } \nu+\mu > 0. \tag{5.6}$$

Let $k_0, k_1, \ldots, k_p$ be the elements of $\{ k \in D_{12}+D_{21} \ ; \ k > 0 \}$

in decreasing order . Then clearly $a_{\nu}^{(12)}a_{\mu}^{(21)} \equiv 0$ for the unique $a_{\nu}^{(12)}$, $a_{\mu}^{(21)}$ with $\nu+\mu = k_0$. Suppose now that we have already proved that $a_{\nu}^{(12)}\, a_{\mu}^{(21)} \equiv 0$ for $\nu+\mu = k_j$ where $j < p$, and consider the equation

$$\sum_{\nu+\mu=k_{j+1}} a_{\nu}^{(12)}\, a_{\mu}^{(21)} \equiv 0 \quad . \tag{5.7}$$

Let $a_{\nu_0}^{(12)}\, a_{\mu_0}^{(21)}$ be the term with minimal μ (and maximal ν) . If we multiply (5.7) by $a_{\nu_0}^{(12)}$ all other terms become $\equiv 0$ by the inductive hypothesis and thus we obtain $(a_{\nu_0}^{(12)})^2 a_{\mu_0}^{(21)} \equiv 0$. Multiply this with $a_{\mu_0}^{(21)}$ and use that in general $f \equiv 0$ if $f^2 \equiv 0$ .. Then we get $a_{\nu_0}^{(12)}\, a_{\mu_0}^{(21)} \equiv 0$. Now let $a_{\nu_1}^{(12)}\, a_{\mu_1}^{(21)}$ be the remaining term in (5.7) with minimal μ and repeat the same procedure... . This gives finally that $a_{\nu}^{(12)} a_{\mu}^{(21)} \equiv 0$ for all ν, μ with $\nu+\mu = k_{j+1}$ and our inductive proof of (5.6) is complete

Using that $a^{(12)} = a^{(21)} = 1$ where $\operatorname{Im} b_1$ and $\operatorname{Im} b_2$ vanish we get from (5.6) that $a_{\nu}^{(12)} \equiv 0$, $a_{\mu}^{(21)} \equiv 0$ for $\nu, \mu > 0$. By symmetry we also have that $a_{\nu}^{(12)}$ and $a_{\mu}^{(21)}$ are $\mathcal{O}((\operatorname{Im} b_2(x))^N)$ for all $N \in \mathbb{Z}_+$, $\nu, \mu > 0$, and this shows that all terms with $\nu, \mu > 0$ in the asymptotic expansion of $a^{(12)}$ and $a^{(21)}$ can be eliminated without destroying (5.3). This completes the proof of the lemma.

Remark 5.3. The lemma is false if we replace S_{ac}^{∞} by $S_{1,0}^{\infty}$ or any larger space of symbols. In fact, let $n = 1$ and put

$$\Phi(x) = e^{-1/x^2} \quad , \quad x \in \mathbb{R} \, ,$$

$$b_1(x) = i\,\Phi(x) \quad , \quad b_2(x) = i\,\Phi(x) + (\Phi(x))^2 \; .$$

Then we have (5.3) with

$$a^{(12)}(x,t) = \chi(t(\Phi(x))^2)\ e^{-it(\Phi(x))^2}$$

$$a^{(21)}(x,t) = \chi(t(\Phi(x))^2)\ e^{it(\Phi(x))^2}$$

where $\chi \in C_0^\infty(\mathbb{R})$ is any function , equal to 1 near the origin . For (5.3) is then equivalent to $e^{-t\Phi}(1-\chi(t\Phi^2)) \sim 0$ which follows from the fact that $t\Phi^2 > c \Rightarrow t\Phi > (ct)^{1/2}$. Since we have local estimates of the form

$$|D_x^\alpha \Phi(x)| \leq C_{\alpha,\delta}(\Phi(x))^{1-\delta}, \quad x \in \mathbb{R} ,$$

for all α and $\delta > 0$, it is easy to see that $a^{(12)}$ and $a^{(21)}$ belong to

$$\bigcap_{\varepsilon>0} S^0_{1,\varepsilon}(\mathbb{R} \times \mathbb{R}_+) \subset \bigcap_{\varepsilon>0} S^\varepsilon_{1,0}(\mathbb{R} \times \mathbb{R}_+) \quad .$$

However b_1 and b_2 are not equivalent in the sense of Lemma 5.2 so this gives a counter example when symbols of type 1,0 are allowed .

We can now prove the necessity in Theorem 4.2 for the case of classical (or almost classical) symbols. Let $\varphi(x,\theta)$ be a regular phase function of positive type, defined in a conic neighbourhood of a point (x_0,θ_0) in $\mathbb{R}^n \times (\mathbb{R}^N \setminus \{0\})$ which is also in $C_{\varphi\mathbb{R}}$. Let $\rho_0 = (x_0, \xi_0) \in \Lambda_{\varphi\mathbb{R}}$ be the corresponding point and assume that the coordinates x in the base are chosen so that Λ_φ is represented locally by

$$\tilde{x} = \tilde{x}(\tilde{\xi}) = \partial H(\tilde{\xi})/\partial\tilde{\xi} \quad , \tilde{\xi} \in C^n , \tag{5.8}$$

where $H(\tilde{\xi}) = \langle\tilde{x}(\tilde{\xi}),\tilde{\xi}\rangle$ is almost analytic , positively homogeneous of degree 1 in $\mathbb{R}^n$ and $\operatorname{Im}H(\xi) \leq 0$, $\xi \in \mathbb{R}^n$.

In the proof of Theorem 4.2 , we have seen that if V is a sufficiently small conic neighbourhood of ξ_0 , then the functions of the form

$$\alpha(\xi) \quad e^{-iH(\xi)}\Big|_V \quad , \alpha \in S^{\infty}_{ac}(\mathbb{R}^n) \ ,$$

and the functions of the form

$$\widehat{I(a,\varphi)}(\xi)\Big|_V \quad , \qquad a \in S^{\infty}_{ac}(\mathbb{R}^n \times \mathbb{R}^N) \quad ,$$

define the same function spaces modulo $S^{-\infty}$. Here $\widehat{I(a,\varphi)}$ is the Fourier transform of

$$I(a,\varphi) \quad = \quad \int e^{i\,\varphi(x,\theta)}\, a(x,\theta) \quad d\theta \quad .$$

Now let ψ be another regular positive phase function and suppose that φ and ψ are equivalent near $\xi_0 \in \Lambda_{\varphi\mathbb{R}} \cap \Lambda_{\psi\mathbb{R}}$ for almost classical symbols. We can choose the local coordinates x such that Λ_φ is still of the form (5.8) and such that Λ_ψ is represented by

$$\tilde{x} \;=\; \partial \tilde{G}(\tilde{\xi}) \,/\partial \tilde{\xi} \,, \; \tilde{\xi} \in \mathbb{C}^n \;, \tag{5.9}$$

where G has the same properties as H . The equivalence of φ and ψ now implies that modulo $S^{-\infty}$ the function spaces

$$S^{\infty}_{ac}(\mathbb{R}^n) \; e^{-iH(\xi)}\Big|_V \qquad \text{and} \qquad S^{\infty}_{ac}(\mathbb{R}^n) \; e^{-iG(\xi)}\Big|_V$$

are the same. Passing to polar coordinates and applying Lemma 5.2, we get

$$|H(\xi) - G(\xi)| \leq C_k \min(\, (\operatorname{Im} H(\xi))^k, (\operatorname{Im} G(\xi))^k)$$

for all $k > 0$ when ξ is in a small real neighbourhood of ξ_0.

By Lemma 1.5 we have the same estimate for $\partial H(\xi)/\partial\xi - \partial G(\xi)/\partial\xi$ and since by the homogeneity we can estimate Im H (and Im G) by Im $\partial H(\xi)/\partial\xi$ (and Im $\partial G(\xi)/\partial\xi$) it follows then from Proposition 1.7 that Λ_φ and Λ_ψ are equivalent near ρ_0. Thus we have proved

Theorem 5.4. Suppose that $\varphi(x,\theta)$, $\psi(x,w)$ are equivalent for classical (or almost classical) symbols near the point $\rho_0 \in \Lambda_{\varphi\mathbb{R}} \cap \Lambda_{\psi\mathbb{R}}$. Then Λ_φ and Λ_ψ are equivalent near ρ_0.

Of course it is easy to construct counter examples to this theorem when symbols of type 1,0 or of type $1-\delta,\delta$ are allowed. This is an immediate application of Remark 5.3.

6. The principal symbol.

In this section we shall define a principal symbol for elements of $I_c^m(X,\Lambda)$. In the case of real phase functions we know (cf [3]) that the principal symbol is a section of the tensor product of the line bundle of densities of order 1/2 on Λ and of the so called Maslov line bundle, which has transition functions of the form i^ν, $\nu \in \mathbb{Z}$.

In the complex case we would like to define the principal symbol as a density of order 1/2 on Λ. Unfortunately it is impossible to define holomorphic (or almost analytic) densities of order 1/2 on complex manifolds, so that the transition functions for arbitrary changes of local coordinates are continuous under small perturbations. To avoid this difficulty we shall restrict suitably the set of local coordinate systems.

Considering first the linearized situation, we let $\tilde{M}$ be the complexification of a real symplectic vector space M of dimension $2n$, with symplectic bilinear form σ. Assume that we have a fixed real Lagrangean plane $F \subset M$ with complexification $\tilde{F}$.

We say that a Lagrangean plane $\Lambda \subset \tilde{M}$ is positive semi-definite (definite) if $\operatorname{Im} \sigma(u,\bar{u}) \geq 0$ $(\operatorname{Im} \sigma(u,\bar{u}) > 0)$ for all $u \in \Lambda$, $u \neq 0$. Similarly we define negative semi-definite and definite Lagrangean planes. We denote by $\mathcal{L}^-$ the set of negative definite Lagrangean planes.

Clearly if $L \in \mathcal{L}^-$, then L is transversal to all positive semi-definite Lagrangean planes and for all real linear symplectic coordinates (x,ξ) in M, L takes the form $\tilde{\xi} = B\tilde{x}$, where B is symmetric and $\operatorname{Im} B < 0$. The set of such matrices will be denoted by $\overset{\circ}{S}{}^-$ and by S^- we denote the set of symmetric matrices B with $\operatorname{Im} B \leq 0$.

Definition 6.1. Let $\Lambda \subset \tilde{M}$ be a positive semi-definite Lagrangean plane. We say that a basis $e = (e_1, \ldots, e_n)$ for Λ is admissible if there is a basis $f = (f_1, \ldots, f_n)$ in F and a plane $L \in \mathcal{L}^-$ such that e_j is the projection of f_j along L for all j. (Note that L is transversal to both Λ and $\tilde{F}$.) We write

$$e = E(f,L) = E_\Lambda(f,L)$$

and denote by $\mathcal{B}(\Lambda)$ the set of admissible bases for Λ, equipped with the product topology from Λ^n.

Note that in the representation $e = E(f,L)$, the f and L are in general not unique. For instance if $\Lambda = \tilde{F}$, we can take L arbitrary and $\mathcal{B}(\tilde{F})$ becomes the set of real bases in F. In the following proposition and its proof, Λ denotes some positive semi-definite Lagrangean plane.

Proposition 6.2. $\mathcal{B}(\Lambda)$ is the union of two disjoint arc-wise connected subsets in such a way that two bases $e = E(f,L)$, $e' = E(f',L') \in \mathcal{B}(\Lambda)$ belong to the same subset if and only if $f, f' \in \mathcal{B}(\tilde{F})$ have the same orientation. Moreover there exists a unique function $s = s_\Lambda : \mathcal{B}(\Lambda) \times \mathcal{B}(\Lambda) \longrightarrow \mathbb{C} \setminus 0$ with the following properties:

For every compact set $K \subset \mathcal{B}(\tilde{F}) \times \mathcal{L}^-$, $s_\Lambda(e,e')$ is a continuous function of e, e' and Λ, when $e, e' \in E_\Lambda(K)$. More precisely, if $e_\nu, e'_\nu \in E_{\Lambda_\nu}(K)$, $\nu = 1,2,\ldots$ and $e_\nu \longrightarrow e \in E_\Lambda(K)$, $e'_\nu \longrightarrow e' \in E_\Lambda(K)$, (and $\Lambda_\nu \longrightarrow \Lambda$) then $s_{\Lambda_\nu}(e_\nu, e'_\nu) \longrightarrow s_\Lambda(e,e')$. (6.1)

If we put $e/e' = e_1 \wedge \ldots \wedge e_n / e'_1 \wedge \ldots \wedge e'_n$, (6.2)
$e, e' \in \mathcal{B}(\Lambda)$, then $s^2(e,e') = \pm\, e/e'$ where the plus sign

is valid precisely when f and f' have the same orientation. Here $e = E(f,L)$, $e' = E(f',L')$.

If $e,e',e'' \in \mathcal{B}(\Lambda)$ then $s(e,e')s(e',e'') = s(e,e'')$. (6.3)

If $e = E(f,L)$, $e' = E(f',L) \in \mathcal{B}(\Lambda)$ (with the <u>same</u> $L \in \mathcal{L}^-$), then $s(e,e') > 0$. (6.4)

<u>Proof</u>. The uniqueness of the function s_Λ is a trivial consequence of the other properties, and (6.3) follows from the statement about $\mathcal{B}(\Lambda)$ and from (6.1),(6.2) and (6.4).

Let Λ be a positive semi-definite Lagrangean plane and choose symplectic linear coordinates (x,ξ) in M so that F is given by $x = 0$ and Λ is given by $\tilde{x} = A\tilde{\xi}$ where $A \in S^-$. Let $L \in \mathcal{L}^-$ be of the form $\tilde{\xi} = B\tilde{x}$ where $B \in \mathring{S}^-$. If $(0,\tilde{\xi}_0) \in \tilde{F}$ and $(\tilde{x},\tilde{\xi}) \in \Lambda$ is the projection of $(0,\tilde{\xi}_0)$ along L, we have

$$\begin{cases} \tilde{x} = A\tilde{\xi} \\ \tilde{\xi} = B\tilde{x} + \tilde{\xi}_0 \end{cases}$$

which implies that $\tilde{\xi} = (I - BA)^{-1}\tilde{\xi}_0$. Thus the set of projections : $\tilde{F} \longrightarrow \Lambda$ along planes in $\mathcal{L}^-$ is in one to one correspondence with matrices of the form $(I - BA)^{-1}$ where $B \in \mathring{S}^-$. Therefore $\mathcal{B}(\Lambda)$ can be identified with the set of matrices of the form

$$C = C_A(B,R) = (I - BA)^{-1}R, \quad B \in \mathring{S}^-, \quad R \in GL(n,\mathbb{R}).$$

Now suppose that

$$(I - BA)^{-1}R = (I - B_1A)^{-1}R_1, \quad B,B_1 \in \mathring{S}^-, \quad R,R_1 \in GL(n,\mathbb{R}) \qquad (6.5)$$

For $0 \leq t \leq 1$ we put

$$B_t = (1-t)B + tB_1 \in \mathring{S}^-$$

$$R_t = (I - B_tA)(I - B_1A)^{-1}R_1 = (1-t)R + tR_1 \in GL(n,\mathbb{R}) .$$

Then

$$(I - BA)^{-1}R = (I - B_tA)^{-1}R_t = (I - B_1A)^{-1}R_1 \qquad (6.6)$$

and we conclude from this homotopy that $\det R / \det R_1 > 0$. This shows that $\mathcal{B}(\Lambda)$ splits into two disjoint arcwise connected subsets, determined by the orientation of $f \in \mathcal{B}(\bar{F})$ in the representation $e = E(f,L)$. (Note that $\mathring{S}^-$ is contractible and that $GL(n,\mathbb{R})$ has two components.)

Now define $(\det(I - BA)^{-1})^{1/2}$, $B \in \mathring{S}^-$, as the branch of the square root which is deformed into a positive real number when B is deformed into $-iI$ in $\mathring{S}^-$ and A is deformed into $-iI$ in S^-. Put

$$p(A,B,R) = (\det(I - BA)^{-1})^{1/2} \, |\det R|^{1/2} , \quad B \in \mathring{S}^- , \quad R \in GL(n,\mathbb{R}).$$

Of course p is continuous, so it follows from the homotopy (6.6) that p is constant on the fibers of the mapping

$$\mathcal{G} : S^- \times \mathring{S}^- \times GL(n,\mathbb{R}) \ni (A,B,R) \longrightarrow (A,(I - BA)^{-1}R) \in S^- \times GL(n,\mathbb{C}).$$

Thus we can define a function

$$S_A(C) = p(A,B,R) , \text{ where } C = (I - BA)^{-1}R .$$

By an elementary topological argument it follows that for every compact set $K \subset S^- \times \mathring{S}^- \times GL(n,\mathbb{R})$, the map

$$\mathcal{G}(K) \ni (A,C) \longrightarrow S_A(C) \in \mathbb{C}$$

is continuous.

Now let $e = E_\Lambda(f,L)$, $e' = E_\Lambda(f',L') \in \mathcal{B}(\Lambda)$ be represented by $(I - BA)^{-1}R$ and $(I - B'A)^{-1}R'$ respectively. Then

$$e/e' = \det (I - BA)^{-1}R \, / \, \det (I - B'A)^{-1}R'$$

so it is natural to define $s_\Lambda(e,e')$ (for Λ in a small neighbourhood of a fixed Lagrangean plane $\tilde{x} = A_0 \tilde{\xi}$) by

$$s_\Lambda(e,e') = S_A((I - BA)^{-1}R) \, / \, S_A((I - B'A)^{-1}R').$$

Then (6.1) - (6.4) follow immediately and since the function $s_\Lambda(e,e')$ is unique with these properties , the proof is complete.

For compact sets $K \subset \mathcal{B}(\tilde{F}) \times \mathcal{L}^-$ we shall write $\mathcal{B}_K(\Lambda)$ in the following instead of $E_\Lambda(K)$.

Now let $\Lambda \subset (T^*(X) \setminus 0)^\sim$ be a positive closed conic Lagrangean manifold, where X is a paracompact C^∞ manifold of dimension n . If $\rho \in \Lambda_{\mathbb{R}}$ we define $\mathcal{B}(T_\rho(\Lambda))$ as above , taking $M = T_\rho(T^*(X))$ and taking for F the tangent space of the fiber. For every choice of local coordinates in X we get natural linear symplectic coordinates in $T_\rho(T^*(X))$ and thus an identification of $T_\rho(T^*(X))$ and $T_\mu(T^*(X))$ if $\rho, \mu \in \Lambda_{\mathbb{R}}$ are sufficiently close. If we therefore consider a section

$$\Lambda_{\mathbb{R}} \ni \rho \longrightarrow e(\rho) \in \mathcal{B}(T_\rho(\Lambda)) ,$$

it makes sense to say that $e(\rho)$ belongs locally (with respect to ρ) to $\mathcal{B}_K(T_\rho(\Lambda))$ for some compact set $K \subset \mathcal{B}(\tilde{F}) \times \mathcal{L}^-$.

<u>Definition 6.3</u>. Let $\lambda_1, \ldots, \lambda_n$ be almost analytic functions on Λ , defined in some complex neighbourhood $U^\lambda = U^{\lambda_1, \ldots, \lambda_n}$ of some real point. We say that $\lambda_1, \ldots, \lambda_n$ are

admissible coordinates on Λ if

1^0 $d\lambda_1, \ldots, d\lambda_n$ are linearly independent over $\mathbb{C}$ at the real points,

2^0 $(\delta\lambda_1, \ldots, \delta\lambda_n)$ belongs locally to $\mathcal{B}_K(T_\rho(\Lambda))$ with respect to $\rho \in U^\lambda \cap \Lambda_{\mathbb{R}}$ for some compact set $K \subset \mathcal{B}(\tilde{F}) \times \mathcal{L}^-$. Here $(\delta\lambda_1, \ldots, \delta\lambda_n)$ is the dual basis of $(d\lambda_1, \ldots, d\lambda_n)$ in $T_\rho(\Lambda)^*$.

We shall see below that one can always find admissible coordinates locally.Now let $U^\mu = U^{\mu_1, \ldots, \mu_n}$ be another system of admissible coordinates.By Proposition 6.2 , we know that

$$s((\delta\lambda_1, \ldots, \delta\lambda_n), (\delta\mu_1, \ldots, \delta\mu_n)) = s(\delta\lambda, \delta\mu)$$

is a continuous function on $U^\lambda \cap U^\mu \cap \Lambda_{\mathbb{R}}$ and that the square of this function is

$$\pm \frac{d(\mu)}{d(\lambda)} = \pm \det\left[(\partial\mu_j/\partial\lambda_k)_{j,k}\right]$$

(where $\partial\mu_j/\partial\lambda_k$ is defined by : $d\mu_j = \sum_k (\partial\mu_j/\partial\lambda_k) d\lambda_k +$ $+ \sum_k (\partial\mu_j/\partial\bar{\lambda}_k) d\bar{\lambda}_k$). It is then clear that $s(\delta\lambda, \delta\mu)$ has a unique almost analytic extension (up to equivalence) defined in a small complex neighbourhood in Λ of $U^\lambda \cap U^\mu \cap \Lambda_{\mathbb{R}}$ which satisfies

$$(s_{\lambda,\mu})^2 \sim \pm \frac{d(\mu)}{d(\lambda)} .$$

Then clearly we have

$$s_{\lambda,\lambda} \sim 1 \ , \quad s_{\lambda,\mu}\, s_{\mu,\omega} \sim s_{\lambda,\omega} \tag{6.7}$$

and we also see that $s_{\lambda,\mu}$ are continuous under small perturbations of λ, μ for which $\delta\lambda, \delta\mu$ stay locally in $\mathcal{B}_K(T_\rho(\Lambda))$ with respect to $\rho \in \Lambda_{\mathbb{R}}$ for some compact set $K \subset \mathcal{B}(\tilde{F}) \times \mathcal{L}^-$.

We now define an almost analytic " Maslov " line bundle $\mathcal{L}$ on Λ as the family of admissible coordinate systems U^λ on Λ with transition functions $s_{\lambda,\mu}$. A section $f \in \Gamma(\Lambda; \mathcal{L})$ is then given by an almost analytic function f_λ on U^λ for all λ such that $f_\lambda \sim s_{\lambda,\mu} f_\mu$ for all λ and μ.

Suppose that $\lambda = (\lambda_1, \ldots, \lambda_n)$ are admissible coordinates near the point $\rho \in \Lambda_{\mathbb{R}}$. For $t \in \mathbb{R}_+$ we denote by t also the natural multiplication $t: \Lambda \to \Lambda$. Then $(t^{-1})^* \lambda_j = \lambda_j \circ t^{-1}$ are also admissible coordinates near the point $t\rho$. In fact, suppose that we choose the real coordinates x in X so that $T_\rho(\Lambda)$ is of the form $\tilde{x} = A\tilde{\xi}$ (for the induced coordinates in $T_\rho(T^*(X))$), then $T_{t\rho}(\Lambda)$ is of the form $\tilde{x} = A\tilde{\xi}/t$. Thus if $(\delta\lambda_1, \ldots, \delta\lambda_n) \in \mathcal{B}(T_\rho(\Lambda))$ is given as the projection of the real basis $(f_1, \ldots, f_n)$ in the fiber along the plane $\tilde{\xi} = B\tilde{x}$, then $(\delta(t^{-1})^*\lambda_1, \ldots, \delta(t^{-1})^*\lambda_n) = (t_*\delta\lambda_1, \ldots, t_*\delta\lambda_n)$ in $T_{t\rho}(\Lambda)$ is the projection of $(tf_1, \ldots, tf_n)$ along the plane $\tilde{\xi} = tB\tilde{x}$.

We say that $f \in \Gamma(\Lambda; \mathcal{L})$ is homogeneous of degree m if we have

$$f_{t^*\lambda} \sim t^m \; t^*(f_\lambda) \qquad \text{near} \quad \rho \tag{6.8}$$

for all $\rho \in \Lambda_{\mathbb{R}}$, $t \in \mathbb{R}_+$ and all admissible coordinates λ defined near $t\rho$. Note that the property (6.8) is invariant under changes of admissible coordinates for it is easy to verify that

$$s_{t^*\lambda, t^*_\mu} = t^* s_{\lambda,\mu} \quad , t \in \mathbb{R}_+ .$$

We denote by $\Gamma^m(\Lambda ; \mathcal{L})$ the space of equivalence classes of homogeneous sections of degree m .

The following theorem is the main result here and the rest of this section will be devoted its proof.

Theorem 6.4. Let $\Lambda \subset (T^*(X) \setminus 0)^{\sim}$ be a closed conic positive Lagrangean manifold. Then there is a "natural" linear bijection

$$\mathcal{P} : \Gamma^{m+n/4}(\Lambda ; \mathcal{L}) \longrightarrow I^m_c(X,\Lambda)/ I^{m-1}_c(X,\Lambda) .$$

If $A \in I^m_c(X,\Lambda)$ and $[A]$ denotes the image in I^m_c/I^{m-1}_c we define the principal symbol of A as $\mathcal{P}^{-1}([A])$.

The difficulty in the proof of Theorem 6.4 is actually the local construction of $\mathcal{P}$. We let $\rho_0 \in \Lambda$ be a real point and we choose some real coordinates $x_1,\dots,x_n$ near the projection of ρ_0. Let $\varphi(x,\theta)$, $x \in \mathbb{R}^n$, $\theta \in \mathbb{R}^N \setminus \{0\}$, be a positive regular phase function defined in a conic neighbourhood of a point (x_0,θ_0) and generating Λ near ρ_0 (with respect to the coordinates chosen above). To φ we shall associate a certain non-vanishing section $\sqrt{d_\varphi}$ in $\Gamma^{N/2}(\Lambda;\mathcal{L})$ which will be the "square root" of a certain almost analytic n-form on Λ . By definition, an almost analytic p-form on an almost analytic manifold is defined by local representatives of the form $\sum a_k\, df_{1,k} \wedge \dots \wedge df_{p,k}$ where the a_k and $f_{j,k}$ are almost analytic and the evident equivalence relations should be satisfied.

In general let M_j , $j = 1,2$, be real manifolds of dimension m_j and suppose that M_1 is a submanifold of M_2. Let $\omega_2 = v_1 \wedge \ldots \wedge v_{m_2-m_1}$ be a (m_2-m_1)-form on M_2 such that the 1-forms v_j vanish on $T(M_1)$ at every point of M_1. Then if ω_1 is a m_1-form on M_1 , we can define a unique m_2-form $\omega_1 \wedge \omega_2$ at the points of M_1 , by putting

$$\omega_1 \wedge \omega_2 = \Omega_1 \wedge \omega_2$$

where Ω_1 is an arbitrary local extension to M_2 of ω_1. For almost analytic manifolds and forms we have of course the analogous result with the only difference that this time the form $\omega_1 \wedge \omega_2$ is only defined up to equivalence.

Let $\tilde{\varphi}$ be an almost analytic extension of φ . With the remark above we now define $d_{\tilde{\varphi}}$ as the almost analytic n-form on $C_{\tilde{\varphi}}$ which satisfies

$$d_{\tilde{\varphi}} \wedge d(\partial\tilde{\varphi}/\partial\tilde{\theta}_1) \wedge \ldots \wedge d(\partial\tilde{\varphi}/\partial\tilde{\theta}_N) \sim i^{n+N} d\tilde{x}_1 \wedge \ldots \wedge d\tilde{x}_n \wedge d\tilde{\theta}_1 \wedge \ldots \wedge d\tilde{\theta}_N \quad (6.9)$$

at the points of $C_{\tilde{\varphi}}$. (The form $d_{\tilde{\varphi}}$ should be compared with Leray's "forme-résidu" in [10].) The form $d_{\tilde{\varphi}}$ exists and is uniquely determined by φ up to equivalence (motivating the notation d_φ in stead of $d_{\tilde{\varphi}}$). In fact , let $\lambda_1, \ldots, \lambda_n$ be almost analytic coordinates on $C_{\tilde{\varphi}}$ and set $d_{\tilde{\varphi}} \sim a(\lambda)\, d\lambda_1 \wedge \ldots \wedge d\lambda_n$. Denote by λ_j also some almost analytic extensions as in Lemma 1.10 , then for (6.9) to hold it is necessary and sufficient that

$$a(\lambda) \sim \left[\det \frac{1}{i}\begin{pmatrix} \partial\lambda/\partial\tilde{x} & \partial\lambda/\partial\tilde{\theta} \\ \partial^2\tilde{\varphi}/\partial\tilde{\theta}\partial\tilde{x} & \partial^2\tilde{\varphi}/\partial\tilde{\theta}^2 \end{pmatrix}\right]^{-1}$$

Since we have a local identification of Λ and $C_{\tilde{\varphi}}$ (cf (3.23)) we can also consider d_φ as a n-form on Λ , defined in a conic neighbourhood of ρ_0.

Now let $\psi \in C^\infty(\mathbb{R}^n)$ with $\operatorname{Im} \psi''_{xx} < 0$. Let $\xi = (\xi_1, \dots, \xi_n)$ be the dual coordinates to the coordinates $x = (x_1, \dots, x_n)$ chosen above . Then the restriction to Λ of

$$\tau = \tilde{\xi} - \partial\tilde{\psi}/\partial\tilde{x} = (\tilde{\xi}_1 - \partial\tilde{\psi}/\partial\tilde{x}_1, \dots, \tilde{\xi}_n - \partial\tilde{\psi}/\partial\tilde{x}_n) \qquad (6.10)$$

(where $\tilde{\psi}$ is some almost analytic extension of ψ) are admissible coordinates on Λ , for at a real point $\rho \in \Lambda$, the basis $(\delta\tau_1, \dots, \delta\tau_n)$ in $T_\rho(\Lambda)$ is precisely the projection of the basis $(\delta\xi_1, \dots, \delta\xi_n)$ in $T_\rho(T^*_{\pi(\rho)}(X))$ along the negative definite Lagrangean plane $t_{\tilde{\xi}} = \psi''_{xx} t_{\tilde{x}}$ (in the induced coordinates in $T_\rho(T^*(X))$). On the other hand $\tilde{\xi}_j \sim \partial\tilde{\varphi}/\partial\tilde{x}_j$ on Λ so in the coordinates (6.10) d_φ takes the form

$$d_\varphi \sim \left[\det \frac{1}{i}\begin{pmatrix} \tilde{\varphi}''_{\tilde{x}\tilde{x}} - \tilde{\psi}''_{\tilde{x}\tilde{x}} & \tilde{\varphi}''_{\tilde{x}\tilde{\theta}} \\ \tilde{\varphi}''_{\tilde{\theta}\tilde{x}} & \tilde{\varphi}''_{\tilde{\theta}\tilde{\theta}} \end{pmatrix}\right]^{-1} d\tau_1 \wedge \dots \wedge d\tau_n \qquad (6.11)$$

We now define $\sqrt{d_\varphi} \in \Gamma(\Lambda; \mathcal{L})$ by giving the value

$$(\sqrt{d_\varphi})_\tau \sim \left[\det \frac{1}{i}\begin{pmatrix} \tilde{\varphi}''_{\tilde{x}\tilde{x}} - \tilde{\psi}''_{\tilde{x}\tilde{x}} & \tilde{\varphi}''_{\tilde{x}\tilde{\theta}} \\ \tilde{\varphi}''_{\tilde{\theta}\tilde{x}} & \tilde{\varphi}''_{\tilde{\theta}\tilde{\theta}} \end{pmatrix}\right]^{-1/2} \qquad (6.12)$$

for the admissible coordinates τ in (6.10) .Here the branch of the root should be chosen as in Theorem 2.3 . This definition does not depend on the choice of ψ for if ψ' is another such function and τ' are the corresponding admissible coordinates on Λ , then the square of $(\sqrt{d_\varphi})_\tau / (\sqrt{d_\varphi})_{\tau'}$ is the square of the corresponding transition function in $\mathcal{L}$ because of (6.11) and the fact that d_φ is invariant .By a deformation from ψ to ψ' we see that $(\sqrt{d_\varphi})_\tau/(\sqrt{d_\varphi})_{\tau'} \sim s_{\tau,\tau'}$ so it follows that the definition (6.12) is independent of the choice of ψ. (On the other hand $\sqrt{d_\varphi}$ depends on the choice of local coordinates x as we shall see soon .)

To prove that $\sqrt{d_\varphi}$ is homogeneous of degree $N/2$, we let $t \in \mathbb{R}_+$ and we let τ be admissible coordinates of the form (6.10) near the point $t\rho_0 = (x_0, t\xi_0)$. Then

$$t^*\tau = t\tilde{\xi} - \partial\tilde{\psi}(\tilde{x})/\partial\tilde{x} = t(\tilde{\xi} - t^{-1}\partial\tilde{\psi}/\partial\tilde{x})$$

are admissible coordinates near ρ_0. In general if $f \in \Gamma(\Lambda;\mathcal{L})$ we have

$$f_{t\tau} \sim t^{-n/2} f_\tau$$

for all $t > 0$ and all admissible coordinates τ. Thus in our case we get

$$(\sqrt{d_\varphi})_{t^*\tau} \sim t^{-n/2} \left[\det \frac{1}{i}\begin{pmatrix} \tilde{\varphi}''_{\tilde{x}\tilde{x}}(\tilde{x},\tilde{\theta}) - t^{-1}\tilde{\psi}''_{\tilde{x}\tilde{x}}(\tilde{x}) & \tilde{\varphi}''_{\tilde{x}\tilde{\theta}}(\tilde{x},\tilde{\theta}) \\ \tilde{\varphi}''_{\tilde{\theta}\tilde{x}}(\tilde{x},\tilde{\theta}) & \tilde{\varphi}''_{\tilde{\theta}\tilde{\theta}}(\tilde{x},\tilde{\theta}) \end{pmatrix}\right]^{-1/2}.$$

On the other hand

$$t^*(\sqrt{d_\varphi})_\tau \sim \left[\det \frac{1}{i}\begin{pmatrix} \tilde{\varphi}''_{\tilde{x}\tilde{x}}(\tilde{x},t\tilde{\theta}) - \tilde{\psi}''_{\tilde{x}\tilde{x}}(\tilde{x}) & \tilde{\varphi}''_{\tilde{x}\tilde{\theta}}(\tilde{x},t\tilde{\theta}) \\ \tilde{\varphi}''_{\tilde{\theta}\tilde{x}}(\tilde{x},t\tilde{\theta}) & \tilde{\varphi}''_{\tilde{\theta}\tilde{\theta}}(\tilde{x},t\tilde{\theta}) \end{pmatrix}\right]^{-1/2}$$

$$\sim t^{-(n-N)/2}\left[\det \frac{1}{i}\begin{pmatrix} \tilde{\varphi}''_{\tilde{x}\tilde{x}}(\tilde{x},\tilde{\theta}) - t^{-1}\tilde{\psi}''_{\tilde{x}\tilde{x}}(\tilde{x}) & \tilde{\varphi}''_{\tilde{x}\tilde{\theta}}(\tilde{x},\tilde{\theta}) \\ \tilde{\varphi}''_{\tilde{\theta}\tilde{x}}(\tilde{x},\tilde{\theta}) & \tilde{\varphi}''_{\tilde{\theta}\tilde{\theta}}(\tilde{x},\tilde{\theta}) \end{pmatrix}\right]^{-1/2}.$$

Comparing with (6.8) we see that $\sqrt{d_\varphi} \in \Gamma^{N/2}(\Lambda;\mathcal{L})$.

Next we consider the effect of changes of coordinates in X. Let $y = (y_1, \dots, y_n)$ be some new local coordinates in X near the projection of ρ_0 and put $\varphi_1(y,\theta) = \varphi(x,\theta)$ so that φ_1 also generates Λ near ρ_0. Comparing the definitions of d_φ and d_{φ_1} we get

$$d_{\varphi_1} \sim \frac{d(y)}{d(x)}\, d_\varphi \quad \text{on } \Lambda . \tag{6.13}$$

We now state that

$$\sqrt{d_{\varphi_1}} \sim \left|\frac{d(\tilde{y})}{d(\tilde{x})}\right|^{1/2} \sqrt{d_\varphi} \tag{6.14}$$

where $\left|\frac{d(\tilde{y})}{d(\tilde{x})}\right|^{1/2}$ denotes an almost analytic extension of $\left|\frac{d(y)}{d(x)}\right|^{1/2}$, defined in the real domain . In fact , it is easy to verify (6.14) in the case when $(y_1, \dots, y_n)$ are obtained from $(x_1, \dots, x_n)$ by permuting the first two coordinates. In the case when $\frac{d(y)}{d(x)} > 0$, (6.14) follows from (6.13) by a simple deformation argument . In general , (6.14) follows by composition of these two cases .

We can now give a local definition of $\mathcal{P}$ in Theorem 6.4. Take $s \in \Gamma^{m+n/4}(\Lambda;\mathcal{L})$ with support in a small conic neighbourhood of ρ_0 . Choose local coordinates $x_1, \dots x_n$ and a positive regular phase function φ in $C^\infty(\mathbb{R}^n \times \mathbb{R}^N)$ which generates Λ close to ρ_0 in these coordinates .Then there is an almost analytic function a on Λ , homogeneous of degree $m+n/4 - N/2$ and unique up to equivalence such that

$$s \sim a \sqrt{d_\varphi} \quad .$$

We consider a as a function on $C_{\tilde{\varphi}}$ and let $A(\tilde{x},\tilde{\theta})$ be a homogeneous almost analytic extension of a to $\mathbb{C}^n \times \mathbb{C}^N$ as in Lemma 1.10 with support in a small conic neighbourhood of (x_0,θ_0) ; the point corresponding to ρ_0. We now define $\mathcal{P}(s) \in I_c^m(X,\Lambda)/I_c^{m-1}(X,\Lambda)$ as the element , given in the coordinates $x_1,\dots,x_n$ by the distribution

$$(2\pi)^{-(n+2N)/4} \int e^{i\,\varphi(x,\theta)} A(x,\theta)\, d\theta \quad .$$

Of course we have to verify that this definition does not depend on the choice of local coordinates x , phase function φ , or extension A . More precisely ,to show that $\mathcal{P}$ is well defined it suffices to verify

1^0 For given local coordinates $x_1,\dots,x_n$ and phase function φ , $\mathcal{P}(s)$ is independent of the choice of extension $A(x,\theta)$ as above.

2^0 $\mathcal{P}(s)$ does not change if we replace the local coordinates x by some new local coordinates y and replace $\varphi(x,\theta)$ by $\varphi_1(y,\theta) \overset{\text{def.}}{=} \varphi(x,\theta)$.

3^0 For some suitable coordinates $x_1,\dots,x_n$ in X it is true that $\mathcal{P}(s)$ does not change if we replace the phase function $\varphi(x,\theta)$ by some new phase function $\varphi_1(x,w)$, $w \in \mathbb{R}^M$.

It follows from a partial integration and the remark after Lemma 1.10 (cf Proposition 1.2.5 in [3]) that if

$$A(x,\theta) \sim A_m(x,\theta) + A_{m-1}(x,\theta) + \dots$$

is a classical symbol and $A_m \big|_{C_\varphi} \sim 0$, then

$$\int e^{i\varphi(x,\theta)} A(x,\theta)\, d\theta \equiv \int e^{i\varphi(x,\theta)} B(x,\theta)\, d\theta \qquad \text{mod. } C^\infty(\mathbb{R}^n)$$

for some $B \in S_c^{m-1}(\mathbb{R}^n \times \mathbb{R}^N)$. This proves 1^0.

To prove 2^0 , let $y_1,\dots,y_n$ be some new coordinates in X . Then from (6.14) we get

$$s \sim a\sqrt{d_\varphi} \sim a \left|\frac{d(\tilde{y})}{d(\tilde{x})}\right|^{-1/2} \sqrt{d_{\varphi_1}} \quad ,$$

so in the coordinates $y_1,\dots,y_n$ we get

$$\int e^{i\varphi_1(y,\theta)}\, A(x(y),\theta) \left|\frac{d(y)}{d(x)}\right|^{-1/2} d\theta$$

as a "new" value for $\mathcal{P}(s)$. Recalling that the elements of $I_c^m(X,\Lambda)$ are densities of order 1/2 , we see that 2^0 is valid.

In order to prove 3^0 , we choose our local coordinates x , so that Λ is represented by an almost analytic equation

$$\tilde{x} = H'_{\tilde{\xi}}(\tilde{\xi})$$

near ρ_0. For the corresponding phase functions $\varphi(x,\theta)$ and $\varphi_1(x,w)$ this means that Hess φ and Hess φ_1 are non-degenerate , as we have seen in section 3. Let

$$\tau = \tilde{\xi} - \partial\tilde{\Psi}(\tilde{x})/\partial\tilde{x}$$

be admissible coordinates on Λ as before and put

$$\tau_s = \tilde{\xi} - s\,\partial\tilde{\Psi}/\partial\tilde{x} \quad , \; 0 \leq s \leq 1 \; .$$

By the usual continuity arguments we get from (6.11) and the invariance of d_{φ} that

$$\left[\det \frac{1}{i}\begin{pmatrix} \tilde{\varphi}''_{\tilde{x}\tilde{x}} & \tilde{\varphi}''_{\tilde{x}\tilde{\theta}} \\ \tilde{\varphi}''_{\tilde{\theta}\tilde{x}} & \tilde{\varphi}''_{\tilde{\theta}\tilde{\theta}} \end{pmatrix}\right]^{-1/2} \sim \left(\frac{d(\tilde{\tau})}{d(\tilde{\xi})}\right)^{1/2} \left[\det \frac{1}{i}\begin{pmatrix} \tilde{\varphi}''_{\tilde{x}\tilde{x}} - \tilde{\psi}''_{\tilde{x}\tilde{x}} & \tilde{\varphi}''_{\tilde{x}\tilde{\theta}} \\ \tilde{\varphi}''_{\tilde{\theta}\tilde{x}} & \tilde{\varphi}''_{\tilde{\theta}\tilde{\theta}} \end{pmatrix}\right]^{-1/2} \tag{6.15}$$

on $C_{\tilde{\varphi}}$. Here $\frac{d(\tilde{\tau})}{d(\tilde{\xi})}$ is the functional determinant between almost analytic coordinates on $C_{\tilde{\varphi}}$, and the square root is the branch obtained by continuity from $s = 1$ under the curve τ_s above.(Of course we have the same formula for $\varphi_1(x,w)$.)

Now let $s \in \Gamma^{m+n/4}(\Lambda; \mathcal{L})$ and write

$$s \sim a\sqrt{d_{\varphi}} \sim a_1 \sqrt{d_{\varphi_1}} \ .$$

It then follows from (6.15) that

$$a\,(\det i^{-1}\mathrm{Hess}\ \tilde{\varphi}\,)^{-1/2} \sim a_1(\det i^{-1}\mathrm{Hess}\ \tilde{\varphi}_1)^{-1/2} \tag{6.16}$$

on Λ . Let $A(\tilde{x},\tilde{\theta})$ and $A_1(\tilde{x},\tilde{w})$ be extensions as above of a and a_1 respectively and put

$$\mathcal{P}(s) = (2\pi)^{-(n+2N)/4} \int e^{i\,\varphi(x,\theta)}\,A(x,\theta)\,d\theta$$

$$\mathcal{P}_1(s) = (2\pi)^{-(n+2M)/4} \int e^{i\,\varphi_1(x,w)}\,A_1(x,w)\,dw \ .$$

It then follows from (6.16) and the proof of Theorem 4.2 (in particular (4.7)) that

$$\mathcal{P}(s) - \mathcal{P}_1(s) \in I_c^{m-1}(X,\Lambda) \ .$$

This completes the proof that $\mathcal{P}$ is locally well defined . By a partition of unity it is now immediate to get a global definition and it is a simple consequence of formula (4.7) that the map

$\mathcal{P}$ is surjective , and using Lemma 5.1 it also follows that $\mathcal{P}$ is injective . (Cf the proof of Theorem 5.4.) This completes the proof of the theorem .

7. Products of Fourier integral operators.

Let X and Y be paracompact C^∞ manifolds. If $C \subset T^*(X) \times T^*(Y) = T^*(X \times Y)$ is an arbitrary submanifold, we define C' as $\{(x,\xi,y,-\eta);\ (x,\xi,y,\eta) \in C\}$. Clearly there is a natural way to extend this definition when $C \subset (T^*(X \times Y))^\sim$ is an arbitrary almost analytic manifold. (C' is then well defined up to equivalence.)

Definition 7.1. We say that $C \subset (T^*(X \times Y) \setminus 0)^\sim$ is a positive canonical relation if $C' \subset (T^*(X \times Y) \setminus 0)^\sim$ is a closed conic positive Lagrangean manifold and if $C_{\mathbb{R}} \subset (T^*(X) \setminus 0) \times (T^*(Y) \setminus 0)$. (Recall from section 4 that an almost analytic manifold $\Lambda \subset \tilde{M}$ is said to be closed if $\Lambda_{\mathbb{R}} \subset M$ is a closed set.) If in addition C' is strictly positive (Def. 3.3) then C is also called strictly positive.

Let $A \in I^m_{1-\delta}(X \times Y, \Lambda)$, where $\delta < 1/2$ and where $\Lambda \subset (T^*(X \times Y) \setminus 0)^\sim$ is a closed conic positive Lagrangean manifold. Then A is the distribution kernel of a continuous operator $C_0^\infty(Y; \Omega_{1/2}) \longrightarrow \mathcal{D}'(X; \Omega_{1/2})$ which we also denote by A. Recall that $C_0^\infty(Y; \Omega_{1/2})$ is the space of compactly supported C^∞ densities of order 1/2 on Y and that $\mathcal{D}'(X; \Omega_{1/2})$ is the corresponding space of distribution densities on X. (We shall call A a Fourier integral operator.) Note that $WF'(A) \subset C_{\mathbb{R}}$ if $C = \Lambda'$. Thus if C is a canonical relation, it follows from the results in [3, section 2.5], that A is continuous $C_0^\infty(Y; \Omega_{1/2}) \longrightarrow C^\infty(X; \Omega_{1/2})$ and can be extended to a continuous operator $\mathcal{E}'(Y; \Omega_{1/2}) \longrightarrow \mathcal{D}'(X; \Omega_{1/2})$.

Now let X, Y, Z be paracompact C^∞ manifolds of dimensions n_X, n_Y and n_Z respectively and assume that

$$A_1 \in I^{m_1}_{1-\delta}(X \times Y, \Lambda_1) \ , \quad A_2 \in I^{m_2}_{1-\delta}(Y \times Z, \Lambda_2) \ , \quad (\delta < 1/2)$$

are properly supported operators where $C_j = \Lambda_j'$ are positive canonical relations. Then the product $A_1 \circ A_2$ is well defined and from [3, section 2.5] it follows that

$$WF'(A_1 \circ A_2) \subset WF'(A_1) \circ WF'(A_2) \subset C_{1\mathbb{R}} \circ C_{2\mathbb{R}} \ , \tag{7.1}$$

where $C_{1\mathbb{R}}$ and $C_{2\mathbb{R}}$ are regarded as relations and $C_{1\mathbb{R}} \circ C_{2\mathbb{R}}$ denotes the products of the relations. We shall give a sufficient condition for $A_1 \circ A_2$ to be a Fourier integral operator.

In general if B is a set, we put $\mathrm{diag}(B) = \{(b,b) \ ; b \in B\}$. Put $\Delta = T^*(X) \times \mathrm{diag}(T^*(Y)) \times T^*(Z)$ and let $\tilde{\Delta} \subset (T^*(X \times Y \times Y \times Z))^\sim$ be its almost analytic complexification, well defined up to equivalence by the conditions : $\tilde{\Delta}_{\mathbb{R}} = \Delta$ and $\dim_{\mathbb{R}} \tilde{\Delta} = 2\dim_{\mathbb{R}} \Delta$. We introduce the following condition :

(7.2)

(a) $C_1 \times C_2$ and $\tilde{\Delta}$ intersect transversally at the points of $(C_{1\mathbb{R}} \times C_{2\mathbb{R}}) \cap \Delta$.

(b) The natural projection

$$(C_{1\mathbb{R}} \times C_{2\mathbb{R}}) \longrightarrow (T^*(X) \setminus 0) \times (T^*(Z) \setminus 0)$$

is injective and proper.

<u>Proposition 7.1</u>. If (7.2) is valid we can define in a natural way (up to equivalence) a positive canonical relation $C_1 \circ C_2 \subset (T^*(X \times Z) \setminus 0)^\sim$ and we have $(C_1 \circ C_2)_{\mathbb{R}} = C_{1\mathbb{R}} \circ C_{2\mathbb{R}}$.

<u>Proof</u>. Let $\rho_1 = (x_0, \xi_0, y_0, \eta_0) \in C_{1\mathbb{R}}$, $\rho_2 = (y_0, \eta_0, z_0, \zeta_0) \in C_{2\mathbb{R}}$

and choose local coordinates x, y, z near x_0, y_0 and z_0 respectively. For the corresponding coordinates in the cotangent spaces we identify $C_1, C_2, \tilde{\Delta}$ with some local representatives. Then we have

<u>Lemma 7.2</u> In a neighbourhood of (ρ_1, ρ_2) the map

$$(C_1 \times C_2) \cap \tilde{\Delta} \ni (\tilde{x}, \tilde{\xi}, \tilde{y}, \tilde{\eta}, \tilde{y}, \tilde{\eta}, \tilde{z}, \tilde{\zeta}) \to (\tilde{x}, \tilde{\xi}, \tilde{z}, \tilde{\zeta}) \in \mathbb{C}^{2n_X + 2n_Z} \tag{7.3}$$

has an injective differential and for some constant C we have the inequality

$$|\mathrm{Im}\,(\tilde{x}, \tilde{\xi}, \tilde{y}, \tilde{\eta}, \tilde{y}, \tilde{\eta}, \tilde{z}, \tilde{\zeta})|^4 \leq C\,|\mathrm{Im}(\tilde{x}, \tilde{\xi}, \tilde{z}, \tilde{\zeta})| \tag{7.4}$$

for the map (7.3).

Admitting the lemma for a moment, we define $C_1 \circ C_2$ as the almost analytic manifold with $(C_1 \circ C_2)_{\mathbb{R}} = C_{1\mathbb{R}} \circ C_{2\mathbb{R}}$ and with local representatives being the images of the maps (7.3) for different choices of (ρ_1, ρ_2) as above and different choices of local coordinates. It is clear from (7.2b), Lemma 7.2 and Lemma 1.11 that this definition makes sense. Note that the maps (7.3) give a natural identification of $C_1 \circ C_2$ with $(C_1 \times C_2) \cap \tilde{\Delta}$. That $C_1 \circ C_2$ is a positive canonical relation will follow from the proof of Lemma 7.2.

<u>Proof of Lemma 7.2</u>. Let $N^*(\mathrm{diag}(Y)) \subset T^*(Y \times Y)$ be the normal bundle of $\mathrm{diag}(Y)$ and define $(N^*)^\sim \subset (T^*(X \times Y \times Y \times Z))^\sim$ as the almost analytic complexification of $N^* = T^*(X) \times N^*(\mathrm{diag}(Y)) \times T^*(Z)$. Then the condition (7.2) is equivalent to

(7.2)'

(a) $\Lambda_1 \times \Lambda_2$ and $(N^*)^\sim$ intersect transversally at the points of $(\Lambda_{1\mathbb{R}} \times \Lambda_{2\mathbb{R}}) \cap N^*$.

(b) The natural projection $(\Lambda_{1\mathbb{R}} \times \Lambda_{2\mathbb{R}}) \cap N^*$ $\longrightarrow (T^*(X)\setminus 0) \times (T^*(Z)\setminus 0)$ is injective and proper.

We now take positive regular phase functions $\varphi_1(x,y,\theta)$, $\theta \in \mathbb{R}^{N_1}$, and $\varphi_2(y,z,\sigma)$, $\sigma \in \mathbb{R}^{N_2}$, which generate Λ_1 and Λ_2 near $\rho_1' = (x_0,\xi_0,y_0,-\eta_0)$ and $\rho_2' = (y_0,\eta_0,z_0,-\zeta_0)$ respectively and denote by $\tilde{\varphi}_1$ and $\tilde{\varphi}_2$ some almost analytic extensions. Condition (7.2'a) means that the map

$$\Lambda_1 \times \Lambda_2 \ni (\tilde{x},\tilde{\xi},\tilde{y}',\tilde{\eta}',\tilde{y}'',\tilde{\eta}'',\tilde{z},\tilde{\zeta}) \longrightarrow (\tilde{y}'-\tilde{y}'',\tilde{\eta}'+\tilde{\eta}'') \in \mathbb{C}^{2n_Y}$$

has a surjective differential near (ρ_1', ρ_2'). Equivalently the map

$$C_{\tilde{\varphi}_1} \times C_{\tilde{\varphi}_2} \ni ((\tilde{x},\tilde{y}',\tilde{\theta}),(\tilde{y}'',\tilde{z},\tilde{\sigma})) \longrightarrow (\tilde{y}'-\tilde{y}'',\partial\tilde{\varphi}_1/\partial\tilde{y}' + \partial\tilde{\varphi}_2/\partial\tilde{y}'') \in \mathbb{C}^{2n_Y}$$

has a surjective differential near $(x_0,y_0,\theta_0,y_0,z_0,\sigma_0)$. Since φ_1 and φ_2 are regular phase functions it follows that the map

$$\mathbb{C}^{n_X+2n_Y+n_Z+N_1+N_2} \ni (\tilde{x},\tilde{y}',\tilde{\theta},\tilde{y}'',\tilde{z},\tilde{\sigma}) \longrightarrow$$

$$(\tilde{y}'-\tilde{y}'',\partial\tilde{\varphi}_1/\partial\tilde{y}' + \partial\tilde{\varphi}_2/\partial\tilde{y}'', \partial\tilde{\varphi}_1/\partial\tilde{\theta}, \partial\tilde{\varphi}_2/\partial\tilde{\sigma}) \in \mathbb{C}^{2n_Y+N_1+N_2}$$

has a surjective differential and thus finally that the map

$$\mathbb{C}^{n_X+n_Y+n_Z+N_1+N_2} \ni (\tilde{x},\tilde{y},\tilde{z},\tilde{\theta},\tilde{\sigma}) \longrightarrow \tag{7.5}$$

$$(\partial\tilde{\varphi}_1/\partial\tilde{y} + \partial\tilde{\varphi}_2/\partial\tilde{y}, \partial\tilde{\varphi}_1/\partial\tilde{\theta}, \partial\tilde{\varphi}_2/\partial\tilde{\sigma}) \in \mathbb{C}^{n_Y+N_1+N_2}$$

has a surjective differential. Thus apart from the homogeneity condition, the function $\varphi(x,z,(y,\theta,\sigma)) = \varphi_1(x,y,\theta) + \varphi_2(y,z,\sigma)$ is a positive regular phase function near $(x_0,z_0,y_0,\theta_0,\sigma_0)$ with (y,θ,σ) as fiber variables. Following [3,p.175] we introduce the new fiber variable

$$\omega = ((\theta^2 + \sigma^2)^{1/2} y, \theta, \sigma)$$

where $\theta^2 = \sum \theta_j^2$, $\sigma^2 = \sum \sigma_j^2$. Then $\Phi(x,z,\omega) = \varphi_1(x,y,\theta) +$

$\varphi_2(y,z,\sigma)$ is a positive regular phase function and we have an identification of $C_{\tilde{\Phi}}$ and

$$C^0_{\tilde{\varphi}} = \{((\tilde{x},\tilde{y}',\tilde{\theta}),(\tilde{y}'',\tilde{z},\tilde{\sigma})) \in C_{\varphi_1} \times C_{\varphi_2} \; ; \; \tilde{y}'=\tilde{y}'' \; , \partial\tilde{\varphi}_1/\partial\tilde{y}' + \partial\tilde{\varphi}_2/\partial\tilde{y}'' = 0\}.$$

Now the map

$$(\Lambda_1 \times \Lambda_2) \cap (N^*)^{\sim} \ni (\tilde{x},\tilde{\xi},\tilde{y},-\tilde{\eta},\tilde{y},\tilde{\eta},\tilde{z}.\tilde{\zeta}) \longrightarrow (\tilde{x},\tilde{\xi},\tilde{z},\tilde{\sigma}) \in \mathbb{C}^{2n_X+2n_Z} \tag{7.6}$$

can be factored as

$$(\Lambda_1 \times \Lambda_2) \cap (N^*)^{\sim} \xrightarrow{\alpha} C^0_{\tilde{\varphi}} \xrightarrow{\approx} C_{\tilde{\Phi}} \xrightarrow{\beta} \Lambda_{\tilde{\Phi}} \hookrightarrow \mathbb{C}^{2n_X+2n_Z} ,$$

where β is the map

$$C_{\tilde{\Phi}} \ni (\tilde{x},\tilde{z},\tilde{\omega}) \longrightarrow (\tilde{x},\tilde{z}, \tilde{\Phi}'_{(\tilde{x},\tilde{z})})$$

considered in section 3, and where α is induced by the maps $C_{\tilde{\varphi}_j} \longrightarrow \Lambda_j = \Lambda_{\varphi_j}$ of the same type . Thus we can apply the inequality (3.23) to α and β and obtain the inequality

$$|\mathrm{Im}(\tilde{x},\tilde{\xi},\tilde{y},-\tilde{\eta},\tilde{y},\tilde{\eta},\tilde{z},\tilde{\zeta})|^4 \leq C \; |\mathrm{Im}(\tilde{x},\tilde{\xi},\tilde{z},\tilde{\zeta})| \tag{7.7}$$

for the map (7.6) . This proves the lemma for it also follows from the factorization that (7.6) has an injective differential.

Now it is also clear that $C_1 \circ C_2$ is a positive canonical relation for $(C_1 \circ C_2)'$ is locally generated by phase functions of the type $\overline{\Phi}(x,z,\omega)$ constructed above (and $(C_1 \circ C_2)_{\mathbb{R}}$ is closed by condition (7.2b)). This completes the proof of Proposition 7.1.

<u>Theorem 7.3</u>. Let $C_1 \subset (T^*(X \times Y) \smallsetminus 0)^{\sim}$, $C_2 \subset (T^*(Y \times Z) \smallsetminus 0)^{\sim}$, be positive canonical relations such that (7.2) is valid. Suppose that $A_1 \in I^{m_1}_{1-\delta}(X \times Y, C_1')$ and $A_2 \in I^{m_2}_{1-\delta}(Y \times Z, C_2')$ are properly

supported where $\delta < 1/2$. Then $A_1 \circ A_2 \in I_{1-\delta}^{m_1+m_2}(X \times Z, (C_1 \circ C_2)')$.
The same result is valid in the case of classical symbols.

Proof. By partitions of unity in the cotangent spaces we can reduce the proof to the case when $WF'(A_1)$ and $WF'(A_2)$ are contained in small conic neighbourhoods of some points $(x_0, \xi_0, y_0', \eta_0') \in C_{1\mathbb{R}}$ and $(y_0'', \eta_0'', z_0, \zeta_0) \in C_{2\mathbb{R}}$ respectively .Of course we can also assume that $(y_0', \eta_0') = (y_0'', \eta_0'')$. After having introduced local coordinates x, y, z in X, Y, Z , we can therefore assume that

$$A_1 v(x) = (2\pi)^{-(n_X+n_Y+2N_1)/4} \iint e^{i\varphi_1(x,y,\theta)} a_1(x,y,\theta)\, v(y)\, dy\, d\theta$$

$$A_2 u(y) = (2\pi)^{-(n_Y+n_Z+2N_2)/4} \iint e^{i\varphi_2(y,z,\sigma)} a_2(y,z,\sigma)\, u(z)\, dz\, d\sigma ,$$

where φ_1, φ_2 are phase functions as above and where

$$a_1 \in S_{1-\delta}^{m_1+(n_X+n_Y-2N_1)/4}(\mathbb{R}^{n_X+n_Y} \times \mathbb{R}^{N_1})$$

and

$$a_2 \in S_{1-\delta}^{m_2+(n_Y+n_Z-2N_2)/4}(\mathbb{R}^{n_Y+n_Z} \times \mathbb{R}^{N_2})$$

have supports in small conic sets .As in [3,chapter 4] one can then prove that $A_1 \circ A_2$ is modulo C^∞ given by the integral

$$A = (2\pi)^{-(n_X+n_Z+2N)/4} \int e^{i\Phi(x,z,\omega)} b(x,z,\omega)\, d\omega$$

with Φ and ω as introduced above , with $N = N_1+N_2+n_Y$ and $b(x,z,\omega) \in S_{1-\delta}^{m_1+m_2+(n_X+n_Z-2N)/4}(\mathbb{R}^{n_X+n_Z} \times \mathbb{R}^N)$ of the form

$$b(x,z,\omega) = \chi(\theta,\sigma)\, a_1(x,y,\theta)\, a_2(y,z,\sigma)\, (\theta^2 + \sigma^2)^{-n_Y/2} .$$

Here χ is positively homogeneous of degree 0 , $\chi = 1$ in a neighbourhood of the point (θ_0, σ_0) and vanishes outside a domain of the form $c_1|\theta| \leq |\sigma| \leq c_2|\theta|$, where $c_j > 0$. This proves the theorem.

In the case of classical symbols, the principal symbol of $A_1 \circ A_2$ is clearly given by some bilinear map

$$\gamma : \Gamma^{m_1'}(C_1'; \mathcal{L}) \times \Gamma^{m_2'}(C_2'; \mathcal{L}) \longrightarrow \Gamma^{m'}((C_1 \circ C_2)'; \mathcal{L}),$$

where $m_1' = m_1 + (n_X + n_Y)/4$, $m_2' = m_2 + (n_Y + n_Z)/4$ and $m' = m_1 + m_2 + (n_X + n_Z)/4$. We are now going to describe γ explicitly, modulo a factor i^{ν}, $\nu \in \mathbb{Z}$. Note that in general the square α^2 of a section $\alpha \in \Gamma(\Lambda; \mathcal{L})$ defines up to the sign (and locally) an almost analytic form on Λ of maximal degree. Thus if α_1, α_2, α are the principal symbols of A_1, A_2, $A_1 \circ A_2$ respectively, it suffices to describe α^2 modulo the sign as a function of α_1^2 and α_2^2.

In general, let $\mathcal{N}$ be a submanifold of some manifold $\mathcal{M}$ and let ω be a differential form on $\mathcal{M}$. We shall use the following terminology: We can consider ω as a form <u>on</u> $\mathcal{N}$, which is just the pullback $i^*\omega$ where $i : \mathcal{N} \hookrightarrow \mathcal{M}$ is the inclusion map. We can also consider ω as a form <u>along</u> $\mathcal{N}$, which, strictly spoken, is the equivalence class of ω with respect to the equivalence relation : $\omega_1 \sim \omega_2$ if $\omega_1 - \omega_2$ vanishes at each point of $\mathcal{N}$. In the almost analytic case we shall use the analogous terminology.

<u>Lemma 7.4</u>. Along the normal bundle $N^*(\mathrm{diag}(Y)) \subset T^*(Y \times Y)$ we have an invariant $2n_Y$ - form ω, defined by the forms

$$d(y_1' - y_2'') \wedge \ldots \wedge d(y_{n_Y}' - y_{n_Y}'') \wedge d(\eta_1' + \eta_2') \wedge \ldots \wedge d(\eta_{n_Y}'' + \eta_{n_Y}'')$$

(more shortly written as : $d(y' - y'') \wedge d(\eta' + \eta'')$)

for each choice of local coordinates $y_1, \dots y_{n_Y}$ in Y .

<u>Proof</u>. On $T^*(Y) \times T^*(Y)$ we have the invariant form $\omega_1 = dy'' \wedge d\eta''$, which is the pullback of the invariant volume form on $T^*(Y)$ under the projection : $T^*(Y) \times T^*(Y) \longrightarrow T^*(Y)$ onto the second factor. We also have the invariant volume form $\omega_2 = dy' \wedge d\eta' \wedge dy'' \wedge d\eta''$ on $T^*(Y) \times T^*(Y)$. Put $\omega' = d(y'-y'') \wedge d(\eta'+\eta'')$ for some choice of local coordinates in Y . Then :

1^0. Every first order factor in the expression for ω' vanishes on $N^*(\mathrm{diag}(Y))$.

2^0 $\omega_2 = \omega' \wedge \omega_1$ along $N^*(\mathrm{diag}(Y))$.

These properties determine ω' uniquely along $N^*(\mathrm{diag}(Y))$, so the lemma follows .

With ω as in the lemma let Ω be the well defined form on $T^*(X) \times T^*(Y) \times T^*(Y) \times T^*(Z)$ along N^* which is the pullback of ω under the projection $T^*(X) \times T^*(Y) \times T^*(Y) \times T^*(Z) \longrightarrow T^*(Y) \times T^*(Y)$. By Ω we also denote some almost analytic extension. Then Ω can be written locally as an exterior product of first order forms vanishing on $(N^*)^\sim$.

<u>Theorem 7.5</u>. Under the conditions in Theorem 7.3 , let $A_1 \in I_c^{m_1}(X \times Y, C_1')$, $A_2 \in I_c^{m_2}(Y \times Z, C_2')$, $A = A_1 \circ A_2 \in I_c^{m_1+m_2}(X \times Z, (C_1 \circ C_2)')$ with principal symbols α_1, α_2 and α respectively. Identifying $(C_1 \circ C_2)'$ with $(C_1' \times C_2') \cap (N^*)^\sim$, we have the following equivalence of forms on $C_1' \times C_2'$ along $(C_1' \times C_2') \cap (N^*)^\sim$:

$$\alpha_1^2 \wedge \alpha_2^2 \sim \pm \alpha^2 \wedge \Omega ,$$

where $\alpha^2 \wedge \Omega$ is well defined along $(C_1' \times C_2') \cap (N^*)^\sim$ by the general remark in the proof of Theorem 6.4 .(The same general remark will be used constantly in the proof below.)

Proof . This is clearly a local statement ,so we can take local coordinates x,y,z in X, Y, Z respectively and introduce phase functions $\varphi_1(x,y,\theta)$, $\varphi_2(y,z,\sigma)$ and $\Phi(x,z,\omega)$ as above. Examinating the proof of Theorem 7.3 , we see that it suffices to prove that (when $\tilde{\varphi}_j$ and $\tilde{\Phi}$ are almost analytic extensions)

$$d_{\tilde{\varphi}_1} \wedge d_{\tilde{\varphi}_2} \sim \pm (\tilde{\theta}^2 + \tilde{\sigma}^2)^{-n_Y} d_{\tilde{\Phi}} \wedge \Omega \tag{7.8}$$

on $C_1' \times C_2'$ along $(C_1' \times C_2') \cap (N^*)^\sim$, where the forms $d_{\tilde{\varphi}_j}$ and $d_{\tilde{\Phi}}$ are defined in the preceding section.By the definition of $d_{\tilde{\varphi}_1}$, $d_{\tilde{\varphi}_2}$ we have

$$d_{\tilde{\varphi}_1} \wedge d_{\tilde{\varphi}_2} \wedge d(\partial\tilde{\varphi}_1/\partial\tilde{\theta}) \wedge d(\partial\tilde{\varphi}_2/\partial\tilde{\sigma}) \sim \pm d\tilde{x} \wedge d\tilde{y}' \wedge d\tilde{y}'' \wedge d\tilde{z} \wedge d\tilde{\theta} \wedge d\tilde{\sigma} \tag{7.9}$$

along $C_{\tilde{\varphi}_1} \times C_{\tilde{\varphi}_2}$ in $\mathbb{C}^{n_X + 2n_Y + n_Z + N_1 + N_2}$. For $d_{\tilde{\Phi}}$ we have on the other hand:

$$d_{\tilde{\Phi}} \wedge d(\partial\tilde{\Phi}/\partial\tilde{\omega}) \sim d\tilde{x} \wedge d\tilde{z} \wedge d\tilde{\omega} \tag{7.10}$$

along $C_{\tilde{\Phi}}$. If we recall that $\Phi(x,z,\omega) = \varphi_1(x,y,\theta) + \varphi_2(y,z,\sigma)$ and that $\omega = ((\theta^2 + \sigma^2)^{1/2} y, \theta, \sigma)$, we get from (7.10) that

$$\begin{aligned} & d_{\tilde{\Phi}} \wedge d(\partial\tilde{\varphi}_1/\partial\tilde{y} + \partial\tilde{\varphi}_2/\partial\tilde{y}) \wedge d(\partial\tilde{\varphi}_1/\partial\tilde{\theta}) \wedge d(\partial\tilde{\varphi}_2/\partial\tilde{\sigma}) \\ & \sim \pm (\tilde{\theta}^2 + \tilde{\sigma}^2)^{n_Y} d\tilde{x} \wedge d\tilde{y} \wedge d\tilde{z} \wedge d\tilde{\theta} \wedge d\tilde{\sigma} \end{aligned} \tag{7.11}$$

along the surface

$$C_{\tilde{\Phi}} = \{ (\tilde{x},\tilde{y},\tilde{z},\tilde{\theta},\tilde{\sigma}) \; ; \; \partial\tilde{\varphi}_1/\partial\tilde{y} + \partial\tilde{\varphi}_2/\partial\tilde{y} = 0 \; , \; \partial\tilde{\varphi}_1/\partial\tilde{\theta} = 0, \; \partial\tilde{\varphi}_2/\partial\tilde{\sigma} = 0 \}$$

which is naturally identified with C_{Φ} . Now we identify $\mathbb{C}^{n_X+n_Y+n_Z+N_1+N_2}$ with the subspace $\tilde{y}' = \tilde{y}''$ in $\mathbb{C}^{n_X+2n_Y+n_Z+N_1+N_2}$ and correspondingly we identify $C_{\tilde{\Phi}}$ with

$$C^0_{\tilde{\Phi}} = \{ (\tilde{x},\tilde{y}',\tilde{\theta},\tilde{y}'',\tilde{z},\tilde{\sigma}) \in C_{\tilde{\varphi}_1} \times C_{\tilde{\varphi}_2} \; ; \; \tilde{y}'=\tilde{y}'' \; , \; \partial\tilde{\varphi}_1/\partial\tilde{y}' + \partial\tilde{\varphi}_2/\partial\tilde{y}'' = 0 \}$$

Then we get from (7.11)

$$d_{\tilde{\Phi}} \wedge d(\partial\tilde{\varphi}_1/\partial\tilde{y}' + \partial\tilde{\varphi}_2/\partial\tilde{y}'') \wedge d(\partial\tilde{\varphi}_1/\partial\tilde{\theta}) \wedge d(\partial\tilde{\varphi}_2/\partial\tilde{\sigma}) \qquad (7.12)$$

$$\sim \pm (\tilde{\theta}^2 + \tilde{\sigma}^2)^{n_Y} \, d\tilde{x} \wedge d\tilde{y}' \wedge d\tilde{z} \wedge d\tilde{\theta} \wedge d\tilde{\sigma}$$

along $C^0_{\tilde{\Phi}}$ in the subspace $\tilde{y}' = \tilde{y}''$ in $\mathbb{C}^{n_X+2n_Y+n_Z+N_1+N_2}$. By the remark in the proof of Theorem 6.4 , we can extend (7.12) to $\mathbb{C}^{n_X+2n_Y+n_Z+N_1+N_2}$ if we multiply by $d(\tilde{y}'-\tilde{y}'')$ at the same time:

$$d_{\tilde{\Phi}} \wedge d(\tilde{y}'-\tilde{y}'') \wedge d(\partial\tilde{\varphi}_1/\partial\tilde{y}' + \partial\tilde{\varphi}_2/\partial\tilde{y}'') \wedge d(\partial\tilde{\varphi}_1/\partial\tilde{\theta}) \wedge d(\partial\tilde{\varphi}_2/\partial\tilde{\sigma}) \qquad (7.13)$$

$$\sim \pm (\tilde{\theta}^2 + \tilde{\sigma}^2)^{n_Y} \, d\tilde{x} \wedge d\tilde{y}' \wedge d(\tilde{y}'-\tilde{y}'') \wedge d\tilde{z} \wedge d\tilde{\theta} \wedge d\tilde{\sigma}$$

along $C^0_{\tilde{\Phi}}$ in $\mathbb{C}^{n_X+2n_Y+n_Z+N_1+N_2}$. Now $d\tilde{y}' \wedge d(\tilde{y}'-\tilde{y}'') = \pm d\tilde{y}' \wedge d\tilde{y}''$ so comparation of (7.13) and (7.9) gives

$$d_{\tilde{\Phi}} \wedge d(\tilde{y}'-\tilde{y}'') \wedge d(\partial\tilde{\varphi}_1/\partial\tilde{y}' + \partial\tilde{\varphi}_2/\partial\tilde{y}'') \sim \pm (\tilde{\theta}^2+\tilde{\sigma}^2)^{n_Y} \, d_{\tilde{\varphi}_1} \wedge d_{\tilde{\varphi}_2} \qquad (7.14)$$

on $C_{\tilde{\varphi}_1} \times C_{\tilde{\varphi}_2}$ along $C^0_{\tilde{\Phi}}$. Considering this equation on $C_1' \times C_2'$ along $(C_1' \times C_2') \cap (N^*)^{\sim}$ instead , we get precisely (7.8) because

$$d(\tilde{y}'-\tilde{y}'') \wedge d(\partial\tilde{\varphi}_1/\partial\tilde{y}' + \partial\tilde{\varphi}_2/\partial\tilde{y}'') = d(\tilde{y}'-\tilde{y}'') \wedge d(\tilde{\eta}' + \tilde{\eta}'') = \Omega$$

on $C_1' \times C_2'$ along $(C_1' \times C_2') \cap (N^*)^{\sim}$. This completes the proof.

8. Two applications .

Let X be a paracompact C^{∞} manifold of dimension n and let $P \in L^{m}(X)$ be a properly supported pseudo-differential operator with principal symbol $p \in C^{\infty}(T^{*}(X) \setminus 0)$, positively homogeneous of degree m . We shall assume that P operates on densities of order $1/2$ .Let Σ be the set where p vanishes and suppose that the Poisson bracket $\{p,\bar{p}\}$ satisfies

$$i^{-1}\{p,\bar{p}\} \neq 0 \quad \text{on } \Sigma \tag{8.1}$$

(Such operators have been much studied by Hörmander and others; we refer to [2] for further references .) From (8.1) it follows that $d(\mathrm{Re}\, p)$ and $d(\mathrm{Im}\, p)$ are linearly independent on Σ , so Σ is a closed conic submanifold of $T^{*}(X) \setminus 0$ of co-dimension 2 . According to the sign of $i^{-1}\{p,\bar{p}\}$ we split Σ as a union $\Sigma = \Sigma^{+} \cup \Sigma^{-}$. In Duistermaat - Sjöstrand [2] it was shown that there exist properly supported operators

$$F \,,\; F^{+} \,,\; F^{-} \;:\; \mathcal{D}'(X;\Omega_{1/2}) \longrightarrow \mathcal{D}'(X;\Omega_{1/2})$$

uniquely determined modulo $L^{-\infty}(X)$ with the following properties:

$$F^{+} \text{ and } F^{-} \text{ are continuous } H_{s}^{loc} \longrightarrow H_{s}^{loc} \text{ and} \tag{8.2}$$
$$F \text{ is continuous } H_{s}^{loc} \longrightarrow H_{s+m-1/2}^{loc} \text{ for all } s \in \mathbb{R} .$$

$$F^{+} + FP \equiv I \mod L^{-\infty} \tag{8.3}$$

$$F^{-} + PF \equiv I \mod L^{-\infty} \tag{8.4}$$

$$(F^{\pm})^{*} \equiv F^{\pm} \mod L^{-\infty}, \text{ where the adjoints are} \tag{8.5}$$

taken with respect to the scalar product

$\int u \bar{v}$, $u, v \in C_0^\infty(X; \Omega_{1/2})$.

$$WF'(F) = \operatorname{diag}(T^*(X) \setminus 0) = \{(\rho,\rho) \ ; \rho \in T^*(X) \setminus 0\} \tag{8.6}$$

$$WF'(F^{\pm}) = \operatorname{diag}(\Sigma^{\pm}) \ . \tag{8.7}$$

F^+ and F^- can be regarded as some kind of approximate orthogonal projections on the kernel and the cokernel of P respectively. It was conjectured in [2] that F^+ and F^- are Fourier integral operators with complex phase functions and we can now prove this, using the result about composition, established in section 7 . Since F^- is essentially of the same type as F^+ we shall concentrate the attention on F^+ . Our proof will use the construction in [2] .

If $\rho_0 \in \Sigma^+$ there is a homogeneous canonical transformation $\mathcal{H}: T^*(X) \setminus 0 \longrightarrow T^*(\mathbb{R}^n) \setminus 0$, defined in a conic neighbourhood of ρ_0 and such that

$$p \circ \mathcal{H}^{-1} = a(x,\xi) \ (\xi_n + i x_n \xi_{n-1}) \tag{8.8}$$

where a is a non-vanishing factor . If $\tilde{\rho}_0 = \mathcal{H}(\rho_0) = (x_0', 0, \xi_0', 0)$ it follows from (8.1) that $\xi_{0,n-1} < 0$. For the operator $\tilde{P} = D_n + i x_n D_{n-1}$ the construction was made microlocally near $\tilde{\rho}_0$ in [2] . From that construction it follows that the operator corresponding to F^+ is given by

$$\tilde{F}^+ u(x) = (2\pi)^{1-n} \iint e^{i\langle x'-y', \xi'\rangle + (x_n^2 + y_n^2)\xi_{n-1}/2} \psi(\xi') u(y) \, dy \, d\xi' \tag{8.9}$$

where $\psi \in S^{1/2}_{1,0}(\mathbb{R}^{n-1})$ has its support in some closed cone in $\mathbb{R}^{n-1} \setminus \{0\}$ which is contained in the domain $\xi_{n-1} < 0$ .The phase function

$$\varphi(x,y,\xi') = \langle x'-y', \xi'\rangle - i\,(x_n^2 + y_n^2)\,\xi_{n-1}/2$$

is regular and of positive type in the domain $\xi_{n-1} < 0$, and if we let Λ'_φ be the corresponding canonical relation, it is easy to see that

$$\Lambda'_{\varphi\mathbb{R}} = \mathrm{diag}\,(\Gamma^+) \qquad \text{close to} \quad (\tilde{\rho}_0, \tilde{\rho}_0)$$

where $\Gamma^+ = \{(x,\xi) \in \mathbb{R}^n \times \mathbb{R}^n \ ; \ x_n = \xi_n = 0\}$. Moreover we see that $\xi_n + ix_n\xi_{n-1}$ vanishes on Λ'_φ as a function on the first factor of $(T^*(\mathbb{R}^n)\setminus 0)^\sim \times (T^*(\mathbb{R}^n)\setminus 0)^\sim$ as well as a function on the second factor. Since we are now in the holomorphic category , it is clear that Λ'_φ is the flow out of $\mathrm{diag}(\tilde{\Gamma}^+)$ along the two commuting Hamilton fields $(H_{\tilde{\xi}_n + i\tilde{x}_n\tilde{\xi}_{n-1}}, 0)$ and $(0, H_{\tilde{\xi}_n + i\tilde{x}_n\tilde{\xi}_{n-1}})$, where

$$\tilde{\Gamma}^+ = \{(\tilde{x},\tilde{\xi}) \in \mathbb{C}^{2n} \ ; \ \tilde{x}_n = \tilde{\xi}_n = 0 \ , \ |\mathrm{Im}\,(\tilde{x},\tilde{\xi})| \ \text{small}\} \ .$$

Clearly $\tilde{F}^+ \in I_1^0(\mathbb{R}^n \times \mathbb{R}^n, \Lambda_\varphi)$ and we also see that Λ_φ is a strictly positive Lagrangean manifold .

By the construction in [2], F^+ is microlocally of the form

$$F^+ = A\,\tilde{F}^+ B \qquad \text{near } \rho_0 \ , \tag{8.10}$$

where $A \in I_1^0(X \times \mathbb{R}^n, (\mathrm{graph}\,\kappa^{-1})')$ and $B \in I_1^0(\mathbb{R}^n \times X, (\mathrm{graph}\,\kappa)')$. It is therefore clear by the composition results in section 7 , that F^+ is a Fourier integral operator of order 0 . The corresponding canonical relation Λ' is trivial to describe: By (8.8) we can always choose local symplectic coordinates (x,ξ) in $T^*(X)\setminus 0$ so that $p(x,\xi)$

$= a(x,\xi)(\xi_n + ix_n\xi_{n-1})$ and so that Σ^+ is given by $x_n = \xi_n = 0$. If $\tilde{\Sigma}^+$ is an almost analytic complexification of Σ^+ , then Λ' (for the chosen coordinates) is the flow out of diag ($\tilde{\Sigma}^+$) along $(H_{\tilde{\xi}_n+i\tilde{x}_n\tilde{\xi}_{n-1}}, 0)$ and $(0,H_{\tilde{\xi}_n+i\tilde{x}_n\tilde{\xi}_{n-1}})$. Summing up we have

<u>Theorem 8.1</u>. With Λ' as above we have $F^+\in I_1^0(X \times X,\Lambda)$.

We next consider a situation which has been previously studied locally by Hörmander [5] . The global result we shall present here has also been more or less conjectured in [5]. Let $P \in L^1(X)$ be a properly supported classical pseudo-differential operator on a paracompact manifold X of dimension n . Let $p \in C^\infty(T^*(X)\setminus 0)$ be the homogeneous principal symbol and assume that

$$\mathbb{R} \ni t \longrightarrow \gamma(t) \in T^*(X)\setminus 0$$

is a bicharacteristic strip such that the following conditions are satisfied :

$$\partial\gamma(t)/\partial t = H_p(\gamma(t)) \neq 0 \quad , \quad \forall\, t \, , \tag{8.11}$$

$$p\circ\gamma = 0 \, , \tag{8.12}$$

the map $\mathbb{R} \ni t \longrightarrow \pi\circ\gamma(t) \in X$ is proper , where (8.13)
$\pi : T^*(X)\setminus 0 \longrightarrow X$ is the natural projection ,

$$d_\xi p \neq 0 \quad \text{on } \gamma \, . \tag{8.14}$$

It follows that the cone $\Gamma \subset T^*(X)\setminus 0$ generated by

γ is a smooth closed submanifold . We shall give a rather implicit global sufficient condition for the existence of an element $u \in \mathcal{D}'(X; \Omega_{1/2})$ such that $WF(u) = \Gamma$ and $Pu \in C^\infty$. The proof will only be sketched here . A more detailed but somewhat different proof will be given elsewhere.

Write $\gamma(t) = (x_t, \xi_t)$. By the Malgrange preparation theorem there are local coordinates in X such that

$$p(x,\xi) = a(x,\xi)(\xi_n - ib(x, \xi'))$$

in a conic neighbourhood of (x_0, ξ_0) , where $a \neq 0$ and $\text{grad}(\text{Re } b) = 0$ at (x_0, ξ_0). Let $\Lambda_0' \subset \mathbb{C}^{n-1} \times (\mathbb{C}^{n-1} \setminus \{0\})$ be a strictly positive conic Lagrangean manifold with $\Lambda'_{0\mathbb{R}} = \{(x_0', \lambda \xi_0') ; \lambda \in \mathbb{R}_+\}$ and define an almost analytic manifold $\Lambda_0 \subset (T^*(X) \setminus 0)^\sim$ by the local representative (where b also denotes some almost analytic extension)

$$\{(\tilde{x}', x_{0n}, \tilde{\xi}', ib(\tilde{x}', x_{0n}, \tilde{\xi}')) ; (\tilde{x}', \tilde{\xi}') \in \Lambda_0' \} .$$

Then $\Lambda_{0\mathbb{R}} = \{(x_0, \lambda \xi_0) ; \lambda \in \mathbb{R}_+\}$ and if σ and p also denote almost analytic extensions , we have $\sigma|_{\Lambda_0} \sim 0$ and $p|_{\Lambda_0} \sim 0$. Moreover $i^{-1}\sigma(u,\bar{u}) > 0$ for all u in $T_{(x_0, \xi_0)}(\Lambda_0) \setminus (T_{(x_0,\xi_0)}(\Lambda_{0\mathbb{R}}))^\sim$ and H_p is not tangential to Λ_0 at (x_0, ξ_0) .

The Hamilton field H_p induces in a natural way linear canonical transformations

$$A_t : T_{\gamma(0)}(\widetilde{T^*(X)}) \longrightarrow T_{\gamma(t)}(\widetilde{T^*(X)}) , \quad \forall\, t \in \mathbb{R} .$$

These transformations are obtained by "integrating the almost analytic Hamilton field H_p in $(T^*(X)\setminus 0)^{\sim}$ from 0 to t " and then taking the differential at $\gamma(0)$ of this mapping. (The flow Φ_t of an almost analytic vectorfield $\sum_1^n a_j(z)\,\partial/\partial z_j$ in $\mathbb{C}^n$ is defined by

$$\partial\Phi_{st}(z)/\partial s = t\,a(\Phi_{st}(z))\ , \quad 0\leq s\leq 1\ , \quad t\in\mathbb{C}\ , \quad |t| \text{ small}$$

$$\Phi_0(z) = z\ , \qquad a = (a_1,\dots,a_n)\ ,\ z\in\mathbb{C}^n\ .\ )$$

We do not enter into the details of this construction , we just notice that the transformations A_t are in general not real, except in the case when p is real-valued . Put $\Gamma_t = A_t(T_{\gamma(0)}(\Lambda_0))$ and introduce the following condition

Λ_0 above can be chosen so that for all $t\in\mathbb{R}$: (8.15)

1^0 $\Gamma_{t\mathbb{R}} = A_t(\Gamma_{0\mathbb{R}})$

2^0 $i^{-1}\sigma(u,\bar{u}) > 0$ for all $u\in\Gamma_t\setminus\widetilde{\Gamma_{t\mathbb{R}}}$.

Note that (8.15) is a global condition for 1^0 and 2^0 are obviously satisfied for small t . When Hess(Im p) = 0 along γ , then the transformations A_t are real and (8.15) is satisfied .

<u>Theorem 8.2</u> . When the condition (8.15) is satisfied there exists a strictly positive conic Lagrangean manifold $\Lambda\subset(T^*(X)\setminus 0)^{\sim}$ with $\Lambda_{\mathbb{R}}=\Gamma$ and such that $p|_{\Lambda}\sim 0$. Moreover there exists $u\in I_c^0(X,\Lambda)$ with non-vanishing principal symbol such that $Pu\in C^{\infty}$. In particular $WF(u) = \Gamma$.

Sketch of the proof. The condition (8.15) implies that we can define an almost analytic manifold $\Lambda \subset (T^*(X) \setminus 0)^\sim$ as the "flow out" of Λ_0 along the almost analytic vectorfield H_p in $(T^*(X) \setminus 0)^\sim$. We shall not enter into the details of this construction . It actually follows that $\Lambda_{\mathbb{R}} = \Gamma$, that $T_{\gamma(t)}(\Lambda)$ is spanned by Γ_t and H_p and that $p|_\Lambda \sim 0$, $\sigma|_\Lambda \sim 0$, where p and σ also denote almost analytic extensions. Since Λ also turns out to be conic , it is clear that Λ is a strictly positive conic Lagrangean manifold.

For the construction of u we need some preparations. Let v be a real vectorfield on some real manifold . If a is a density of order 1/2 we can define its Lie derivative $\mathcal{L}_v(a)$ and if we write $a(\lambda) = u(\lambda)\sqrt{d\lambda_1 \dots d\lambda_n}$ for local coordinates $\lambda_1, \dots, \lambda_n$ we get

$$\mathcal{L}_v(a) = (v(u) + 2^{-1}\sum \partial v_j/\partial\lambda_j)\sqrt{d\lambda_1 \dots d\lambda_n} . \tag{8.16}$$

Here we have written

$$v = \sum v_j(\lambda)\,\partial/\partial\lambda_j . \tag{8.17}$$

Now let $\Lambda \subset (T^*(X) \setminus 0)^\sim$ be a positive conic Lagrangean manifold. An almost analytic vectorfield v on Λ is then defined by giving operators of the form (8.17) for every choice of local admissible coordinates $\lambda_1, \dots, \lambda_n$. Here the v_j should be almost analytic and of course we should have the obvious equivalence relations between different representatives of the form (8.17) under changes of admissible coordinates. (Our definition generalizes of course immediately to the case of arbitrary almost analytic manifolds .)

If $a \in \Gamma(\Lambda;\mathcal{L})$ is a section of the Maslov line bundle <u>we define</u> $\mathcal{L}_v(a) \in \Gamma(\Lambda;\mathcal{L})$ for local representatives by the formula (8.16) , where now $\lambda_1,\ldots,\lambda_n$ are admissible coordinates and $\sqrt{d\lambda_1 \ldots d\lambda_n}$ denotes the section of $\mathcal{L}$ taking the value 1 for these coordinates . It is easy to check that the definition is independent of the choice of the λ_j .

Suppose that $p \in C^{\infty}(T^*(X) \setminus 0)$ and denote by p also an almost analytic extension . If $p|_{\Lambda} \sim 0$ it is easy to verify that the almost analytic vectorfield

$$H_p = \sum \frac{\partial p}{\partial \xi_j} \frac{\partial}{\partial x_j} - \frac{\partial p}{\partial x_j} \frac{\partial}{\partial \xi_j}$$

on $(T^*(X) \setminus 0)^{\sim}$, induces an almost analytic vectorfield on Λ. We can now formulate

<u>Proposition 8.3</u>. Let $P \in L^m(X)$ be a properly supported classical operator with principal symbol p and let $u \in I_c^k(X,\Lambda)$, where $\Lambda \subset (T^*(X) \setminus 0)^{\sim}$ is a closed conic positive Lagrangean manifold such that $p|_{\Lambda} \sim 0$, if p also denotes some almost analytic extension. Then $Pu \in I_c^{m+k-1}(X,\Lambda)$ and if $a \in \Gamma^{n/4+k}(\Lambda;\mathcal{L})$ is the principal symbol of u , then Pu has the principal symbol $i^{-1}\mathcal{L}_{H_p} a + ca$. Here c is the subprincipal symbol of P , defined in local coordinates by

$$c(x,\xi) = p_{m-1}(x,\xi) - (2i)^{-1} \sum_1^n \partial^2 p_m(x,\xi)/\partial x_j \partial \xi_j$$

where $\sigma_P \sim p_m(x,\xi) + p_{m-1}(x,\xi) + \ldots$ is the asymptotic expansion of the symbol in these coordinates .

This proposition has been proved by Hörmander-Duistermaat [1] in the case of real phase functions . The proof in the complex case is almost the same so we omit it .

Returning to the situation in Theorem 8.2 (m=1) we look for $u_0 \in I_c^0(X,\Lambda)$ with principal symbol $a_0 \in \Gamma^{n/4}(\Lambda;\mathcal{L})$ such that

$$(i^{-1}\mathcal{L}_{H_p} + c)\, a_0 \sim 0 . \tag{8.18}$$

H_p is a non-vanishing real vector field on $\Lambda_{\mathbb{R}}$ whose integral curves are obtained from γ by multiplication by some positive number in the fiber variable . Since $\mathbb{R} \ni t \longrightarrow \pi \circ \gamma(t) \in X$ is a proper map it is rather easy to construct a non-vanishing solution a_0 of (8.18) by considering Taylor expansions at $\Lambda_{\mathbb{R}}$ in the complex directions of Λ. For the corresponding $u_0 \in I_c^0$ we have $Pu_0 \in I_c^{m-2}(X,\Lambda)$ and if $v_{m-2} \in \Gamma^{m-2+n/4}$ is the principal symbol we can find $a_{-1} \in \Gamma^{-1+n/4}$ such that

$$(i^{-1}\mathcal{L}_{H_p} + c)\, a_{-1} \sim - v_{m-2} .$$

Then if $u_{-1} \in I_c^{-1}(X,\Lambda)$ has principal symbol a_{-1} it follows that

$$P(u_0+u_{-1}) \in I_c^{m-3}(X,\Lambda) .$$

By repeating this construction we obtain the desired $u \in I_c^0(X,\Lambda)$ as an asymptotic sum

$$u \sim u_0 + u_{-1} + \dots \quad , \; u_j \in I_c^{-j}(X,\Lambda) ,$$

and this completes the sketch of the proof of Theorem 8.2 .

References

1. Duistermaat,J.J. and Hörmander,L., Fourier integral operators II. Acta Math.,128(1972),183-269.

2. Duistermaat,J.J. and Sjöstrand,J., A global construction for pseudo-differential operators with non-involutive characteristics.Inventiones math.,20(1973),209-225.

3. Hörmander,L.,Fourier integral operators I.Acta Math.,127(1971), 79-183.

4. Hörmander,L.,Lecture notes at the Nordic Summer School of Mathematics ,1969.

5. Hörmander,L.,On the existence and the regularity of solutions of linear pseudo-differential operators.Enseignement Math., 17(1971),99-163.

6. Hörmander,L.,Pseudo-differential operators and hypoelliptic equations.Amer.Math.Soc.Symp.on Singular Integral Operators, 1966,138-183.

7. Hörmander,L.,Pseudo-differential operators and non-elliptic boundary problems.Ann.Math.,83(1966),129-209.

8. Kucherenko,V.V.,Hamilton-Jacobi equations in a complex non-analytic situation.Dokl.Akad.Nauk SSSR,213(1973),1021-1024.

9. Kucherenko,V.V.,Maslov's canonical operator on a germ of complex,almost analytic manifold.Dokl.Akad.Nauk SSSR,213(1973), 1251-1254.

10. Leray,J.,Le calcul différentiel et intégral sur une variété analytique complexe(Problème de Cauchy,III).Bull.Soc.math. France,87(1959),81-180.

11. Maslov,V.,The characteristics of pseudo-differential operators and difference schemes.Actes Congrès Intern.Math.Nice 1970, Tome 2,755-769.

12. Nirenberg,L.,A proof of the Malgrange preparation theorem.Proc. Liverpool Singularities Symp. I,Dept.pure Math. Univ. Liverpool 1969-1970,(1971),97-105.

13. Nirenberg,L. and Treves,F.,On local solvability of linear partial differential equations.Part I.Comm.Pure Appl.Math., 23(1970),1-38.

14. Wells,R.O.Jr,Compact real submanifolds of a complex manifold with non-degenerate holomorphic tangent bundles.Math.Ann., 179(1969),123-129.

ON A PROBLEM OF HANS LEWY

L. Nirenberg *

1. Consider a linear first order partial differential operator with C^∞ complex coefficients in a neighbourhood of the origin in R^3

(1) $$P = \sum_1^3 a^j(x) \frac{\partial}{\partial x^j} , \qquad \sum |a^j| > 0 .$$

Around the same time that Lewy presented his famous example of a "nonsolvable" operator P i.e. one for which, for general f in C^∞, the equation

(2) $$Pw = f$$

has no solution in any open set, he posed the following:

Question 1 Do homogeneous equations

(3) $$Pw = 0$$

always have nonconstant solutions?

* This research was supported by NSF Grant NSF-GP-37069X.

This talk is a report on the paper [3] in which an example is constructed for which $w \equiv$ constant is the only C^1 solution in a neighbourhood of the origin.

The question arose in connection with some investigations of Lewy [1] on boundary behaviour of holomorphic functions of two complex variables (z^1, z^2). In $\mathbb{C}^2$ let Ω be a domain with smooth boundary $\partial\Omega$ (the question is purely local so we consider Ω and $\partial\Omega$ in a neighbourhood of some point on $\partial\Omega$). There is a smooth real function $\rho(z^1, z^2)$ with grad $\rho \neq 0$ such that Ω corresponds, locally, to $\rho < 0$. In Ω consider a holomorphic function w, i.e. satisfying the Cauchy-Riemann equations

$$w_{\overline{z^j}} = 0 , \qquad j = 1,2;$$

here $z^j = x^j + i\, y^j$, $w_{\overline{z^j}} = \frac{1}{2}\left(\frac{\partial}{\partial x^j} + i \frac{\partial}{\partial y^j}\right) w$,
$w_{z^j} = \frac{1}{2}\left(\frac{\partial}{\partial x^j} - i \frac{\partial}{\partial y^j}\right) w$, and which is C^∞ in $\Omega \cup \overline{\Omega}$ (locally).
The restriction of w to $\partial\Omega$ then satisfies the "induced or restricted Cauchy Riemann equations"

$$(4) \qquad Pw = \left(\rho_{\overline{z^1}} \frac{\partial}{\partial \overline{z^2}} - \rho_{\overline{z^2}} \frac{\partial}{\partial \overline{z^1}}\right) w = 0 .$$

P is a well defined first order operator acting on functions defined on $\partial\Omega$, and so is of the form (1) in terms of local coordinates on $\partial\Omega$.

In [1] Lewy proved the following local analogue of Hartog's theorem.

<u>Theorem</u> A. Let w be a C^1 function defined on $\partial\Omega$ in a neighbourhood of some point $z_0 = (z_0^1, z_0^2)$ satisfying $Pw = 0$, where P is the operator (4).

If Ω is strongly pseudo-convex at z_0 then there is a neighbourhood U of z_0 in $\mathbb{C}^2$ such that w admits a holomorphic continuous extension into $U \cap \Omega$.

The condition of strong pseudo-convexity may be expressed in terms of P (here $\bar{P}$ is the operator $\sum \overline{a^j} \frac{\partial}{\partial x^j}$ in case $P = \sum a^j \frac{\partial}{\partial x^j}$)

(5) $P, \bar{P}$ and $[P, \bar{P}] = P\bar{P} - \bar{P}P$ are linearly independent on $\partial\Omega$.

Lewy then asked

<u>Question 2</u>. Does every equation $Pw = 0$, where P satisfies (5), arise locally from a strongly pseudo-convex domain Ω in $\mathbb{C}^2$ as the restriction of the Cauchy-Riemann equations to $\partial\Omega$?

The answer (as he showed) is yes, provided (and this is clearly necessary) there exist locally at least two solutions z^1, z^2 of $Pw = 0$ whose gradients are linearly independent over the complex field. For if $z^1(x)$, $z^2(x)$ are two such solutions then the set of points S in $\mathbb{C}^2$ that they fill out is parametrized by three real variables and, using the independence of the gradients, and (5), it is not hard to see that S is a smooth hypersurface in $\mathbb{C}^2$. Furthermore one verifies that $Pw = 0$ is a nonvanishing multiple of the restricted Cauchy-Riemann equatios on S as given in (4). This answer leads naturally to Question 1.

There is a higher dimensional analogue of Question 1 which arises in the following way: If Ω is a smooth hypersurface in $\mathbb{C}^n$ given by $\rho = 0$, with, say, $\rho_{z^n} \neq 0$ then the restriction to $\partial\Omega$ of any holomorphic function w in Ω which is smooth in $\Omega \cup \partial\Omega$, satisfies a system of (n-1) induced Cauchy-Riemann equations:

$$P_j w = \left(\rho_{\bar{z}^n} \frac{\partial}{\partial \bar{z}^j} - \rho_{\bar{z}^j} \frac{\partial}{\partial \bar{z}^n}\right) w = 0 , \qquad j=1,\ldots,n-1.$$

The operators P_j are well defined on $\partial\Omega$ and, for a suitable choice of real vector field T on $\partial\Omega$,

(5)' $P_1,\ldots,P_{n-1}$, $\bar{P}_1,\ldots,\bar{P}_{n-1}$, T span all homogeneous first order operators and the commutator of any two of the P_j is a linear combination of $P_1,\ldots,P_{n-1}$.

The condition of strong pseudo-convexity analogous to (5) is the following: we have

$$[P_j,\bar{P}_k] = i\, c_{jk}\, T \bmod (P_1,\ldots,P_{n-1}, \bar{P}_1,\ldots,\bar{P}_{n-1})$$

for some Hermitian matrix valued function c_{jk}; strong pseudo-convexity means

(5)" c_{jk} is positive definite.

Question 1 now takes the form

Question1'. In a neighbourhood of the origin in R^{2n-1}, given n-1 linear operators

$$P_j = \sum_1^{2n-1} a_j^{\alpha}(x) \frac{\partial}{\partial x^{\alpha}} , \qquad j = 1,\ldots,n-1,$$

satisfying (5)' and (5)" with c_{jk} positive definite, do there exist nontrivial solutions w of the system

$$P_j w = 0 , \qquad j = 1,\ldots,n-1 ?$$

For $n = 2$, as we see in this talk, the answer is no. For $n > 2$, in particular for $n > 3$ there is reason to believe that the situation is quite different, and the answer may be yes. M. Kuranishi seems to be close to proving this assuming that c_{jk} is nonsingular.

2. The example in [3] is somewhat complicated and instead of going into its details we shall describe a simpler two dimensional example taken from section 2 of [2]. The corresponding arguments for the three dimensional example will only be sketched.

In a neighbourhood of the origin in the x,y plane consider the operator

$$(6) \qquad P = \frac{\partial}{\partial x} + i(x + \phi(x,y)) \frac{\partial}{\partial y}$$

where ϕ is a C^∞ nonnegative function which is even in x and satisfies

$$\phi(x,y) < |x| , \quad \text{for} \quad x \neq 0 .$$

Because of this condition the real and imaginary parts of P are linearly independent for $x \neq 0$ (i.e. P is elliptic) and hence, (a nontrivial fact!), in a neighbourhood of any point not on the y axis there are many nontrivial solutions of $Pw = 0$. In particular one may introduce new local coordinates $\zeta = \xi + i\eta$ about such a point so that P takes the form

$$P = \lambda(\xi,\eta) \frac{1}{2} \left(\frac{\partial}{\partial \xi} + i \frac{\partial}{\partial \eta}\right) , \qquad \lambda \neq 0,$$

i.e., a solution is simply a holomorphic function of some new local coordinate ζ.

We shall now specify further conditions on the function ϕ ; in $x > 0$ it is positive inside a sequence of nonoverlapping closed discs $D_j^{m,n}$, $m,n,j = 1,2,...$ and vanishes outside their union. The discs satisfy

(i) the ordinates of the centers of $D_j^{m,n}$ equal $1/n$

(ii) $\frac{1}{m} < x < \frac{1}{m-1}$ for any (x,y) in $D_j^{m,n}$, $j = 1,2,...$

(iii) The abscissae of the centres of $D_j^{m,n}$ decrease to $1/m$ as $j \to \infty$.

It is easy to construct such discs, and then to construct a function ϕ as described above; ϕ is then to be extended to $x < 0$ as even in x.

Theorem 1. With ϕ chosen as above, if w is a distribution solution of $Pw = 0$ in a neighbourhood of the origin then $w \equiv$ constant.

The proof is elementary but a bit tricky. Observe first that since in $x \neq 0$ w is a holomorphic function of new local variables, it follows that w is in C^∞ for $x \neq 0$.

Proof: Step 1. Decompose w as a sum of functions

$$w = u + v$$

with u and v respectively odd and even in x. The even part of the equation then reads

$$u_x + ix\, u_y = -i\phi\, v_y \tag{7}$$

In $x \geq 0$ if we set $s = x^2/2$ we find that outside the discs, i.e. where ϕ vanishes, we have on dividing by x,

$$u_s + i\, u_y = 0$$

i.e. u is a holomorphic function of $x^2/2 + i\,y$. Since u vanishes for $s = 0$ and $y < 0$ it follows that $u \equiv 0$ outside the discs, for the complement of the discs is connected. In particular u vanishes on the boundary of each $D_j^{m,n}$.

Step 2. We claim that for $m,n = 1,2,\ldots$

$$v_y\left(\frac{1}{m}, \frac{1}{n}\right) = 0 . \tag{8}$$

Suppose to the contrary, that $v_y(1/m,1/n) \neq 0$ for some m,n. Integrating the equation (7) over the disc $D_j^{m,n} = D$ we find by Green's theorem

$$0 = \iint_D (u_x + ix\, u_y)\, dx\, dy = -i \iint_D \phi\, v_y\, dx\, dy. \tag{9}$$

However, for j large, in $D_j^{m,n}$, $\arg v_y(x,y)$ is close to $\arg v_y(1/m,1/n)$ and the same is true for $\arg \phi\, v_y$; but then (9) is impossible. Thus (8) holds.

Step 3. Keeping m fixed let $n \to \infty$. It follows from (8) that all y derivatives of v vanish at $(1/m,0)$. We claim that, in fact, <u>all derivatives</u> of v vanish at $(1/m,0)$. This is because we may simply compute all these derivatives by repeated differentiations of the odd part of the equation $Pw = 0$:

$$v_x + i\, v_y = -i\phi\, u_y .$$

We conclude that all derivatives of w vanish at $(1/m,0)$.

Step 4. As we noted earlier in a neighbourhood of $(1/m,0)$ the equation $Pw = 0$ simply asserts that w is a holomorphic function of some new local coordinates. But since all its derivatives vanish there it follows that $w \equiv$ constant in some neighbourhood of $(1/m,0)$. By a simple (analytic extension) argument, we may infer that $w \equiv$ constant in $x > 0$. Similarly $w \equiv$ constant in $x < 0$. It is an exercise to conclude that a distribution solution with these properties is identically constant in a full neighbourhood of the origin.

3. We turn now to the example of [3] in R^3 with real coordinates x,y,t; set

$$x + i y = r e^{i\theta} .$$

The example is a slight modification of Lewy's original example (2) in which $P = \partial/\partial \bar{z} + iz\, \partial/\partial t$. It is of the form

$$(10) \qquad P = \frac{\partial}{\partial \bar{z}} + iz \frac{\partial}{\partial t} + z\phi \frac{\partial}{\partial t} + z\psi \frac{\partial}{\partial \theta} ,$$

here $\partial/\partial\theta = i(z\, \partial/\partial z - \bar{z}\, \partial/\partial \bar{z})$; ϕ and ψ are suitably chosen C^∞ functions which are <u>real analytic</u> in θ for $r > 0$, and vanish of infinite order on the t axis.

The functions ϕ and ψ are complicated, as in the two dimensional example. The discs $D_j^{m,n}$ in that example in which $\phi > 0$, are replaced by similar discs rotated about the t axis. That is, the functions ϕ, ψ have their supports in infinitely many nonoverlapping closed tori about the t axis. With suitably chosen ϕ, ψ one proves that if w is a C^1 solution of $Pw = 0$ (P given in (10)) in a neighbourhood of the origin then $w \equiv$ constant.

The proof follows the arguments of the two dimensional example. The decomposition of step 1 is now the infinite sum decomposition of w in its Fourier series representation in θ: $w = \sum_j w_j e^{ij\theta}$. One then obtains an infinite system of equations for the w_j, but these are treated as in steps 1,2,3 and we end up with the result that:

(11) all derivatives of w vanish on the circles: $x^2 + y^2 = \frac{1}{m^2}$, $t = 0$, for sufficiently large integers m.

The difficulty in completing the argument is that there is no analogue of step 4. We wish to infer from (11) that $w \equiv$ constant. This leads to the following open problem on unique continuation:

Problem 1. Let w be a smooth solution in a domain G in R^3 of

$$Pw = \sum_1^3 a^j(x) \frac{\partial w}{\partial x^j} = 0$$

with P satisfying (5). If w vanishes in an open set in G does it follow that $w \equiv 0$ in G ?

In some cases the answer is yes, in particular if P arises as the restricted Cauchy-Riemann equations on a boundary $\partial\Omega$ of a domain Ω in $\mathbb{C}^2$. This is easy to see with the aid of Lewy's Theorem A. Thus the answer to problem 1 is yes provided there exist two linearly independent solutions of $Pw = 0$.

This does not seem a useful approach in our problem since we are trying to show just the opposite: that $Pw = 0$ has only trivial solutions. However because the coefficients of the operator P are analytic in one of the variables θ, we claim that in a neighbourhood of any point off the t axis there are many nontrivial solutions of $Pw = 0$. This nontrivial fact may be proved with the aid of the complex Frobenius theorem (theorem on integrability of almost complex structures). Using this one may complete the proof (see [3] for details), that $w \equiv$ const.

I wish to conclude with an open problem which is related to the result just stated. Consider vector valued functions $w(t,\theta) = (w^1,\dots,w^N)$, w^j complex, of the variables

$$t = (t_1,\dots,t_n) \in R^n \ , \qquad \theta = (\theta_1,\dots,\theta_k) \in R^k \ ,$$

and a system of N equations in a neighbourhood of the origin in $R^n \times R^k$:

$$(12) \qquad Pw = \sum_1^n a_j(t,\theta) \frac{\partial w}{\partial t_j} + \sum_1^k b_\alpha(t,\theta) \frac{\partial w}{\partial \theta_\alpha} = f(t,\theta) \ .$$

Here a_j, b_α are $N\times N$ matrices, and f is an N-vector.

Problem 2. Assume that the coefficients (i.e. their real and imaginary parts), as well as the components of f, are real analytic in θ. Assume also that the operator

$$\sum_1^n a_j(t,\theta) \frac{\partial}{\partial t_j}$$

is elliptic in the t variables for each θ, i.e.

$\sum_1^n a_j(t,\theta)\tau_j$ is nonsingular for $\tau = (\tau_1,\dots,\tau_n) \in R^n \setminus 0$.

Does there exist a local solution of (12) which is analytic in θ ?

If we extend θ to complex values in $\mathbb{C}^k$ and extend the coefficients, and f, as holomorphic in θ we see that we are really concerned with one overdetermined elliptic system:

$$Pw = f$$

$$Q_\alpha w = \frac{\partial w}{\partial \bar{\theta}_\alpha} = 0$$

This is a special case of the general question, posed by D. C. Spencer for overdetermined elliptic systems.

In case $N = 1$, $n = 2$ this corresponds to the complex Frobenius theorem and the answer is yes. But for $N = 2$, $n = 2$, $k = 1$ the answer is not known.

Bibliography

[1] H. Lewy, On the local character of the solutions of an atypical linear differential equation in three variables and a related theorem for regular functions of two complex variables. Ann. of Math. 64 (1956), pp. 514-522.

[2] L. Nirenberg, Lectures on linear partial differential equations, Conf. Board of Math. Sci., Reg. Conf. Series in Math. No. 17, Amer. Math. Soc. 1973.

[3] On a question of Hans Lewy. Uspekhi Mat. Nauk Vol. 29, 2 (176), (1974), pp. 241-251.

On structures of L^2-well-posed mixed problems for hyperbolic operators

Taira SHIROTA
Departement of Mathematics,
Hokkaido University and Nice University.

1. Introduction and results.

Let P be a x_0-strictly hyperbolic $2p \times 2p$-system of partial differential operators of the first order defined over a C^∞-cylinder $R^1 \times \Omega \subset R^{n+1}$, where Ω is a C^∞-domain in R^n. Let B be a $p \times 2p$ system of functions defined on the boundary Γ of $R^1 \times \Omega$. We consider the following mixed problems under certain conditions:
for some positive T

$$(P,\ B,\ \Omega)\begin{cases} P(X, D)u = f & \text{for } X \in R^1 \times \Omega \text{ and } T \geqq x_0 \geqq 0, \\ B(X)u = g & \text{for } X \in \Gamma \text{ and } T \geqq x_0 \geqq 0, \\ \text{and} & \\ u = h & \text{for } X \in \{x_0 = 0\} \times \Omega \end{cases}$$

where $\sqrt{-1}\, D = (\frac{\partial}{\partial x_0}, \cdots, \frac{\partial}{\partial x_n})$.

For the sake of simplicity of descriptions and by the hyperbolicity of P, we may only consider the case where $\Omega = \{x_n > 0\}$ and we assume here the following conditions:

(I) α). The coefficients of P are real and those of P and B belong to $C^\infty(R^1 \times \bar{\Omega})$ and constant outside some compact set of $R^1 \times \bar{\Omega}$.

β). For the operator P, it satisfies the # condition with respect to Γ, i.e., for fixed (X, τ, σ) there exist at most real double roots λ of $|P^0|(X, \tau, \sigma, \lambda) = 0$, where $X \in \Gamma$ and (τ, σ, λ) is the covector of $X = (x_0, (x_1, \cdots, x_{n-1}), x_n)$. Furthermore we assume that there is at most one such real double root for fixed such (X, τ, σ). Finally it is non-characteristic with respect to Γ and is normal, i.e., $|P^0|(X, 0, \sigma, \lambda) \neq 0$, where $|P^0|$ is the determinant of the principal part of P and $(\sigma, \lambda) \neq 0$.

γ). The p-row vectors of $B(X)$ are linearly independent for each $X \in \Gamma$.

In general theoretic point of view, the above conditions are somewhat strong, but our restrictions are nutural and essential in studie for hyperbolic mixed problems.

Nowaday, though investigations about Cauchy problems for hyperbolic operators are extensively developed, but there are only few results about mixed problems as a general theoretic point of view. That is Kreiss' and Sakamoto's results ([5], [9]) who obtain the existence of solutions of mixed problems under so called uniform Lopatinsky condition, but there are not any essential results about singulartities of solutions for mixed poblems. The reasons why it is so, we have already known, are the appearances of singularities of phase funtions and amplitude functions at the begining of the investigations, i.e., the existences of glancing waves. However the solutions of mixed problems of other types exist and investigated by Lax and Phillips in the pont of view of scattering theory ([7]). Therefore as a general theoretic point of view, it seems to me to be interesting to consider systematic investigations of existence theorems for mixed problems, bearing in mind the appearances of such singulartities.

We have already defined the L^2-well-posedness for problems (P, B, Ω) with constant coefficients cases ([1]) such that for any data $f \in H^1((-\infty, T) \times \Omega)$ with

$$\langle \text{supp } f, e_0 \rangle \geqq 0 \text{ and } g = h = 0,$$

there exists a unique solution u with the following properties:

$$u \in H^1((-\infty, T) \times \Omega), \quad \langle \text{supp } u, e_0 \rangle \geqq 0$$

and for some positive C

$$\|u\|_0 ((0, T) \times \Omega) \leqq C \|f\|_0((0, T) \times \Omega)$$

where the norm is the usual Sobolev's one. Then we obtain the necessary and sufficient condition for L^2-well-posedness in the case of constant coefficients and remark that even if it is variable coefficients case, the L^2-well-posedness implies the L^2-well-posedness for the freezing problems at any point of the boundary Γ ([2]). Therefore in order to investigate the L^2-well-posedness in variable coefficients case, we may assume the following condition:

(II). The freezing problems at each boundary point are L^2-well-posed.

Then our question is whether our original problem is L^2-well-posed or not. Here we remark that for constant coefficients problems, our L^2-well-posedness implies that the speeds of propagation of solutions for any direction are not faster than that for Cauchy problems ([11]), so it seems to me that our L^2-well-posedness is not so strong and not so weak, where we define the propagation speed in the direction $\xi \in S^{n-1}$ is the minimum s such that setting $t = x_0$

$$\sup < \text{supp of solution } u(x_0, x_1, \cdots x_n), \xi>$$
$$\leqq st + \sup < \text{supp of the data } u(0, x_1 \cdots x_n), \xi>$$

for any $t > 0$ and for any solution u. But we remark that if we consider the stability of problem (P, B, Ω), it will be better to consider more wide class of well posedness. Therefore I am afraid that our L^2-well-posed class of problems (P, B, Ω) is somewhat narrow.

Here we shall state that under certain conditions our question is answered affirmatively. But unfortunately, we can't construct any parametrics and any Riemann functions in our general theoretic point of view even if it is the most elementary case, as it is previously stated. Thus our method used here is the classical energy one, but the condition (II) is just equevalent to certain relations of the coefficients of B, i.e., the condition (II) will be able to control the coefficients of B, generalized reflection coefficients or coupling coefficients for (P^0, B, Ω) so that we will get informations about structures of our problems. Therefore our interests are to know how these relations are able to imply the L^2-well-posedness for our problem (P, B, Ω) in variable coefficients cases.

Furthermore we remark that for certain cases the coupling coefficients indicate us the behaviours of the propagations of singularities of solutions.

Now we set the additional assumptions in a neighbourhood of points (X_0, τ_0, σ_0) from which the singularities mentioned above occur:

(III) α). If the Lopatinsky determinant $R(X_0, \tau_0, \sigma_0)$ for (P^0, B, Ω) is vanshing for real point (X_0, τ_0, σ_0) such that there are no real double root λ of $|P^0|(X_0, \tau_0, \sigma_0, \lambda) = 0$, then

$$R(X_0, \tau_0 - i\gamma, \sigma_0) \geqq O(\gamma^1) \quad \text{as } \gamma > 0 \text{ and } \gamma \to 0.$$

Futhermore if there is at least one real simple root $\lambda(X_0, \tau_0, \sigma_0)$, the zeros of $R(X, \tau, \sigma)$ in some neighbourhood $U(X_0, \tau_0, \sigma_0)$ is in the set $\{\gamma=0\}$.

β) i). Let (X_0, τ_0, σ_0) be the real point such that there is a real double root λ of $|P^0|(X_0, \tau_0, \sigma_0, \lambda) = 0$.
If $R(X_0, \tau_0, \sigma_0) = 0$, we assume that

$$|R(X_0, \tau_0 - i\gamma, \sigma_0)| \geqq O(\gamma^{\frac{1}{2}}) \quad \text{as } \gamma > 0 \text{ and } \gamma \to 0.$$

Then we can decompose R as follows: for any (X, τ, σ) in some neighbourhood $U(X_0, \sigma_0) \times U(\tau_0)$ with $\operatorname{Im} \tau = \gamma < 0$

$$R(X, \tau, \sigma) = c(X, \tau, \sigma)(\zeta^{\frac{1}{2}} - D(X, \sigma)).$$

Where $C(X, \tau, \sigma) \neq 0, \sqrt{1} = 1$ and ζ is defined as follows:

$$\zeta = \tau - \tau_0(X, \sigma),$$

$$\tau_0 = \tau_0(X_0, \sigma_0)$$

and $\zeta = 0$ implies that $|P^0|(X, \tau, \sigma, \lambda) = 0$ has the real double root $\lambda(X, \tau, \sigma)$.

ii). We assume that for any $(X', \sigma) \in U(X_0, \sigma_0) \cap \Gamma \times R^{n-1}$ and for some constant $C > 0$

$$- \operatorname{Re} D(X', \sigma) \geqq C(\operatorname{Im} D(X', \sigma))^2 \qquad \text{(a)}$$

or

$$\operatorname{Im} D(X', \sigma) \geqq C(\operatorname{Re} D(X', \sigma))^2 \qquad \text{(b)}$$

according as for $\tau_0 > 0$ the normal surface cutted by $X = X_0$ and $\sigma = \sigma_0$ is convex or concave with respect to τ at $(\tau_0, \lambda(X_0, \tau_0, \sigma_0))$ respectively.

γ). In the case β) ii) (a), if there is at least one real simple root λ of $|P^0|(X_0, \tau_0, \sigma_0, \lambda) = 0$, we assume that

the rank of the Hessian of $\underset{(x', \sigma)}{\operatorname{Re} D(X', \sigma)}$ at its zeros
$=$ the codim of $\operatorname{Re}(X', \sigma) = 0$ in R^{2n-1},

where we preassume that $\operatorname{Re} D(X', \sigma) = 0$ is regular.
In the case β) ii) (b) we replace $\operatorname{Re} D(X', \sigma)$ by $\operatorname{Im} D(X', \sigma)$ in the above condition.

Then we have the following

<u>Theorem.</u> Under the assumptions (I), (II) and (III) our problem (P, B, Ω) is L^2-well-posed. Furthermore for any $u \in H_1((-\infty, T) \times \Omega)$ with $\operatorname{supp} u \subset [0\ T] \times \Omega$ and for some $C > 0$ the following a priori estimate is valid:

$$C\|u\|_0((0, T) \times \Omega) \leqq \|Pu\|_0((0, T) \times \Omega) + |Bu|_{\frac{1}{2}}((0, T) \times \dot{\Omega}).$$

Here we use certain modification of Kreiss' consideration ([5]) with together our result about reflection coefficients ([1]). In particular we make use of the micro-localizations of the characterization for L^2-well-posed mixed problems of order two ([2]).

2. The outline of the proof of Theorem.

First of all we shall give the definitions of coupling coefficients for problem (P^0, B, Ω). Then for a fixed real point (X_0, τ_0, σ_0) from (I) we can find an infinitely differentialble and non-singular matrix $S(X, \tau, \sigma)$ of homogeneous order zero (with respect to (τ, σ)), defined in a neighbourhood $U(X_0) \times U(\tau_0, \sigma_0)$ with $\operatorname{Im} \tau \leqq 0$ such that

$$S^{-1} \cdot P^0 \cdot S = E\lambda - M(X, \tau, \sigma)$$

$$M = \begin{bmatrix} \lambda_I^+ & & & & \\ & \lambda_I^- & & & \\ & & M_{II}' & & \\ & & & M_{III}^+ & \\ & & & & M_{III}^- \end{bmatrix} .$$

Where, if it exists, $\lambda_I^{\pm} = \begin{pmatrix} \lambda_1^{\pm} & & \\ & \ddots & \\ & & \lambda_{\ell-1}^{\pm} \end{pmatrix}$ $(I=\{1,\cdots,\ell-1\})$

whose elements $\lambda_i^{\pm}$ $(i \in I)$ are real and continous,

$$M_{II}'(X, \tau, \sigma) = M_{II}(X, \zeta, \sigma)$$

$$= \begin{bmatrix} \lambda_{II}^+(X, 0, \sigma), & \Lambda_\gamma \\ 0 & , \lambda_{II}^-(X, 0, \sigma) \end{bmatrix} + \zeta \begin{bmatrix} P_{11}, & P_{12} \\ P_{21}, & P_{22} \end{bmatrix} ,$$

$\lambda_{II}^{\pm}(X, \zeta, \sigma)$ $(II=\{\ell\})$ are eigenvalues of M_{II}' with positive imaginary part and negative one, when $\operatorname{Im} \tau = \operatorname{Im} \zeta < 0$, repectively and it is a real double root at (X_0, τ_0, σ_0), if it is present. Here $\Lambda_\gamma = (|\tau|^2 + |\sigma|^2)$ for $\operatorname{Im} \tau = -\gamma$. Furthermore $M_{III}^{\pm}$ $(III=\{\ell+1,\cdots,p\})$ are $(p-\ell) \times (p-\ell)$ matrix whose eigenvalues $\lambda_i^{\pm}$ $(i \in III)$ have imaginary parts and negative ones at (X_0, τ_0, σ_0) respectively.

Let $S_{II}(X, \zeta, \sigma) = \begin{bmatrix} 1 & 0 \\ S_{21} & 1 \end{bmatrix}$, where it is homogeneous of order 0 and $S_{21}(X, \zeta, \sigma) = (\lambda_{II}^+(X, \zeta, \sigma) - \lambda_{II}^+(X, 0, \sigma) - \zeta P_{11}(X, \zeta, \sigma)) \cdot$

$$\cdot (\Lambda_\gamma + \zeta P_{12}(X, \zeta, \sigma))^{-1}.$$

Let

$$S_1 = \begin{bmatrix} E_I & & \\ & S_{II} & \\ & & E_{III} \end{bmatrix},$$ where E_I, E_{III} are $2(\ell-1) \times 2(\ell-1)$, $2(p-\ell) \times 2(p-\ell)$ identity matrices resptectively. Then

$$(S\cdot S_1)^{-1}\cdot P^0\cdot(S\cdot S_1)$$

$$= E\lambda - \begin{bmatrix} \lambda_I^+ & & & & \\ & \lambda_I^- & & & \\ & & \begin{bmatrix} \lambda_{II}^+ & \alpha \\ 0 & \lambda_{II}^- \end{bmatrix} & & \\ & & & M_{III}^+ & \\ & & & & M_{III}^- \end{bmatrix}.$$

Here $\alpha(X, \zeta, \sigma)$ is a function of homogeneous order 1 such that it is real for $\zeta \geqq 0$ and $\alpha(X, 0, \sigma) = \Lambda_\gamma$. Furthermore we remark that $S\cdot S_1$ is non-singular and continuous, but not infinitely differentiable, when $\mathrm{Im}\ \tau \leqq 0$.

Definition. Let S' be a matrix of homogeneous order 0 which is non-singular and continuous when $\mathrm{Im}\ \tau \leqq 0$. If $S'^{-1}\cdot P^0\cdot S'$ becomes the same form as described above, setting

$$B\cdot S' = (V_I^+, V_I^-, V_{II}^+, V_{II}^-, V_{III}^+, V_{III}^-),$$

we call the components b_{ij} of the matrix

$$(V_I^+, V_{II}^+, V_{III}^+)^{-1}\cdot(V_I^-, V_{II}^-, V_{III}^-)$$

coupling coefficients with respect to S' for problem (P^0, B, Ω) and the determinant of $(V_I^+, V_{II}^+, V_{III}^+)$ Lopatinsky's one.

Here we must remark that if S'' is a matrix such that it is non-singular and continuous when $\mathrm{Im}\,\tau < 0$ and $(S'')^{-1}\cdot P^0\cdot S''$ becomes a generalized diagonal form with $\alpha(X, \zeta, \sigma) \equiv 0$, then we may say that the corresponding components of the matrix are reflection coefficients with respect to S'' for problem (P^0, B, Ω). But we use here only the conception of coupling coefficients because it is more natural than that of the reflection coefficients above defined.

Now from (I) we see the following

Lemma 2.1. ([1],[2]) If the coefficients of P, B are constant and (P, B, Ω) is L^2-well-posed, then for any real (τ_0, σ_0) there exists a constant $C > 0$ such that for any $i,j = 1\cdots p$

$$|b_{ij}(\tau_0-i\gamma, \sigma_0)| \leqq C|\mathrm{Im}\,\lambda_i^+(\tau_0-i\gamma,\sigma_0)|^{\frac{1}{2}}\cdot|\mathrm{Im}\,\lambda_j^-(\tau_0-i\gamma,\sigma_0)|^{\frac{1}{2}}\cdot\gamma^{-1}$$

as $\gamma > 0$ and $\gamma \to 0$.

Applying the above lemma to our freezing problem, we see the following lemmas about micro-localized problems over a conical neighbourhood $U(X_0) \times U(\tau_0,\sigma_0)$ of fixed points (X_0, τ_0, σ_0) in order to obtain a priori estimate.
From (II) we have only to consider the cases α), β) in (III) where for some $c > 0$ $|\tau| < c\,|\sigma|$ and τ is real, because in the other cases and for any $X \in \Gamma$ $R(X, \tau, \sigma) \neq 0$ ([11]). Hereafter we assume always the conditions(I), (II) and (III).

A). First of all we consider the case where the set II is empty and let $I = \{1,\cdots\ell\}$. Then from (III) α) we see that for $X \in \Gamma \cap U(X_0)$, $(\tau, \sigma) \in U(\tau_0, \sigma_0)$

$$R(X, \tau, \sigma) = C_0(X, \tau, \sigma)(\tau-\nu(X, \sigma))$$

Where $C_0(X_0, \tau_0 \sigma_0) \neq 0$, $\mathrm{Im}\, \nu(X, \sigma) \geqq 0$ and they are infinitely differentiable.

Lemma 2.2. (i). There exist indeces $j, k \in \mathrm{III}$ such that for any $(X, \tau, \sigma) \in U(X_0) \times U(\tau_0, \sigma_0)$ with Im

the vectors $\{V_I^+, V_{\ell+1}^+, \cdots, V_{j-1}^+, V_k^-, V_{j+1}^+, \cdots, V_p^+\}$ are linearly independent.

Let $j \in \mathrm{III}$ in (i) be $\ell+1$, for simplicity, and donote by $L(\cdot)$ the linear space spaned by the vectors in the parentheses. Then for $(X, (\tau, \sigma) \in U(X_0) \times U(\tau_0, \sigma_0)$ with $\tau = \nu(X, \sigma)$

(ii). $V_{\ell+1}^+(X, \tau, \sigma) \in L(V_{\ell+2}^+, \cdots, V_p^+)$,

(iii). $V_j^-(X, \tau, \sigma) \in L(V_I^+, V_{\ell+2}^+, \cdots, V_p^+)$ for any $j \in I$

and

(iv). If the set I is not empty, $\nu(X, \sigma)$ is real.

Let $B^+(X, \tau, \sigma) = (V_I^+, V_{\mathrm{III}}^+)$.

Then

$$R(X, \tau, \sigma) = \det\ (V_I^+, V_{\mathrm{III}}^+) \quad \text{for} \quad X \in \Gamma.$$

From the above lemma we obtain

Lemma 2.3. (i). There exist smooth scalar functions $a_i(X, \tau, \sigma)$ of homogeneous order 0 and a smooth, p-vector valued function $a(X, \tau, \sigma)$ of homogeneous order -1 for $X \in \Gamma$ such that

$$V_{\ell+1}^+ = a_{\ell+2} V_{\ell+2}^+ + \cdots + a_p V_p^+ + (\tau - \nu(X, \sigma))a,$$

(ii). $C \equiv (V_I^+, a\Lambda_\gamma, V_{\ell+2}^+, \cdots, V_p^+)$ is non-singular

and

(iii). There exists a smooth $p \times \ell$ matrix $C_I(X, \tau, \sigma)$ of homogeneous order 0 such that $V_I^-(X, \tau, \sigma) = B^+ \cdot C_I$.

Then from Lemma 2.3 (i) and (ii) we see that

$$B^+ \cdot \left[\begin{array}{c|ccc} E_I^+ & & 0 & \\ \hline & 1 & & 0 \cdots 0 \\ & -a_{\ell+2} & & 1 \cdots 0 \\ & \vdots & & \ddots \\ 0 & -a_p & & 0 \cdots 1 \end{array}\right] = C \cdot \left[\begin{array}{ccc} E_I^+, & 0, & 0 \\ 0, & (\tau-\nu(X,\sigma))\Lambda_\gamma^{-1}, & 0 \\ 0, & 0, & E'^+ \end{array}\right].$$

Where E_I^+ and E'^+ are the (ℓ, ℓ) and $(p-\ell-1,\ p-\ell-1)$ identity matrix respectively.

Now let $|V|_{q,\gamma}^2$ be

$$\int_{R^n} (\gamma^2 + |\xi'|^2)^q \, |\widehat{e^{-\gamma x_0} V}(\xi')|^2 d\xi'$$

for functions or vector-valued functions V defined over Γ and for covectors ξ' of $X' \in \Gamma$. Then from the above equality we see the following

<u>Lemma 2.4.</u> There exist positive constant and γ_0 such that for any $\gamma \geq \gamma_0$ and $V^+(X') = {}^t({}^tV_I^+, {}^tV_{III}^+) \in C_0^\infty(R^n)$,

$$|\psi_1 \circ B^+ \cdot \psi_2 V^+|_{\frac{1}{2},\gamma} \geq C_I(\gamma\, |\psi_2 V_I^+|_{0,\gamma} + \gamma |\psi_2 V^+|_{-\frac{1}{2},\gamma})$$

$$\bmod\ (|V^+|_{-\frac{1}{2},\gamma}).$$

Where ψ_1, ψ_2 are smooth, homogeneous of order 0 such that their supports are contained in a sufficiently small complex conical neighbourhood $U(X_0) \times U(\tau_0, \sigma_0)$ and $\psi_1(X', \tau, \sigma) = 1$ on supp ψ_2.

Finally from Lemma 2.3. (iii), the above inequality and reducing our problem to the problem concerning only on the Γ ([4]), we obtain an a priori estimate for problem (P, B, Ω) in micro-local sense as the above estimate.

B). Next we shall treat the case where II is present. Here we use τ, σ, ζ and τ in symbolic notations as normalized ones, because all functions described below except P_{II} and M'_{II} are considered as ones of homogeneous order 0.

Let (X_0, τ_0, σ_0) be a pont satisfying the condition (III) β) in its neighbourhood $U(X_0) \times U(\tau_0, \sigma_0)$. Let $B \cdot S$ be

$$(V_I^+, V_I^-, V'_{II}, V''_{II}, V_{III}^+, V_{III}^-)$$

and change the variable τ by $\zeta = \tau - \tau_0(X, \sigma)$. Let $\eta = \sqrt{\zeta}$. Then from (III) β) i) we see that for

$$X \in \Gamma \cap U(X_0),\ (\tau, \sigma) \in U(\tau_0, \sigma_0) \quad \text{with} \quad \operatorname{Im} \tau \leqq 0$$

$$\begin{aligned} R(X, \tau, \sigma) &= |V_I^+, V_{II}^+, V_{III}^+|(X, \zeta, \sigma) \\ &= C(X, \eta, \sigma)(\eta - D(X, \sigma)), \end{aligned}$$

and form (III) β) ii) we see that if $R(X, \tau, \sigma) = 0$, then $\zeta = \eta = 0$, where $C(X, \eta, \sigma)$ and $D(X, \sigma)$ are both infinitely differentiable functions.

Let $Q(X, \ , \sigma)$ be $(a_{11} + a_{21} b_{\ell\ell})(a_{12} + a_{22} b_{\ell\ell})^{-1}$

Where $\begin{bmatrix} a_{11} & a_{12} \\ a_{21} & a_{22} \end{bmatrix} = S_{II}^{-1}$ and $b_{\ell\ell} = |V_I^+ \ V_{II}^- \ V_{III}^+| \cdot |V_I^+ \ V_{II}^+ \ V_{III}^+|^{-1}$.

Then $Q(X, \tau, \sigma) = |V_I^+, V_{II}', V_{III}^+| \cdot |V_I^+, V_{II}'', V_{III}^+|^{-1}$.

From (II), (III) β) and Lemma 2.1 we see the following

Lemma 2.5. For $(X, \zeta, \sigma) \in U(X_0) \cap \Gamma \times U(0, \sigma_0)$

i). $|V_I^+, V_{II}'', V_{III}^+| \ (X, \zeta, \sigma) \neq 0$,

ii). $V_{II}'(X, \zeta, \sigma) \in L(V_{III}^+)$,

iii). $V_I^-(X, \zeta, \sigma) \in L(V_I^+, V_{III}^+)$.

where in ii) and iii) we consider only real (X, ζ, σ) such that $\zeta = 0$ and $D(X', 0, \sigma) = 0$, i.e., $R(X, \zeta, \sigma) = 0$ and

iv). $V_{II}' - QV_{II}'' \in L(V_I^+, V_{III}^+)$.

Now let

$$(V_I^+, V_{II}'', V_{III}^+)^{-1}(V_I^-, V_{II}', V_{III}^-)(X, \zeta, \sigma)$$

$$= \begin{pmatrix} K_{II} & K_{I\,II} & K_{I\,III} \\ K_{II\,I} & K_{II\,II} & K_{II\,III} \\ K_{III\,I} & K_{III\,II} & K_{III\,III} \end{pmatrix} (X, \zeta, \sigma),$$

then $K_{II\,II}(X, \zeta, \sigma) = Q(X, \tau, \sigma)$.

To obtain an a priori estimate micro-locally in this case, we may only consider the problem $(\widetilde{S}^{-1}\cdot P\cdot\widetilde{S}, BS, \Omega)$ with the pseudo-differential operator $\widetilde{S}$ whose symbole is homogeneous of order 0 and certain extention of S. Then our boundary conditions become essentially, micro-locally the following form:

$$g - V_{III}^{-}U_{III}^{-} = (V_I^{+}, V_{II}^{''}, V_{III}^{+})(X', D')\begin{pmatrix} U_I^{+} + K_{II}\,U_I^{-} + K_{I\,II}\,U_{II}^{'}, \\ U_{II}^{''} + K_{II\,I}\,U_I^{-} + QU_{II}^{'}, \\ U_{III}^{+} + K_{III\,I}\,U_I^{-} + K_{III\,II}\,U_{II}^{'} \end{pmatrix}.$$

Where V_{III}^{-} and K are pseudo-differential operators whose symbols are $V_{III}^{-}(X', \zeta, \sigma)$ and $K(X', \zeta, \sigma)$ respectively and $u = {}^t({}^tU_I^{+}, {}^tU_I^{-}, U_{II}^{'}, U_{II}^{''}, {}^tU_{III}^{+}, {}^tU_{III}^{-})$.

Hereafter we consider only the case (a) because we can treat the case (b) analogously. Futhermore from Lemma 2.5 and (III) γ) we see the following

Lemma 2.6. Any components k of $K_{I\,II}(X', \zeta, \sigma)$ and $K_{II\,I}(X', \zeta, \sigma)$ are written in the following form in $U(X_0) \times U(\tau_0, \sigma_0)$

$$k(X', \zeta, \sigma) = k_0(X, \sigma) + \zeta k_1(X, \sigma) + O(|\zeta|^2)$$

and

$$|k(X, \sigma)|^2 \leqq C|\mathrm{Re}\, D(X, \sigma)|$$

for some positive C.

Thus we can reduce our problem to the following micro-local one where we consider only over the complex conical neighbourhood $U(X_0) \times U(\tau_0, \sigma_0)$ and $u = {}^t(u', u'') \in C_0^\infty(R^{n+1})$:

$$(P\ ,\ (Q,I),\ \Omega)\begin{cases} P_{II}\psi_2\, u = (D_{x_n} - \widetilde{M}_{II}^{'}(X, D'))\psi_2 u = f & x_n \geqq 0, \\ \psi_2 u'' + \widetilde{Q}(X', D')\psi_2 u' = g & x_n = 0. \end{cases}$$

Then we see that for real ζ, $\lambda_{II}(X, \zeta, \sigma)$ is real and letting $\lambda^{\pm}_{II}(X, \zeta, \sigma) = a(X, \zeta, \sigma) \pm \sqrt{\zeta} \cdot b(X, \zeta, \sigma)$, $\zeta H(X, \zeta, \sigma) = a(X, \zeta, \sigma) - \lambda^{+}_{II}(X, 0, \sigma)$, we obtain that for $\zeta = 0$

$$P_{21}(X, 0, \sigma) = b^2(X, 0, \sigma)$$

and

$$H(X, 0, \sigma) = \frac{1}{2}(P_{11}(X, 0, \sigma) + P_{22}(X, 0, \sigma)).$$

Considering a construction of symmetrizer, by Kreiss' Method, we see that if for real $\zeta \in U(0)$, there exist real functions d_1, d_2 of ζ such that

$$2d_1(-\mathrm{Re}Q(X, \tau, \sigma) + (P_{11} - P_{22})\tfrac{1}{2}\zeta(1 + \zeta P_{12})^{-1})$$

$$+ d_2(|Q|^2(X, \tau, \sigma) + P_{21}\zeta(1 + \zeta P_{12})^{-1}) \geqq 0,$$

then the following micro-lecal estimate holds: for some positive C_I

$$\|\psi_1 \circ (D_{x_n} - M'(X, D'))\psi_2 u\|_{0,\gamma} + |\psi_2 u'' + \psi_1 \circ Q(X', D')\psi_2 u'|_{\frac{1}{2},\gamma}$$

$$\geqq C_1 \gamma \|\psi_2 u\|_{0,\gamma}, \quad \mathrm{mod}\ (|u|_{-\frac{1}{2},\gamma} + \|u\|_{0,\gamma}).$$

Here and in the above problem ψ_1 and ψ_2 are the anologous functions as in Lemma 2.4. and the norne $\| \ \|_{q,\gamma}$ is defined for functions defined over $R^1 \times \Omega$ by the usual method as $| \ |_{q,r}$.

Moreover using Lemma 2.4, Lemma 2.6 and the method deriving the above micro-lecal estimate and gethering them, we see that for some positive constants C_k, γ_k and $u \in H_{k,\gamma}(R^{n+1}_+)$

$$\|Pu\|_{k,\gamma} + |Bu|_{k+\frac{1}{2},\gamma} \geqq C_k \gamma \|U\|_{k,\gamma}$$

for every $\gamma > \gamma_k$ and integer $k \geqq 0$.

To find d_1 and d_2, we remark that the Lopatinsky determinant $R_{II}(X, \tau, \sigma)$ for the problem $(P_{II}, (Q,I), \Omega)$ is written in the following form :

$$Q(X, \tau, \sigma) + (1+\zeta\cdot P_{12})^{-1} (\lambda^+_{II}(X, \zeta, \sigma) - \lambda^+_{II}(X, 0, \sigma)-\zeta\cdot P_{11})$$

$$= A(X, \eta , \sigma)(\eta - D(X,\sigma))$$

where A is a non-vanishing smooth function with respect to (X, η , σ). Setting $A = A (X, \zeta, \sigma) + \eta A_2(X, \zeta, \sigma)$, we see that for real ζ

$$Q = (1 + \zeta P_{12})^{-1}\left(\frac{1}{2}\zeta(P_{11}-P_{22})\right) - \{A_2 D + (1+\zeta\cdot P_{12})^{-1}b\} D + \zeta A_2.$$

Then form (III) β) ii), (a) and the above form of Q, we can find $d_1 \gg 0$, $d_{20} < 0$ and $d_{21} \gg 0$ such that for d_1, $d_2 = d_{20} + \zeta d_{21}$ it satisfies the desired inequality.

Such methods using symmetrizer are very technical, but the function $\sqrt{\zeta}$ is not any suitable symbol of pseudo-differential operators, so we can't use S', S'' and $R (X, \tau, \sigma)$ directly and we must emphasize that there are no other deeper theory and formal calculation which explain this simple technics.

Since a certain dual problem also satisfies our assumption (I), (II) and (III), the a priori estimate also valid for the dual problem and therefore our proof is finished ([6]).

Finally we remark that our conditions are invariant under transformations of Ω preserving the normal diretions at its boundary and hence our theorem is applicable to the problem (P, B, Ω) where Ω has the smooth bounded boundary.

3. Remarks.

A). In the case where p is a single equation of higher order, it is easily reduced to our system with pseudo-differential operators in order to get an a priori estimate and then we obtain the corresponding theorem concerning the existence theorem for solutions. For example we consider the case where the order of $p = 2$ and $B = 1$ or $B = D_{x_n} - c(X)D_{x_0} - \sum_{i=1}^{n-1} b_i(X)D_{x_i}$. If $c(X)$ and $b_i(X)$ are real, then the L^2 -well-posedness for freezing problem is equivalent to the conditions that

i). $R(X', \tau, \sigma) \neq 0$ for $\operatorname{Im} \tau < 0$ and

ii). in the case (III) β) ii), the $D(X', \sigma)$ is real

and

$$- D(X', \sigma) \geqq 0.$$

Furthermore the conditions i) and ii) for any $X' \in \Gamma$ implies that it is L^2-well-posed. In the case where the coefficients of B are complex the L^2-well-posedness with respect to freezing problems implies that $- \operatorname{Re} D(X', \sigma) \geqq 0$. Therefore there are certain gaps between (III) β) ii) (a) and the above condition.

B). The conditions (III) α) and β) mean that the $R(X, \tau, \sigma)$ has simple roots at (X_0, τ_0, σ_0) in certain sense and so it is the simplest case where Lopatinsky determinant is zero.

The condition γ) is necessary in constant coefficients case of second order. In fact for real $c(X)$ and $b_i(X)$ in A) and $p =$ the $D_{x_0}^2 - \Delta$, the D is real and

$$\begin{aligned} &\text{codim. of } D(X', \sigma) = 0 \text{ in } R^{2n-1} \\ &= \text{rank } \operatorname{Hess.}_{(X', \sigma)} D(X', \sigma) \text{ at its zeros} \\ &= \begin{cases} n - 2 & \text{for } c \neq 0 \ (> 0), \\ 0 & \text{for } c = 0 \ \text{(Neumann case).} \end{cases} \end{aligned}$$

So we see that they have quite different structures. ([3]).

C). Finally we remark about the #-condition for p and Γ. It seems to me to be interesting that the following extension of Ludwig's theorem ([8]) about folding caustics.

Let p be x_0-strictly hyperbolic single operator with the conditions (I) α) and β) with respect to Γ.

Let $\theta(X)$ be the function defined over some neighbourhood of $x_n = 0$ such that it is smooth and

$$p^0(X, \text{grad } \theta(X)) = 0 \quad \text{at } x_n = 0.$$

Then the bicharacteristic strips:

$$(y, s) \longrightarrow (X(y, s), \xi(y, s))$$

with respect to the Hamilton field H_{p^0};

$$\dot{x}_\nu = \frac{\partial p^0}{\partial \xi\nu}(X, \xi) \quad \text{and} \quad \dot{\xi}_\nu = -\frac{\partial p^0}{\partial x\nu}(X, \xi) \quad (\nu = 0, 1, \cdots, n)$$

with initial conditions

$$\begin{cases} x_\nu = y_\nu \quad (\nu = 0, 1, \cdots n-1), \\ x_n = 0, \\ \xi_\nu = \dfrac{\partial \theta}{\partial x_\nu} \quad (\nu = 0, \cdots, n) \quad \text{at } s = 0 \end{cases}$$

is defined a Lagrangian submanifold in $R^{n+1} \times R^{n+1}$ and $x_0 = 0$ is folding caustics if and only if

$$\text{i}). \quad \frac{\partial p^0}{\partial \xi_n}(X, \text{grad } \theta(X)) = 0$$

and

$$\text{ii}). \quad \frac{\partial}{\partial x_n}(p^0(X, \text{grad } \theta(X)) \neq 0$$

for $x_n = 0$.

Hereafter we consider only at some neighbourhood of some point X_0 with $x_{0,0} = x_{0,n} = 0$. Then we can construct such $\theta(X)$ using a surface eikonal equation with given data at $x_0 = x_n = 0$ under the condition #.

Lemma 3.1. $p(X, D)$ is decomposed in some conical neighbourhood $U(X_0) \times U(\pm(\operatorname{grad}_{X'} \theta)(X_0))$ such that

$$p(X, D) = q(X, D)\, p_2(X, D) + K$$

where $q(X, D)$ and $p_2(X, D)$ are polynomials with respect to D_{x_n} of order $2(p - 1)$ and 2 respectively and have coefficients whose symbols belong to $S_{1,0}(R^{n+1} \times R^n \backslash 0)$. Furthermore $p_2(X, D)$ and $\theta(X)$ satisfy the conditions 1) and 2) and $p_2^0(X, \operatorname{grad} \theta(X)) = 0$ at $x_n = 0$ (locally). Finally, K is a polymonial with respect to D_{x_n} whose coefficients have vanishing symbols.

Using Lemma 3.1 we obtain the following

Lemma 3.2. For a given formal sum $\tilde{g}_0(X, k) = \sum_{\alpha=0}^{\infty} \tilde{g}_0{}^{\alpha}(X)\, k^{-\alpha}$ at $x_n = 0$ with $\tilde{g}_0{}^{\alpha} \in C_0^{\infty}(U(X_0) \cap \Gamma)$, there exist its extension

$$g_0(X, k) = \sum_{\alpha=0}^{\infty} g_0{}^{\alpha}(X)\, k^{-\alpha}$$

and an other formal sum

$$g_1(X, k) = \sum_{\alpha=0}^{\infty} g_1{}^{\alpha}(X)\, k^{-\alpha}$$

with $g_i{}^{\alpha} \in C^{\infty}(U_1(X_0))$ $(\overline{U_1(X_0)} \subset U(X_0))$ such that the formal sum

$$u = e^{i\tilde{\theta}k}(g_0 V + (k^{-\frac{1}{3}} \cdot i) g_1 V^{(1)})$$

is a uniform asymptotic solution $p\, u = 0$ in $x_n \geq 0$. Where $\tilde{\theta}$ is the certain extension of θ such that for $x_n \geq 0$ $p_2^0(\frac{\partial}{\partial x_i} \tilde{\theta} \pm \sqrt{x_n}\, \delta_{ni}) = 0$ (eikonal equation),

and
$$V = V(k^{\frac{2}{3}} \cdot x_n) = \int e^{i(k^{\frac{2}{3}} \cdot x_n \cdot \theta - 3^{-1} \cdot \theta^3)} d\theta$$
$$V^{(1)} = V^{(1)}(k^{\frac{2}{3}} \cdot x_n)$$

after certain coordinate transformation preserving the surface $x_n = 0$.

Here we remark that for $x_n \geqq 0$ the corresponding transport equations, which are satisfied by

$$G_{\pm}^{\alpha}(X) = \frac{1}{2}(g_0^{\alpha} \pm \sqrt{x_n}\, g_1^{\alpha}) \quad (\alpha = 0, 1, 2, \cdots)$$

under given data at $x_n = 0$, are constructed only for the case where the order of $p = 2$. Therefore we must use the Lemma 3.1.

Finally we hope that our considerations will become certain basis of the investigations about propagations of singularities of solutions of the mixed problems near future in general theoretic point of view for partial differential operators.

Reference

[1] R. Agemi and T. Shirota: On necessary and sufficient conditions for L^2-well-posedness of mixed problems for hyperbolic equations, Jour. Fac Sci. Hokkaido Univ. Ser. I, Vol. 21, 133-151 (1970).

[2] R. Agemi and T. Shirota: On necessary and sufficient conditions for L^2-well-posedness of mixed problems for hyperbolic equations II, ibid, Vol. 22, 137-149 (1972).

[3] T. Okubo and T. Shirota: On structure of certain L^2-well-posed mixed problems for hyperbolic systems of first order (to appear)

[4] L. Hörmander: Pseudo-differential operators and non-elliptic boundary problems, Ann. of Math. 83, 129-2o9 (1966).

[5] H. O. Kreiss: Initial-boundary value problems for hyperbolic systems. Comm. Pure Appl. Math. 23, 277-298 (1970).

[6] K. kubota: Remarks on boundary value problems for hyperbolic equations. Hokkaido Math. J. Vol. II, No.2, 202-213 (1973).

[7] P. D. Lax and R. S. Phillips: Scattering theory, Academic Press, New York, 1967.

[8] D. Ludwig: Uniform asymptotic expansions at a caustic, Comm. Pure Appl. Math., Vol. 19, 215-250 (1966).
D. Ludwig: Uniform asymptotic expansions of the field scattered by a convex object at high frequencies, Comm. Pure Appl. Math., Vol. 20,103-138 (1967).

[9] R. Sakamoto: Mixed problem for hyperbolic equations I, II, J. Math. Kyoto Univ., 10, 349-373, 403-417 (1970).

[10] T. Shirota: On certain L^2-well-posed mixed problems for hyperbolic system of first order, Proc. Japan Acad., Vol. 50, No 2, 143-147 (1974).

[11] T. Shirota: On the propogation speed of hyperbolic operator with mixed boundary conditions, Jour. Fac. Sci., Hokkaido Univ., Ser. I, Vol. 22, 25-31 (1972).

Applications of Fourier distributions with complex phasefunctions

Johannes Sjöstrand

0. Introduction and statement of the results .

We shall present two applications of the calculus developped jointly with A. Melin [4] . The terminology and the notations will be as in that paper . Both the applications will treat the construction of solutions of a homogeneous pseudodifferential equation ; $P(x, D)u \equiv 0 \mod C^{\infty}$, with prescribed wavefront sets . This will be done in two cases ; (1) when P has a real bicharacteristic strip and (2) when P has a real "bicharacteristic leaf" of dimension 2 . In the case (1) the main result has essentially been given in [4 , section 8]. The proof here is different from the one sketched in [4] because we also want to cover the case (2) . Note that in the case (1) Hörmander [2] has studied the situation locally and he has also observed the global difficulty that will force us to introduce the condition (I) below . In the case (2) Duistermaat and Hörmander [1,Th. 7. 4. 1] have obtained a result under the additional assumption that $p(x,\xi) = 0 \Rightarrow \{p, \overline{p}\}(x,\xi) = 0$, where p is the principal symbol and $\{\ ,\ \}$ is the Poisson bracket .

Let X be a paracompact C^{∞} manifold of dimension n and let $P: \mathcal{D}'(X;\Omega_{1/2}) \longrightarrow \mathcal{D}'(X;\Omega_{1/2})$ be a properly supported classical pseudodifferential operator of order 1 , with principal symbol $p \in C^{\infty}(T^*X\setminus 0)$ positively homogeneous of degree 1 . We assume that there is a bicharacteristic strip :

(0.1) $\quad \mathbb{R} \ni t \longmapsto \gamma(t) \in T^*X\setminus 0$

with the following properties :

(0.2) $\quad \gamma'(t) = \mathcal{H}_p(\gamma(t)) \neq 0 \quad ,$

where $\gamma'(t)$ is the directional derivative of γ and $\mathcal{H}_p$ is the Hamilton field of p (which is consequently real at the points of γ).

(0.3) $$p \circ \gamma = 0 .$$

(0.4) The mapping $\mathbb{R} \ni t \mapsto \pi \circ \gamma(t) \in X$ is proper ,

where $\pi : T^*X \setminus 0 \to X$ is the natural projection . From (0.2)-(0.4) and the homogeneity of p it follows that

(0.5) $\frac{\partial}{\partial \lambda}$ and $\gamma'(t)$ are linearly independent ,

where $\frac{\partial}{\partial \lambda}$ is the vectorfield on $T^*X \setminus 0$ in the cone axis direction , uniquely determined by the property that $\frac{\partial}{\partial \lambda}(f) = f$ for all $f \in C^\omega(T^*X \setminus 0)$ which are positively homogeneous of degree 1 .

(0.6) $$\mathbb{R}_+ \gamma = \bigcup_{\lambda \in \mathbb{R}_+} \lambda \gamma$$ is a closed submanifold of $T^*X \setminus 0$,

where $\lambda\gamma = \{(x, \lambda\xi) ; (x,\xi) \in \gamma\}$ and γ is identified with its image in $T^*X \setminus 0$.

In order to construct a singular solution to the homogeneous equation $Pu \equiv 0 \mod. C^\infty$ we shall first construct a strictly positive Lagrangean manifold $\Lambda \subset \tilde{p}^{-1}(0)$ with $\Lambda_{\mathbb{R}} = \mathbb{R}_+ \gamma$. Here $\tilde{p}$ denotes an almost analytic extension to $\widetilde{T^*X \setminus 0}$ of p . For the construction of Λ we shall introduce the global necessary and sufficient condition (I) below . First some notations : Let $\mathcal{M}_k^m(\gamma, T^*X \setminus 0) \subset C^\infty(T^*X \setminus 0)$ be the subspace of functions which vanish on γ to order m and which are positively homogeneous of degree k . If E is a finite dimensional Euclidean space , we let $\mathcal{P}^m(E)$ be the space of complex homogeneous polynomials on E of degree m . Let $\mathcal{F} = T_\gamma(T^*X)/T_\gamma(\mathbb{R}_+\gamma)$ be the normal bundle of $\mathbb{R}_+\gamma$ restricted to γ and define the bundle $\mathcal{P}^m(\mathcal{F})$ over γ in the obvious way . Using Taylors formula we get a natural identification for all $m \in \mathbb{Z}^+ \cup \{0\}$:

$$(0.7) \quad \mathcal{M}_k^m(\gamma, T^*X\setminus 0)/\mathcal{M}_k^{m+1}(\gamma, T^*X\setminus 0) \simeq C^\infty(\gamma; \mathcal{P}^m(\mathcal{F})) .$$

(The right hand side is the space of C^∞ sections of $\mathcal{P}^m(\mathcal{F})$.) Especially for $m = k = 1$ we get

$$(0.8) \quad \mathcal{M}_1^1(\gamma, T^*X\setminus 0)/\mathcal{M}_1^2 \simeq C^\infty(\gamma; \widetilde{\mathcal{N}_\gamma^*}(\mathbb{R}_+\gamma, T^*X)) ,$$

where $\mathcal{N}_\gamma^* = \mathcal{N}_\gamma^*(\mathbb{R}_+\gamma, T^*X)$ is the conormal bundle of $\mathbb{R}_+\gamma$ restricted to γ and " $\sim$ " means that we have taken the complexification .

For every $\rho \in T^*X\setminus 0$ and $u \in \widetilde{T_\rho^*}(T^*X\setminus 0)$ we define $\mathcal{H}_u \in \widetilde{T_\rho}(T^*X\setminus 0)$ by the equation

$$(0.9) \quad \sigma(t, \mathcal{H}_u) = \langle t, u \rangle , \quad \forall\ t \in \widetilde{T_\rho}(T^*X\setminus 0) ,$$

where σ is the symplectic 2-form on $T^*X\setminus 0$, here considered as a bilinear form on $\widetilde{T_\rho}(T^*X\setminus 0)$. If $v \in C^\infty(T^*X\setminus 0)$ we put $\mathcal{H}_v = \mathcal{H}_{dv}$. We have an injective map :

$$(0.10) \quad C^\infty(\gamma; \widetilde{\mathcal{N}_\gamma^*}(\gamma, T^*X\setminus 0)) \ni u \longmapsto \mathcal{H}_u \in C^\infty(\gamma; \widetilde{T_\gamma}(T^*X\setminus 0)) .$$

The image consists of the sections which are everywhere symplectically orthogonal to $\mathbb{R}_+\gamma$.

The differential operator $\mathcal{H}_p$ clearly leaves the spaces $\mathcal{M}_k^m(\gamma, T^*X\setminus 0)$ invariant since $\mathcal{H}_p$ is tangential to γ . In particular from (0.8) we see that $\mathcal{H}_p$ induces a first order differential operator , also denoted by $\mathcal{H}_p$:

$$\mathcal{H}_p : C^\infty(\gamma; \widetilde{\mathcal{N}_\gamma^*}) \longrightarrow C^\infty(\gamma; \widetilde{\mathcal{N}_\gamma^*}) .$$

We now introduce the following condition :

(I) There exist $u_2, \ldots, u_{n-1} \in C^\infty(\gamma; \tilde{\mathcal{N}}_\gamma^*(\mathbb{R}_+\gamma, T^*X))$ such that $\mathcal{H}_p, \mathcal{H}_{u_2}, \ldots, \mathcal{H}_{u_{n-1}}, \widetilde{\frac{\partial}{\partial\lambda}}$ span a positive semidefinite Lagrangean plane $\Lambda_\rho \subset T_\rho(T^*X \setminus 0)$ at every point $\rho \in \gamma$, with $\Lambda_{\rho\,\mathbb{R}} = T_\rho(\mathbb{R}_+\gamma)$, and such that $\mathcal{H}_p(u_j) = 0$ for all j.

Proposition 1. (I) is a necessary and sufficient condition for the existence of a conic, strictly positive Lagrangean manifold $\Lambda \subset \tilde{p}^{-1}(0)$ with $\Lambda_{\mathbb{R}} = \mathbb{R}_+\gamma$.

Theorem 2. Suppose that (I) is satisfied and let Λ be as in Proposition 1. Then there exists $u \in I_c^0(X, \Lambda)$ with non-vanishing principal symbol, such that $Pu \in C^\infty(X; \Omega_{1/2})$. In particular $WF(u) = \mathbb{R}_+\gamma$.

The condition (I) is of course of a very implicit nature. However it is always possible to construct the sections u_j locally (and even globally if we drop the condition that the Lagrangean planes Λ_ρ are positive semidefinite with $\Lambda_{\rho\,\mathbb{R}} = T_\rho(\mathbb{R}_+\gamma)$). Therefore we can always get microlocal versions of Theorem 2. Such a result has been sketched by Hörmander [2].

Proposition 3. Let P be as above and suppose that there exists $t_0 \in \mathbb{R}$ such that

$$(\text{Hess Im } p)_{\gamma(t)} \begin{cases} \leq 0 & \text{for } t \geq t_0 \\ \geq 0 & \text{for } t \leq t_0 \end{cases}.$$

Then the condition (I) is satisfied.

In particular Theorem 2 covers the case when p is real valued (c.f. Duistermaat-Hörmander [1, Th.6.2.1]).

For the second application we assume that P, p, X are as above. Instead of the existence of γ we assume that there exists a closed connected non compact 2-dimensional submanifold $\Gamma \subset T^*X\setminus 0$, countable at infinity, with the following properties.

(0.11) $\mathcal{H}_{\mathrm{Re}\, p}$, $\mathcal{H}_{\mathrm{Im}\, p}$ span $T(\Gamma)$ at every point.

(0.12) $p = 0$ on Γ.

(0.13) The natural projection $\pi : \Gamma \longrightarrow X$ is proper.

This implies that

(0.14) $\frac{\partial}{\partial\lambda} \notin T_\rho(\Gamma)$ for all $\rho \in \Gamma$,

and that

(0.15) $\mathbb{R}_+\Gamma = \bigcup_{\lambda\in\mathbb{R}_+} \lambda\Gamma$ is a closed submanifold of $T^*X\setminus 0$.

As above we define the spaces $\mathcal{M}^m_k(\Gamma, T^*X\setminus 0)$ and the bundle $\mathcal{N}^*_\Gamma(\mathbb{R}_+\Gamma, T^*X\setminus 0)$. Instead of condition (I) we introduce

(II) There exist $u_3, \ldots, u_{n-1} \in C^\infty(\Gamma; \tilde{\mathcal{N}}^*_\Gamma)$ such that $\mathcal{H}_p(u_j) = 0$ for all j and such that $\mathcal{H}_p, \mathcal{H}_{\bar{p}}, \mathcal{H}_{u_3}, \ldots, \mathcal{H}_{u_{n-1}}, \frac{\partial}{\partial\lambda}$ span a positive semidefinite Lagrangean plane $\Lambda_\rho \subset \tilde{T}_\rho(T^*X\setminus 0)$ at every point $\rho \in \Gamma$, with $\Lambda_{\rho\,\mathbb{R}} = T_\rho(\mathbb{R}_+\Gamma)$.

Proposition 4. (II) is a sufficient condition for the existence of a strictly positive conic Lagrangean manifold $\Lambda \subset \tilde{p}^{-1}(0)$ with $\Lambda_{\mathbb{R}} = \mathbb{R}_+\Gamma$. When Γ is simply connected, (II) is also a necessary condition.

In the case when Γ is not simply connected we think that (II) should be replaced by some weaker condition, giving just the tangent space Λ_ξ at each point, but not a particular basis. We also point out that (II) is a global condition (so there are always microlocal versions of Theorem 5 below).

Theorem 5. Suppose that Λ is as in the preceding proposition. Then there exists $u \in I^0_c(X, \Lambda)$ with principal symbol vanishing only on a discrete set of rays in $\mathbb{R}_+\Gamma$, such that $Pu \in C^\infty(X; \Omega_{1/2})$. In particular $WF(u) = \mathbb{R}_+\Gamma$.

Proposition 6. Suppose that Γ is simply connected and that the Poisson bracket $\{p, \bar{p}\} = \mathcal{H}_p(\bar{p})$ belongs to $\mathcal{M}^3_1(\Gamma, T^*X\setminus 0)$. Then (II) is satisfied.

It would of course be interesting to be able to analyze the condition (II) more than what we will do here. Note that Duistermaat - Hörmander [1, Th. 7.4.1] have obtained a result similar to Theorem 5 in the case when $\{p, \bar{p}\}$ vanishes at $p^{-1}(0)$.

The plan of the paper will be the following. In section 1 we make some simple remarks about almost analytic vector fields. In section 2 we establish the surjectivity for certain Cauchy-Riemann type operators on fiberbundles, which is a simple consequence of the results of Malgrange [3, chapter 3]. In section 3 we make the geometric constructions. Essentially we shall concentrate on the case of condition (II) since all the corresponding work in the case of condition (I) is the same or simpler. In section 4 we prove Theorems 2, 5.

I would like to thank J. Chazarain who has patiently helped me to check most of the technical details in the proofs.

1. Almost analytic vectorfields .

Let $\omega \subset \mathbb{C}^n$ be open and let ν be a complex C^∞ vectorfield on ω. By definition we say that ν is almost analytic if $\nu(f)$ is almost analytic and $\nu(\bar{f}) \sim 0$ for all almost analytic functions f on ω. (Here ν is considered as a differential operator .) If ν and μ are vectorfields in ω we write $\nu \sim \mu$ if the corresponding relations hold between the coefficients . Clearly ν is almost analytic if and only if

$$(1.1) \qquad \nu \sim \sum_1^n a_j(z)\frac{\partial}{\partial z_j} ,$$

where the a_j are almost analytic . (Just apply ν to the functions z_j and $\bar{z}_j$.) To every almost analytic vectorfield ν of the form (1.1) we associate a real vectorfield $\hat{\nu}$ given (up to equivalence) by

$$(1.2) \qquad \hat{\nu} \sim \sum_1^n (\mathrm{Re}\, a_j(z))\frac{\partial}{\partial x_j} + (\mathrm{Im}\, a_j(z))\frac{\partial}{\partial y_j} .$$

Then

$$(1.3) \qquad \hat{\nu}(f) \sim \nu(f) \quad \text{for all almost analytic functions } f .$$

The property (1.3) determines $\hat{\nu}$ up to equivalence because if μ is a real vector field in ω and $\mu(f) \sim 0$ for all almost analytic functions f, it follows that $\mu \sim 0$ by taking $f(z) = z_j$ for $j = 1, \ldots, n$.

As an immediate consequence of (1.3) we have

<u>Lemma 1.1.</u> If ν, μ are almost analytic vectorfields , we have the relation $\widehat{[\nu, \mu]} \sim [\hat{\nu}, \hat{\mu}]$ between the commutators .

Note that if ν is almost analytic and $g(z)$ is an almost analytic function then $\widehat{g(z)\nu} \sim \check{g}(z)(\hat{\nu})$, where $\hat{\nu}$ at the right hand side is considered as a section in $T_\omega(\mathbb{C}^n)$ and $\check{g}(z) : T_z(\mathbb{C}^n) \longrightarrow T_z(\mathbb{C}^n)$ denotes the natural multiplication with the complex number $g(z)$.

Let ν be an almost analytic vectorfield in ω . Then for $s \in \mathbb{C}$ we can define the flow $\Psi_{s,t}$ of $\widehat{s\nu}$ as the local diffeomorphism obtained by integrating this field from 0 to $t \in \mathbb{R}$. We then get a mapping $\omega \ni z \longmapsto \Psi_{s,t}(z)$, defined for $s \in \mathbb{C}$, $t \in \mathbb{R}$ when $|st| < s(z)$. Here $s(z)$ is a positive lower half continuous function on ω . We put

$$\Phi_s(z) = \Psi_{s,1}(z) \quad , \quad z \in \omega \quad , \quad s \in \mathbb{C} \quad , \quad |s| < s(z) \quad ,$$

and call this the flow of ν .

<u>Lemma 1.2.</u> For all $N \in \mathbb{Z}^+$, $\omega' \subset\subset \omega$, $0 < s_0 < \inf_{z \in \omega'} s(z)$, there is a constant C such that

$$\left| \frac{\partial}{\partial \bar{s}} \Phi_s(z) \right| + \left| \frac{\partial}{\partial \bar{z}} \Phi_s(z) \right| \leq C |s| \sup_{0 \leq t \leq 1} |\operatorname{Im} \Phi_{ts}(z)|^N$$

when $z \in \omega'$, $s \in \mathbb{C}$, $|s| \leq s_0$.

<u>Proof.</u> Let ν be of the form (1.1) and put $a(z) = (a_1(z), \ldots, a_n(z))$. Then by definition we have

$$(1.4) \qquad \begin{cases} \frac{\partial}{\partial t} \Phi_{ts}(z) = s\, a(\Phi_{ts}(z)) \quad , \quad 0 \leq t \leq 1 \\ \Phi_0(z) = z \end{cases}$$

Applying $\frac{\partial}{\partial \bar{s}}$ we get

$$(1.5) \qquad \begin{cases} \frac{\partial}{\partial t}\left(\frac{\partial}{\partial \bar{s}} \Phi_{ts}(z)\right) - s \frac{\partial a}{\partial z}(\Phi_{ts}(z))\left(\frac{\partial}{\partial \bar{s}} \Phi_{ts}(z)\right) = s \frac{\partial a}{\partial \bar{z}}(\Phi_{ts}(z))\overline{\left(\frac{\partial}{\partial s} \Phi_{ts}(z)\right)} \\ \left(\frac{\partial}{\partial \bar{s}} \Phi_{ts}(z)\right)_{t=0} = 0 \quad . \end{cases}$$

This system for $\frac{\partial}{\partial \bar{s}} \Phi_{ts}$ gives the desired estimate for $\frac{\partial}{\partial \bar{s}} \Phi_s$ if we remember that a is almost analytic . Similarly we can estimate $\frac{\partial}{\partial \bar{z}} \Phi_s$.

This lemma will serve when we want to construct an almost analytic manifold as the flow out of an almost analytic manifold along an almost analytic vectorfield .

We note that modulo equivalence there is a natural bijection between the set of almost analytic vectorfields on ω and the set of complex C^∞ vectorfields on $\omega_{\mathbb{R}}$. Similarly we can identify the complex C^∞ differential p-forms on $\omega_{\mathbb{R}}$ with the almost analytic p-forms on ω , which by definition are sums of terms of the form $a(z)\, d a_1(z) \wedge \ldots \wedge d a_p(z)$, where a , a_j are almost analytic functions . Let ν be an almost analytic vectorfield and let σ be an almost analytic p form . Then we define the Lie derivative $\mathcal{L}_\nu \sigma$ as $\mathcal{L}_{\hat{\nu}} \sigma$; the Lie derivative along the the real field $\hat{\nu}$. It is trivial that if ν' (σ') is the corresponding vectorfield (p-form) on $\omega_{\mathbb{R}}$, then $\mathcal{L}_\nu \sigma$ is the almost analytic extension of $\mathcal{L}_{\nu'} \sigma'$.

Finally we note that the notions above can easily be extended to the case when ω is replaced be the complexification $\tilde{M}$ of a real manifold M . The only possible exception is the definition of the flow Φ_s , but that will not give any difficulties in the particular situation we shall consider below .

2. Some technical preparations .

Let Γ be a connected non-compact C^∞ manifold , countable at infinity of dimension 2 or 1 . Let $\mathcal{E}$ be an N-dimensional complex C^∞ vectorbundle over Γ and assume that we have a first order differential operator $\mathcal{L} : C^\infty(\Gamma ; \mathcal{E}) \longrightarrow C^\infty(\Gamma ; \mathcal{E})$, which for every local trivialization of $\mathcal{E}$ takes the form

$$(2.1) \qquad \mathcal{L} = \begin{pmatrix} L & 0 & \dots & 0 \\ 0 & L & & \vdots \\ \vdots & & & L \\ 0 & \dots & & \end{pmatrix} + \begin{pmatrix} & \vdots & \\ \dots & b_{jk}(x) & \dots \\ & \vdots & \end{pmatrix} ,$$

where L is an elliptic (complex) vectorfield on Γ and the b_{jk} are C^∞ functions . Clearly L is independent of the choice of local trivialization of $\mathcal{E}$ and we have $\mathcal{L}(u\,v) = (Lu)\,v + u\,(\mathcal{L}v)$ for all $u \in C^\infty(\Gamma)$, $v \in C^\infty(\Gamma;\mathcal{E})$. Note that L gives an orientation of Γ . The following result is trivial when $\dim(\Gamma) = 1$ and is an easy consequence of Malgrange [3 , chapter 3] when $\dim(\Gamma) = 2$.

Theorem 2.1. Let $\mathcal{L}$ and Γ be as above . Then

(a) $\mathcal{L}$ is surjective $C^\infty(\Gamma;\mathcal{E}) \longrightarrow C^\infty(\Gamma;\mathcal{E})$,

(b) For any open set $\omega \subset \Gamma$, put $\mathcal{N}_{\mathcal{L}}(\omega;\mathcal{E}) = \{u \in C^\infty(\omega;\mathcal{E})\,;\,\mathcal{L}u=0\}$. Then $\mathcal{N}_{\mathcal{L}}(\Gamma;\mathcal{E})$ is dense in $\mathcal{N}_{\mathcal{L}}(\omega;\mathcal{E})$ (by restriction) if and only if $\Gamma \setminus \omega$ has no compact components .

Proof. We can assume that $\dim(\Gamma) = 2$. Let ${}^t\mathcal{L} : C^\infty(\Gamma;\mathcal{E}') \longrightarrow C^\infty(\Gamma;\mathcal{E}')$ be the adjoint , defined as in Malgrange [3] . Let $\omega \subset \Gamma$ be open , connected and put $\mathcal{N}_\omega = \{u \in C^\infty(\omega;\mathcal{E}')\,;\,{}^t\mathcal{L}\,u = 0\}$. By the results of [3] it suffices to prove that if $u \in \mathcal{N}_\omega$ vanishes on an open subset of ω then $u = 0$. Now ${}^t\mathcal{L}$ is still of the form (2.1) with with L replaced by $-L$ and as in Duistermaat-Hörmander [1] (see also the proof of Lemma 3.2) we can find local coordinates x_1 , x_2 and a C^∞ function $\mu(x)$ such that $-L = e^{-\mu(x)}\dfrac{\partial}{\partial \bar z}$ if $z = x_1 + i\,x_2$. Put $\mathcal{L}' = e^{\mu(x)}\,{}^t\mathcal{L}$ and choose some local trivialization of $\mathcal{E}'$.

Lemma 2.2. Locally we can find a C^{∞} matrix $\mathcal{B}(z)$ with norm $< 1/2$ such that

$$(2.2) \qquad (I + \mathcal{B}(z))\,\mathcal{L}' = \frac{\partial}{\partial \bar{z}} \circ (I + \mathcal{B}(z)) \quad .$$

Proof of the Lemma. We work near the origin in $\mathbb{C}$. We have $\mathcal{L}' = \frac{\partial}{\partial \bar{z}} + \mathcal{A}(z)$ for some C^{∞} matrix $\mathcal{A}$. Let $\chi_{\varepsilon} \in C^{\infty}(\mathbb{C})$ be equal to 1 near the origin with support in $|z| < \varepsilon$ and such that $|\chi_{\varepsilon}(z)| < 1$. Let $K_{\varepsilon}(z) = (\pi z)^{-1}$ for $|z| < 2\varepsilon$ and $= 0$ for $|z| \geq 2\varepsilon$, so that $K_{\varepsilon} \in L^1(\mathbb{C})$ and $\|K_{\varepsilon}\|_{L^1} \leq (4\,\|\chi_{\varepsilon}\mathcal{A}\|_{L^{\infty}})^{-1}$ if $\varepsilon > 0$ is small enough. Define the matrices $\mathcal{B}_{\nu}$ succesively by convolution: $\mathcal{B}_0 = I$, $\mathcal{B}_1 = K_{\varepsilon} * (\chi_{\varepsilon}\mathcal{A})$, .., $\mathcal{B}_{\nu+1} = K_{\varepsilon} * (\mathcal{B}_{\nu}\chi_{\varepsilon}\mathcal{A})$. Then $\|\mathcal{B}_{\nu}\|_{L^{\infty}} \leq 4^{-\nu}$ and

$$(\nu) \qquad \left[\frac{\partial}{\partial \bar{z}}, \mathcal{B}_{\nu+1}\right] = \frac{\partial}{\partial \bar{z}}(\mathcal{B}_{\nu+1}) = \mathcal{B}_{\nu}\mathcal{A} \quad ,$$

in a neighbourhood of the origin, independent of ν. If $\mathcal{B} = \sum_{1}^{\infty} \mathcal{B}_{\nu}$ we get by adding the equations (ν) that

$$(2.3) \qquad \left[\frac{\partial}{\partial \bar{z}}, \mathcal{B}\right] = (I + \mathcal{B})\mathcal{A}$$

near the origin and this is equivalent to (2.2). Since (2.3) is an elliptic equation for $\mathcal{B}$ it is clear that $\mathcal{B}$ is C^{∞} and the lemma follows.

Now suppose that $u \in \mathcal{N}_{\omega}$, where ω is open, connected, and that u vanishes on an open subset. Then locally $(I + \mathcal{B}(z))\,u$ is holomorphic for suitable local coordinates and suitable small matrix $\mathcal{B}$, so it follows that $\mathcal{U} = 0$. This proves Theorem 2.1.

Now let M be a real paracompact C^{∞} manifold and let $\Lambda \subset \tilde{M}$ be an almost analytic manifold. By definition an (N-dimensional) almost analytic vectorbundle $\mathcal{E}$ over Λ is given by a covering of Λ by open subsets U_j and almost analytic transition functions G_{jk}, defined in $U_j \cap U_k$ with values in $Gl(N, \mathbb{C})$ such that $G_{jj} \sim id$, $G_{jk} G_{kl} \sim G_{jl}$. We do not enter into the details. In the following we shall denote by $C^{aa}(\Lambda)$ the space of almost analytic functions on Λ and by $C^{aa}(\Lambda ; \mathcal{E})$ the space of almost analytic sections of $\mathcal{E}$. Note that when $\Lambda = \tilde{M}$ then $\mathcal{E}$ can be identified with the complex vectorbundle $\mathcal{E}|_M$ over M. This is an immediate consequence of the identification $C^{aa}(M) \simeq C^{\infty}(M)$.

Consider the following situation. M and Λ are as above. Assume that $\Gamma = \Lambda_{\mathbb{R}}$ is a closed connected non-compact 2-dimensional submanifold of M which is countable at infinity. Moreover, assume that Λ intersects M in a " proper way " so that we locally have an inequality of the form

$$(2.4) \qquad d(z, \Gamma) \leq C \, |\operatorname{Im} z| \quad , \quad z \in \Lambda \quad ,$$

where $d(z, \Gamma)$ is the distance from z to Γ. Let L be an almost analytic vectorfield on $\tilde{M}$ which is tangential to Λ and $\tilde{\Gamma}$; the almost analytic complexification of Γ. By defintion, this means that the corresponding real field $\hat{L}$ is tangential to Λ and $\tilde{\Gamma}$ to infinite order at the real points. By restriction to M, L is a complex vectorfield which is tangential to Γ. We assume that L is elliptic as a differential operator on Γ. Note that L gives a differential operator : $C^{aa}(\Lambda) \rightarrow C^{aa}(\Lambda)$.

Now let $\mathcal{E}$ be an almost analytic vectorbundle over Λ and let

$\mathcal{L} : C^{aa}(\Lambda ; \mathcal{E}) \longrightarrow C^{aa}(\Lambda ; \mathcal{E})$ be a differential operator taking the form (2.1) (with almost analytic functions b_{jk}) for every local trivialization of $\mathcal{E}$. Then $\mathcal{L}(u\,v) = u\,(\mathcal{L}v) + (Lu)\,v$, $u \in C^{aa}(\Lambda)$, $v \in C^{aa}(\Lambda ; \mathcal{E})$. Our purpose is to extend Theorem 2.1 to this situation .

If E is a complex finite -dimensional vectorspace we denote by $\mathcal{P}^m(E)$, $m \in \mathbf{Z}^+ \cup \{0\}$, the space of homogeneous holomorphic polynomials on E of degree m . Let $\mathcal{F} = T_{\tilde{\Gamma}}(\Lambda) / T(\tilde{\Gamma})$ be the normal bundle of $\tilde{\Gamma}$ in Λ and define the almost analytic bundle $\mathcal{P}^m(\mathcal{F})$ over $\tilde{\Gamma}$ in the obvious way . Let $\mathcal{M}^m(\tilde{\Gamma}, \Lambda ; \mathcal{E}) \subset C^{aa}(\Lambda ; \mathcal{E})$ be the space of sections vanishing to order m at $\tilde{\Gamma}$.

Lemma 2.3. For all $m \in \mathbf{Z}^+ \cup \{0\}$ we have natural isomorphisms :

$$\mathcal{M}^m(\tilde{\Gamma}, \Lambda ; \mathcal{E}) / \mathcal{M}^{m+1}(\tilde{\Gamma}, \Lambda ; \mathcal{E}) \simeq C^{aa}(\tilde{\Gamma} ; \mathcal{E}|_{\tilde{\Gamma}} \otimes \mathcal{P}^m(\mathcal{F})) . \tag{2.5}$$

Proof. We consider first the situation locally and we choose local representatives for all the almost analytic objects . Let $u \in \mathcal{M}^m(\tilde{\Gamma}, \Lambda ; \mathcal{E})$ (where now Λ and $\tilde{\Gamma}$ are concrete manifolds) and let ν be a real vectorfield on Λ . If $x \in \tilde{\Gamma}$ we see that $(\nu^m u)(x)$ is defined without ambiguity and gives a homogeneous polynomial in t with values in $\mathcal{E}_x$ (= the fiber of $\mathcal{E}$) if $t \in \mathcal{F}_x$ is the image of ν at x . ($\mathcal{F}_x = T_x(\Lambda)/T_x(\tilde{\Gamma})$). This local consideration globalizes and gives a map

$$\mathcal{M}^m(\tilde{\Gamma}, \Lambda ; \mathcal{E}) \ni u \longmapsto \hat{u} \in C^{aa}(\tilde{\Gamma} ; \mathcal{E}|_{\tilde{\Gamma}} \otimes \mathcal{P}^m(\mathcal{F})) \tag{2.6}$$

with kernel $\mathcal{M}^{m+1}(\tilde{\Gamma}, \Lambda ; \mathcal{E})$.

To prove the surjectivity of this map we again consider the situation locally and choose local representatives . It is easy to find almost analytic functions $z_1, \dots, z_n$ (locally) on Λ with linearly independent differentials so that $\tilde{\Gamma}$ is given by $z_3 = z_4 = \dots = z_n = 0$. (Here $n = \dim_{\mathbb{C}} \Lambda$.) Let $\lambda : C^{aa}(\tilde{\Gamma}; \mathcal{E}|_{\tilde{\Gamma}}) \longrightarrow C^{aa}(\Lambda; \mathcal{E})$ be a (local) right inverse of the restriction map $\gamma : C^{aa}(\Lambda; \mathcal{E}) \longrightarrow C^{aa}(\tilde{\Gamma}; \mathcal{E}|_{\tilde{\Gamma}})$. Let $v \in C^{aa}(\tilde{\Gamma}; \mathcal{E}|_{\tilde{\Gamma}} \otimes \mathcal{P}^m(\mathcal{F}))$. $\mathcal{P}^m(\mathcal{F})$ is naturally trivialized by the choice of $z_1, \dots, z_n$ and v can therefore be considered as a polynomial $\sum_{|\alpha|=m} a_\alpha(z') z''^{\alpha}$, $z' = (z_1, z_2)$, $z'' = (z_3, \dots, z_n)$ where $a_\alpha \in C^{aa}(\tilde{\Gamma}; \mathcal{E}|_{\tilde{\Gamma}})$. Put $u = \sum_{|\alpha|=m} z''^{\alpha} \lambda(a_\alpha)/m! \in C^{aa}(\Lambda; \mathcal{E})$. Then $\hat{u} = v$ and by an almost analytic partition of unity this construction can be globalized to prove the surjectivity of (2.6) . The lemma is proved.

Note that we also have a natural identification

$$(2.7) \qquad C^{aa}(\tilde{\Gamma}; \mathcal{E}|_{\tilde{\Gamma}} \otimes \mathcal{P}^m(\mathcal{F})) \simeq C^{\infty}(\Gamma; \mathcal{E}|_{\Gamma} \otimes \mathcal{P}^m(\mathcal{F})|_{\Gamma}) .$$

Since L is tangential to $\tilde{\Gamma}$, $\mathcal{L}$ induces an operator $\mathcal{L}_m$ from $\mathcal{M}^m(\tilde{\Gamma}, \Lambda; \mathcal{E}) / \mathcal{M}^{m+1}(\tilde{\Gamma}, \Lambda; \mathcal{E})$ into itself and by (2.6) , (2.7) we can also consider $\mathcal{L}_m$ as a differential operator from $C^{\infty}(\Gamma; \mathcal{E}|_{\Gamma} \otimes \mathcal{P}^m(\mathcal{F})|_{\Gamma})$ into itself , which is easily seen to satisfy the assumptions of Theorem 2.1 . Thus

$$\mathcal{L}_m : \mathcal{M}^m(\tilde{\Gamma}, \Lambda; \mathcal{E}) / \mathcal{M}^{m+1} \longrightarrow \mathcal{M}^m / \mathcal{M}^{m+1}$$

is surjective .

Now let $v \in C^{aa}(\Lambda; \mathcal{E})$. We want to find $u \in C^{aa}(\Lambda; \mathcal{E})$ such that

$$(2.8) \qquad \mathcal{L} u = v \quad .$$

By induction we assume that we have already found $u_N \in C^{aa}(\Lambda ; \mathcal{E})$ such that

$$\mathcal{L} u_N \equiv v \quad \text{mod.} \ \mathcal{M}^N(\tilde{\Gamma}, \Lambda ; \mathcal{E}) .$$

Then by the surjectivity of $\mathcal{L}_N$ there is $w_N \in \mathcal{M}^N(\tilde{\Gamma}, \Lambda ; \mathcal{E})$ such that $\mathcal{L} w_N \equiv v - \mathcal{L} u_N \mod. \mathcal{M}^{N+1}$. Thus if $u_{N+1} = u_N + w_N$, we have

$$\mathcal{L} u_{N+1} \equiv v \quad \text{mod.} \ \mathcal{M}^{N+1}(\tilde{\Gamma}, \Lambda ; \mathcal{E}) .$$

Now by the property (2.4) we can find $u \in C^{aa}(\Lambda ; \mathcal{E})$ such that $u \sim \lim_{N \to \infty} u_N$ in the sence that $u - u_N \in \mathcal{M}^{N+1}(\tilde{\Gamma}, \Lambda ; \mathcal{E})$ for all N. Clearly $\mathcal{L} u = v$ and we have therefore proved the first part of the following theorem :

Theorem 2.4. Let $\mathcal{L}$, Λ, $\tilde{\Gamma}$, $\mathcal{E}$ be as above . Then

(a) $\mathcal{L} : C^{aa}(\Lambda ; \mathcal{E}) \longrightarrow C^{aa}(\Lambda ; \mathcal{E})$ is surjective .

(b) If $\omega \subset \Gamma$ is open and $\Gamma \setminus \omega$ has no compact components , then every $u \in C^{\infty}(\omega ; \mathcal{E}|_{\omega})$, such that $\mathcal{L} u = 0$, can be approximated in the C^{∞} topology by restrictions to ω of sections $u' \in C^{aa}(\Lambda ; \mathcal{E})$ such that $\mathcal{L} u' = 0$.

Proof of (b). Let $u \in C^{\infty}(\omega ; \mathcal{E}|_{\omega})$ be such that $\mathcal{L} u = 0$. By Theorem 2.1 (b) u can be approximated by sections $v \in C^{\infty}(\Gamma ; \mathcal{E}|_{\Gamma})$ such that $\mathcal{L} v = 0$. By the proof of (a) above every such section v can be extended to a section $u' \in C^{aa}(\Lambda ; \mathcal{E})$ such that $\mathcal{L} u' = 0$. The proof is complete .

3. The geometric constructions .

As mentioned in the introduction we shall concentrate on the case of condition (II) , since the corresponding constructions in the case of (I) are the same or simpler .

Proof of the sufficiency in Proposition 4. We place ourselves in the situation of Proposition 4 , so that (0.11) - (0.15) are fullfilled . Assume that $u_3, \dots, u_{n-1} \in C^\infty(\Gamma ; \tilde{\mathcal{N}}_\Gamma^*(\mathbb{R}_+\Gamma , T^*X))$ are sections which satisfy the condition II . For every $\rho \in \Gamma$ we have $\Lambda_{\rho\mathbb{R}} = T_\rho(\mathbb{R}_+\Gamma)$ by condition II so we have the inequality

$$(3.1) \qquad \left| \operatorname{Im}(v + t_3 \mathcal{H}_{u_3} + \dots + t_{n-1}\mathcal{H}_{u_{n-1}} + t\,\tfrac{\partial}{\partial\lambda}) \right|$$
$$\geq C \left(|\operatorname{Im} v| + |t_3| + \dots + |t_{n-1}| + |\operatorname{Im} t| \right) , \quad v \in T_\rho(\widetilde{\Gamma}) , \ t_j , \ t \in \mathbb{C},$$

where C is a positive constant which locally can be taken independent of ρ .

If $v_3 \in \mathcal{M}_1^1(\Gamma , T^*X\setminus 0)$ is a representative of $u_3 \in C^\infty(\Gamma ; \tilde{\mathcal{N}}_\Gamma^*) \simeq \mathcal{M}_1^1(\Gamma, T^*X\setminus 0) / \mathcal{M}_1^2$, we can define an almost analytic manifold $\Gamma_3 \subset \widetilde{T^*X\setminus 0}$ in a natural way as the " flow out " of $\tilde{\Gamma}$ (the complexification of Γ) along $\mathcal{H}_{v_3}$. This is done as follows : Take real local canonical coordinates (x , ξ) in $T^*X\setminus 0$ so that $T^*X\setminus 0$ is locally identified with an open set $W \subset \mathbb{R}^{2n}$. Let $\tilde{W} \subset \mathbb{C}^{2n}$ be open with $\tilde{W}_{\mathbb{R}} = W$ and let $\tilde{v}_3 \in C^{aa}(\tilde{W})$ be an almost analytic extension of $v_3(x, \xi)$. Then $\mathcal{H}_{\tilde{v}_3} = \sum (\partial\tilde{v}_3/\partial\xi_j)\, \partial/\partial x_j - (\partial\tilde{v}_3/\partial x_j)\, \partial/\partial\xi_j$ is an almost analytic vectorfield . If $W' \subset\subset \tilde{W}$ is open , the flow of $\mathcal{H}_{\tilde{v}_3}$ is defined as in section 1 :

$$\Phi_t : W' \longrightarrow \tilde{W} \quad , \quad t \in \mathbb{C} \quad , \quad |t| < \varepsilon_{W'} \quad ,$$

where $\varepsilon_{W'}$ is a small positive number . Let $\tilde{\Gamma}' \subset \tilde{W}$ be a representative of $\tilde{\Gamma}$. Then for $\varepsilon > 0$ sufficiently small , we get from (3.1) that

$$(3.2) \qquad |\operatorname{Im} \Phi_t(x,\xi)| \geq C(|t| + |\operatorname{Im}(x,\xi)|) \,,$$
$$(x,\xi) \in \tilde{\Gamma}' \cap W' \,, \quad |\operatorname{Im}(x,\xi)| < \varepsilon \,, \quad |t| < \varepsilon \,,$$

where C is a positive constant . Put

$$\Gamma_3' = \{ \Phi_t(x,\xi) \, ; (x,\xi) \in \tilde{\Gamma}' \cap W' \, , |\operatorname{Im}(x,\xi)| < \varepsilon \, , \ |t| < \varepsilon \} .$$

It follows easily from (3.2) and Lemma 1.2 (as in the proof of Lemma 1.11 in [4]) that Γ_3' is almost analytic . Moreover , for different choices of local coordinates (x,ξ) , the different Γ_3' obtained in this way are all local representatives of a certain almost analytic manifold $\Gamma_3 \subset \widetilde{T^*X\setminus 0}$.

More generally , if $v_k \in \mathcal{M}_1^1(\Gamma, T^*X\setminus 0)$, $k = 3, \ldots, n-1$ are representatives of the u_k in condition II , we can define an almost analytic manifold $\Gamma_k \subset \widetilde{T^*X\setminus 0}$ as the flow out of $\tilde{\Gamma}$ along succesively $\mathcal{H}_{v_3}, \ldots, \mathcal{H}_{v_k}$. It follows from (3.1) that $\Gamma_{k\mathbb{R}} = \Gamma$.

The manifold Λ will be constructed by this procedure , where the v_k have to be chosen with special care .

<u>Proposition 3.1.</u> There exist $v_3, \ldots, v_{n-1} \in \mathcal{M}_1^1(\Gamma, T^*X\setminus 0)$, being representatives of $u_3, \ldots, u_{n-1}$ respectively , such that $\mathcal{H}_p(v_j) \in \mathcal{M}_1^{\infty}(\Gamma, T^*X\setminus 0) = \bigcap_{m \in \mathbb{Z}^+} \mathcal{M}_1^m(\Gamma, T^*X\setminus 0)$ and such that $v_{k+1} \sim 0$ on Γ_k ; the flow out of $\tilde{\Gamma}$ along $\mathcal{H}_{v_3}, \ldots, \mathcal{H}_{v_k}$ succesively . Here v_{k+1} also denotes some almost analytic extension of v_{k+1} .

In the following we shall often identify C^∞ functions with their almost analytic extensions without changing the notations. If $\Lambda_1 \subset \Lambda_2 \subset \widetilde{T^*X\setminus 0}$ are almost analytic manifolds and Λ_2 is conic, we let $\mathcal{M}^m_k(\Lambda_1, \Lambda_2) \subset C^{aa}(\Lambda_2)$ be the subspace of functions which are positively homogeneous of degree k and which vanish to order m on Λ_1. There is a natural identification $\mathcal{M}^1_1(\Gamma, T^*X\setminus 0) \simeq \mathcal{M}^1_1(\tilde{\Gamma}, \widetilde{T^*X\setminus 0})$.

Proof of Proposition 3.1. The proof will be by induction, so we assume that $v_3, \ldots, v_k$ have already been constructed. We claim that $\mathcal{H}_p$ is tangential to Γ_k in the sense that the corresponding real field $\hat{\mathcal{H}}_p$ is tangential to Γ_k of infinite order at Γ (when local representatives of everything have been chosen). In fact, by Lemma 1.1 and the induction hypothesis :

$$\mathcal{L}_{\hat{\mathcal{H}}_{v_j}}(\hat{\mathcal{H}}_p) = [\hat{\mathcal{H}}_{v_j}, \hat{\mathcal{H}}_p] = \widehat{[\mathcal{H}_{v_j}, \mathcal{H}_p]} = -\widehat{\mathcal{H}_{\mathcal{H}_p(v_j)}}$$

which vanishes to infinite order on $\tilde{\Gamma}$ for $j = 3, \ldots, k$. Thus $\mathcal{H}_p$ is (approximately) invariant under the flows of $\mathcal{H}_{v_j}$, $j = 3, \ldots, k$. Since $\mathcal{H}_p$ is already tangential to $\tilde{\Gamma}$, it remains tangential to $\Gamma_3, \ldots, \Gamma_k$. (Without getting into the details we point out that inequality (3.1) is important here. It tells us that the flows of $\mathcal{H}_{v_j}$ transport all " errors " away from the real domain.)

Now let $v \in \mathcal{M}^1_1(\Gamma, T^*X\setminus 0)$ be a representative for u_{k+1}. Since $\langle \mathcal{H}_{v_j}, dv \rangle = \sigma(\mathcal{H}_{v_j}, \mathcal{H}_v) = 0$ at $\tilde{\Gamma}$ for $j = 3, \ldots, k$, we know that $v|_{\Gamma_k}$ vanishes on $\tilde{\Gamma}$ to order 2. Clearly $v|_{\Gamma_k}$ can be extended to an element $v' \in \mathcal{M}^2_1(\tilde{\Gamma}, \widetilde{T^*X\setminus 0})$ so after substracting v' from v, we can assume that

$$v \in \mathcal{M}^1_1(\Gamma_k, \widetilde{T^*X\setminus 0}) \quad \text{and} \quad \mathcal{H}_p(v) \in \mathcal{M}^2_1(\tilde{\Gamma}, \widetilde{T^*X\setminus 0}) . \tag{3.3}$$

Let $\mathcal{F}_k = T_{\Gamma_k}(\widetilde{T^*X\setminus 0}) / T(\Gamma_k) \oplus (\frac{\partial}{\partial\lambda})$, where $(\frac{\partial}{\partial\lambda})$ denotes the onedimensional almost analytic bundle on Γ_k generated by the field $\frac{\partial}{\partial\lambda}$. As in the introduction and in section 2 we have natural identifications

$$(3.4) \qquad \mathcal{M}_1^m(\Gamma_k, \widetilde{T^*X\setminus 0}) / \mathcal{M}_1^{m+1} \simeq C^{aa}(\Gamma_k ; \mathcal{P}^m(\mathcal{F}_k)), \; m \in \mathbb{Z}^+ .$$

Since $\mathcal{H}_p$ is tangential to Γ_k it induces a differential operator $\mathcal{H}_p : C^{aa}(\Gamma_k ; \mathcal{P}^m(\tilde{\mathcal{F}}_k)) \longrightarrow C^{aa}(\Gamma_k ; \mathcal{P}^m(\tilde{\mathcal{F}}_k))$ which satisfies the assumptions of Theorem 2.4 with Λ replaced by Γ_k and $\mathcal{E}$ by $\mathcal{P}^m(\tilde{\mathcal{F}}_k)$. (Note that (2.4) is satisfied in view of (3.1) .)

Let $u_1 \in \mathcal{M}_1^1(\Gamma_k, \widetilde{T^*X\setminus 0}) / \mathcal{M}_1^2$ be the image of v . Since $\mathcal{H}_p v \in \mathcal{M}_1^2(\tilde{\Gamma}, \widetilde{T^*X\setminus 0})$ by (3.3) , it is clear that $\mathcal{H}_p(u_1)$, considered as a section in $C^{aa}(\Gamma_k ; \mathcal{P}^1(\mathcal{F}_k))$, vanishes on $\tilde{\Gamma}$. By Theorem 2.4 and its proof , there exists $w^1 \in C^{aa}(\Gamma_k ; \mathcal{P}^1(\tilde{\mathcal{F}}_k))$, vanishing on $\tilde{\Gamma}$, such that $\mathcal{H}_p w^1 = -\mathcal{H}_p u_1$. If $v' \in \mathcal{M}_1^1(\Gamma_k, \widetilde{T^*X\setminus 0})$ is a representative for w^1 it is clear that $v' \in \mathcal{M}_1^2(\tilde{\Gamma}, \widetilde{T^*X\setminus 0})$. Thus $v^1 = v + v' \in \mathcal{M}_1^1(\Gamma_k, \widetilde{T^*X\setminus 0})$ is also a representative for u_{k+1} and $\mathcal{H}_p v^1 \in \mathcal{M}_1^2(\Gamma_k, \widetilde{T^*X\setminus 0})$. Applying Theorem 2.4 and (3.4) repeatedly , we get a sequence $v^j \in \mathcal{M}_1^j(\Gamma_k, \widetilde{T^*X\setminus 0})$, $j = 2, 3, \dots$ such that

$$\mathcal{H}_p v^j = -\mathcal{H}_p v^{j-1} \quad \text{mod.} \; \mathcal{M}_1^{j+1}(\Gamma_k, \widetilde{T^*X\setminus 0}) .$$

Now take $v_{k+1} \in \mathcal{M}_1^1(\Gamma_k, \widetilde{T^*X\setminus 0}) \subset \mathcal{M}_1^1(\Gamma, T^*X\setminus 0)$ with Taylor expansion $v_{k+1} \sim \sum_1^\infty v^j$ in the sence that $v_{k+1} - \sum_1^N v^j \in \mathcal{M}_1^{N+1}(\Gamma_k, \widetilde{T^*X\setminus 0})$ for all N . Then $\mathcal{H}_p v_{k+1} \in \mathcal{M}_1^\infty(\Gamma ; T^*X\setminus 0)$ and this completes our induction proof of Proposition 3.1 .

As the manifold Λ in Proposition 4 we now take the conic almost analytic manifold which is generated by Γ_{n-1}. Since the v_j are positively homogeneous of degree 1 it is clear that $\frac{\partial}{\partial \lambda}$ and $\mathcal{H}_{v_j}$ commute. Therefore we could also obtain Λ in the following way: Put $V = \mathbb{R}_+ \Gamma$ and let $\tilde{V} \subset \widetilde{T^*X \setminus 0}$ be the complexification. Let V_k be the flow out of $\tilde{V}$ along succesively $\mathcal{H}_{v_3}, \dots, \mathcal{H}_{v_k}$. Then $\Lambda = V_{n-1}$.

We have to verify that $\sigma|_\Lambda \sim 0$ and $p|_\Lambda \sim 0$, where σ also denotes an almost analytic extension of the symplectic form. To verify the first statement, we note that for any $v \in C^{aa}(\widetilde{T^*X \setminus 0})$, we have $\mathcal{L}_{\hat{\mathcal{H}}_v}(\sigma) \sim 0$. In fact, as noticed in section 1 $\mathcal{L}_{\hat{\mathcal{H}}_v}(\sigma)$ is the almost analytic extension of $\mathcal{L}_{\mathcal{H}_v}(\sigma) = \mathcal{L}_{\mathcal{H}_{\mathrm{Re}\, v}}(\sigma) + i \mathcal{L}_{\mathcal{H}_{\mathrm{Im}\, v}}(\sigma)$ in $T^*X \setminus 0$ which vanishes identically. If Φ_t, $t \in \mathbb{C}$, $|t| < \varepsilon$ are the local diffeomorphisms generated by $\mathcal{H}_v$ in $\widetilde{T^*X \setminus 0}$ (welldefined after choosing local canonical coordinates and local representatives,) it follows that we have locally for all $N \in \mathbb{Z}^+$:

(3.5) $$\left| (\Phi_t^* \sigma)_{(x,\xi)} - \sigma_{(x,\xi)} \right| \leq C_N \sup_{0 \leq s \leq 1} \left| \mathrm{Im}\, \Phi_{st}(x,\xi) \right|^N .$$

We have $\sigma|_{\tilde{V}} \sim 0$ since $\frac{\partial}{\partial \lambda}$, $\mathcal{H}_{\mathrm{Re}\, p}$, $\mathcal{H}_{\mathrm{Im}\, p}$ are symplectically orthogonal at V. Suppose we have already proved that $\sigma|_{V_k} \sim 0$ for some k. Clearly $\mathcal{H}_{v_{k+1}}$ is symplectically orthogonal to V_k since v_{k+1} vanishes on V_k. Since the flow Φ_t corresponding to $\mathcal{H}_{v_{k+1}}$ is approximately a canonical transformation in the sense of (3.5), it follows that $\sigma|_{V_{k+1}} \sim 0$ by (3.1), so by induction we get $\sigma|_\Lambda \sim 0$.

Similarly we can show that $p|_\Lambda \sim 0$. It suffices to use that $\mathcal{H}_{v_k} p = - \mathcal{H}_p v_k \in \mathcal{M}_1^\infty(\Gamma, T^*X \setminus 0)$. This completes the proof of the sufficiency in Proposition 4.

Under the general assumptions (0.11)-(0.15) we have

Lemma 3.2. There exists $b \in C^{\infty}(T^*X\setminus 0)$ positively homogeneous of degree 0, such that $\{q, \bar{q}\} \in \mathcal{M}_1^2(\Gamma, T^*X\setminus 0)$ if $q = e^b p$.

Proof. We know that $\mathcal{H}_{\mathrm{Re}\, p}$, $\mathcal{H}_{\mathrm{Im}\, p}$ span $T(\Gamma)$ at every point so there is an $a \in C^{\infty}(\Gamma)$ such that

$$(3.6) \qquad [\mathcal{H}_p, \mathcal{H}_{\bar{p}}] = a\,\mathcal{H}_p - \bar{a}\,\mathcal{H}_{\bar{p}} \quad \text{at} \quad \Gamma .$$

Let $b \in C^{\infty}(\Gamma)$ be a solution (c.f. Theorem 2.1) of $\mathcal{H}_p \bar{b} = \bar{a}$ and extend b to a homogeneous function on $T^*X\setminus 0$ of degree 0. Since $p = 0$ on Γ it follows that

$$(3.7) \qquad [\mathcal{H}_q, \mathcal{H}_{\bar{q}}] = 0 \quad \text{at} \quad \Gamma ,$$

if $q = e^b p$. Now $[\mathcal{H}_q, \mathcal{H}_{\bar{q}}] = \mathcal{H}_{\{q, \bar{q}\}}$ and since $\{q, \bar{q}\} = 0$ at Γ, it follows that $\{q, \bar{q}\} \in \mathcal{M}_1^2(\Gamma, T^*X\setminus 0)$.
(This argument is copied from Duistermaat-Hörmander [1] .)

Proof of the necessety in Proposition 4 . We assume that (0.11)-(0.15) are satisfied and that $\Lambda \subset p^{-1}(0)$ is a strictly positive conic Lagrangean manifold with $\Lambda_{\mathbb{R}} = \mathbb{R}_+ \Gamma$. For the moment we do not assume that Γ is simply connected . Let q be as in the preceding Lemma .

Lemma 3.3. There exists $v_2 \in \mathcal{M}_1^1(\Gamma, T^*X\setminus 0)$ such that
$v_2 \equiv \bar{q} \mod. \mathcal{M}_1^2(\Gamma, T^*X\setminus 0)$, $v_2|_{\Lambda} \sim 0$ and
$\mathcal{H}_q v_2 \in \mathcal{M}_1^{\infty}(\Gamma, T^*X\setminus 0)$.

Proof. Choose a positive positively homogeneous C^{∞} function $s(x,\xi)$ on $T^*X\setminus 0$ of degree 1 such that $s(x,\xi) = 1$ on Γ and put $S^*X = \{ (x,\xi) \in T^*X\setminus 0 \ ; \ s(x,\xi) = 1 \}$. For every positively homogeneous function $v \in C^{\infty}(T^*X \setminus 0)$ of degree 1, we denote by v' its restriction to S^*X. Then we have $(\mathcal{H}_q v)' = L v' + g v'$, where the vectorfield L on S^*X and the function g are given by the decomposition $\mathcal{H}_q = L + g \frac{\partial}{\partial \lambda}$ at the points of S^*X. If $\Lambda' = \Lambda \cap \widetilde{S^*X}$ we see that L is tangential to Λ' as an almost analytic vectorfield, since $\mathcal{H}_q$ is necessarily tangential to Λ. Let $\tilde{\bar{q}}$ be the almost analytic extension of $\bar{q}$. Then since $\mathcal{H}_{\bar{q}} \in T_\rho(\tilde{\Gamma}) \subset T_\rho(\Lambda)$ at every point $\rho \in \Gamma$, we have $\langle T_\rho(\Lambda), d\tilde{\bar{q}} \rangle = \sigma(T_\rho(\Lambda), \mathcal{H}_{\tilde{\bar{q}}}) = \{0\}$. This means that $\tilde{\bar{q}}|_{\Lambda} \in \mathcal{M}^2_1(\tilde{\Gamma}, \Lambda)$ so we can find $w_1 \in \mathcal{M}^1_1(\Lambda, \widetilde{T^*X \setminus 0})$ with $w_1 \equiv \tilde{\bar{q}}$ mod. $\mathcal{M}^2_1(\Gamma, T^*X \setminus 0)$. Then $w_1' \in \mathcal{M}^1(\Lambda', \widetilde{S^*X})$ and we denote by f_1 its image in $\mathcal{M}^1(\Lambda', \widetilde{S^*X}) / \mathcal{M}^2 \simeq C^{aa}(\Lambda'; \mathcal{P}^1(\mathcal{F}))$, where $\mathcal{F}$ is the normal bundle of Λ'. Using that $\mathcal{H}_q \bar{q} \in \mathcal{M}^2_1(\Gamma, T^*X \setminus 0)$ we see that

(3.8) $\quad (L + g) f_1$ vanishes at $\tilde{\Gamma}$ when considered as a section in $C^{aa}(\Lambda; \mathcal{P}^1(\mathcal{F}))$.

(Here of cource $L + g$ denotes the induced operator in $\mathcal{M}^1 / \mathcal{M}^2 \simeq C^{aa}(\Lambda'; \mathcal{P}^1(\mathcal{F}))$.) Therefore we can apply Theorem 2.4 to construct by succesive approximations a function $w \in \mathcal{M}^1(\Lambda', \widetilde{S^*X})$ with $w \equiv w_1'$ mod. $\mathcal{M}^2(\Gamma, S^*X)$ such that $(L + g) w \in \mathcal{M}^{\infty}(\Gamma, S^*X)$. Let $v_2 \in \mathcal{M}^1_1(\Gamma, T^*X \setminus 0)$ be the homogeneous extension of w. Then the Lemma follows.

<u>Remark 3.4.</u> Without assuming the existence of Λ we can always by a simpler construction find $v_2 \equiv \bar{q}$ mod. $\mathcal{M}^2_1(\Gamma, T^*X \setminus 0)$, such that $\mathcal{H}_q v_2 \in \mathcal{M}^{\infty}_1(\Gamma, T^*X \setminus 0)$.

<u>From now on we assume that Γ is simply connected.</u> We let Λ, q, v_2 be as above. Choose local coordinates x_1, x_2 in Γ so that $\mathcal{H}_{\mathrm{Re}\, q} = \frac{\partial}{\partial x_1}$, $\mathcal{H}_{\mathrm{Im}\, q} = \frac{\partial}{\partial x_2}$ at Γ and let $\Gamma' \subset \Gamma$ be a small quadratic open set with respect to these coordinates. Let $\rho, \mu \in \Gamma'$ and let Φ_t, Ψ_s be the flows in Γ of $\mathcal{H}_{\mathrm{Re}\, q}$ and $\mathcal{H}_{\mathrm{Im}\, q}$ respectively. Then there exist unique small numbers t_0 and s_0 such that $\rho = \mathcal{K}_{\rho,\mu}(\mu)$ if $\mathcal{K}_{\rho,\mu} = \Phi_{t_0} \circ \Psi_{s_0}$. ($t_0$ and s_0 depend on ρ, μ of course.)

The fields $\mathcal{H}_{(q+v_2)/2}$ and $\mathcal{H}_{(q-v_2)/2i}$ commute to infinite order at the points of Γ and coincide with $\mathcal{H}_{\mathrm{Re}\, q}$ and $\mathcal{H}_{\mathrm{Im}\, q}$ respectively at Γ. Let $\tilde{\Phi}_t$, $\tilde{\Psi}_s$ be the corresponding flows in $\widetilde{T^*X\setminus 0}$. Then for all $N \in \mathbf{Z}^+$ we have locally for small $s, t \in \mathbb{R}$ the estimates

$$\left| \tilde{\Phi}_t \circ \tilde{\Psi}_s(x, \xi) - \tilde{\Psi}_s \circ \tilde{\Phi}_t(x, \xi) \right| \leq C_N \, d((x, \xi), \Gamma)^N$$

for $(x, \xi) \in \widetilde{T^*X\setminus 0}$.

Put $\tilde{\mathcal{K}}_{\rho,\mu} = \tilde{\Phi}_{t_0} \circ \tilde{\Psi}_{s_0}$ so that $\tilde{\mathcal{K}}_{\rho,\mu}$ maps a complex neighbourhood of μ onto a complex neighbourhood of ρ. Put $A_{\rho,\mu} = d\tilde{\mathcal{K}}_{\rho,\mu} : T_\mu(\widetilde{T^*X\setminus 0}) \longrightarrow T_\rho(\widetilde{T^*X\setminus 0})$. This is a linear canonic transformation and since $\tilde{\Phi}_t$ and $\tilde{\Psi}_s$ commute to infinite order near Γ, it follows that $A_{\rho,\mu} \circ A_{\mu,\nu} = A_{\rho,\nu}$ if $\rho, \mu, \nu \in \Gamma'$.

Now if $\rho, \mu \in \Gamma$ are arbitrary points we choose a curve γ joining ρ to μ and we put $A_{\rho,\mu} = A_{\rho,\rho_1} \circ A_{\rho_1,\rho_2} \circ \ldots \circ A_{\rho_N,\mu}$ $: T_\mu(\widetilde{T^*X\setminus 0}) \longrightarrow T_\rho(\widetilde{T^*X\setminus 0})$, where ρ_j are points on γ chosen sufficiently densely. Since Γ is simply connected, this definition does not depend on the choice of γ or of the ρ_j. Now $\mathcal{H}_q$ and $\mathcal{H}_{v_2}$ are tangential to Λ since $q|_\Lambda = v_2|_\Lambda = 0$ and it follows that $A_{\rho,\mu}$ maps $T_\mu(\Lambda)$ into $T_\rho(\Lambda)$. It is also easy to see that the sections in $T_\Gamma(T^*X\setminus 0)$ given by $\mathcal{H}_q$, $\mathcal{H}_{\bar{q}}$, $\frac{\partial}{\partial\lambda}$ are invariant under the maps $A_{\rho,\mu}$.

Now let $\mu\in\Gamma$ be a fixed point and let $\mathcal{H}_q, \mathcal{H}_{\bar{q}}, \frac{\partial}{\partial\lambda}, \nu_3, \ldots, \nu_{n-1}$ be a basis for $T_\mu(\Lambda)$. Extend $\nu_3, \ldots, \nu_{n-1}$ to sections in $C^\infty(\Gamma; T_\Gamma(\Lambda))$ by putting $\nu_j(\rho) = A_{\rho,\mu}(\nu_j)$, $\forall\rho\in\Gamma$. Then clearly $\mathcal{H}_q, \mathcal{H}_{\bar{q}}, \frac{\partial}{\partial\lambda}, \nu_3, \ldots, \nu_{n-1}$ span $T_\rho(\Lambda)$ at every point $\rho\in\Gamma$. Let $v_3, \ldots, v_{n-1}\in\mathcal{M}_1^1(\Gamma, T^*X\setminus 0)$ be such that $\mathcal{H}_{v_j} = \nu_j$ at Γ. By the construction, we know that

$$[\mathcal{H}_{(q+v_2)/2}, \mathcal{H}_{v_j}] = \mathcal{L}_{\mathcal{H}_{(q+v_2)/2}}(\mathcal{H}_{v_j}) \quad\text{and}\quad [\mathcal{H}_{(q-v_2)/2i}, \mathcal{H}_{v_j}]$$

vanish at Γ. Thus $[\mathcal{H}_q, \mathcal{H}_{v_j}] = \mathcal{H}_{(\mathcal{H}_q v_j)}$ vanishes at Γ and this implies that $\mathcal{H}_q v_j\in\mathcal{M}_1^2(\Gamma, T^*X\setminus 0)$.

We have now proved that the existence of Λ implies that condition (II) is satisfied with p replaced by $q = e^b p$. To get (II) for p instead, we calculate for $3\le j\le n-1$:

$$\mathcal{H}_p v_j = e^{-b}\mathcal{H}_q v_j + q\,\mathcal{H}_{e^{-b}}(v_j) \equiv w_j p \quad \text{mod.}\ \mathcal{M}_1^2(\Gamma, T^*X\setminus 0),$$

where $w_j\in C^\infty(T^*X\setminus 0)$ is homogeneous of degree 0. Put $v'_j = v_j - f_j p$, where f_j is homogeneous of degree 0 and $f_j|_\Gamma$ is a solution of $\mathcal{H}_p(f_j|_\Gamma) = w_j|_\Gamma$. Then $\mathcal{H}_p(v'_j) \equiv w_j p - p\,\mathcal{H}_p f_j \equiv 0 \ \text{mod}.\ \mathcal{M}_1^2(\Gamma, T^*X\setminus 0)$, so if

$u_j \in \mathcal{M}_1^1(\Gamma, T^*X\setminus 0)/\mathcal{M}_1^2$, $3 \le j \le n-1$ are the images of v_j' it is clear that condition (II) is satisfied . This completes the proof of the necessety in Proposition 4 .

Proof of Proposition 6 . Assume that (0.11) - (0.15) are satisfied and that $\{p, \bar{p}\} \in \mathcal{M}_1^3(\Gamma, T^*X\setminus 0)$, where Γ is simply connected . Let $\mu \in \Gamma$. It is easy to see (c.f. [1]) that there are real symplectic coordinates (x, ξ) in a conic neighbourhood of μ such that $p(x, \xi) = \xi_{n-1} + i\xi_n + a(x, \xi)$ and such that x and ξ are positively homogeneous of degree 0 and 1 respectively . Here a vanishes to second order at $\mu = (x_0, (\xi_0', 0))$, $\xi_0' \in \mathbb{R}^{n-2}\setminus\{0\}$. (We write in general $\mathbb{R}^n \ni x = (x', x'')$, $x' \in \mathbb{R}^{n-2}$, $x'' \in \mathbb{R}^2$.) Let $\Lambda_\mu' \subset T_{(x_0', \xi_0')}(\widetilde{T^*\mathbb{R}^{n-2}})$ be a positive semidefinite Lagrangean plane with $\Lambda'_{\mu\mathbb{R}} = \{\lambda(0, \xi_0') ; \lambda \in \mathbb{R}\}$. Put $\Lambda_\mu =$

$$= \{ (t_x, t_\xi) = (t_x', t_x'', t_\xi', t_\xi'') \ ; (t_x', t_\xi') \in \Lambda_\mu' , \ t_\xi'' = 0 \} .$$

Then $\Lambda_\mu \subset T_\mu(p^{-1}(0))$ is a positive semidefinite Lagrangean plane with $\Lambda_{\mu\mathbb{R}} = T_\mu(\mathbb{R}_+\Gamma)$.

Since $\{p, \bar{p}\} \in \mathcal{M}_1^3(\Gamma, T^*X\setminus 0)$ we can take $q = p$ in Lemma 3.2 and we can find a v_2 as in Remark 3.4 such that $v_2 \equiv \bar{p}$ mod. $\mathcal{M}_1^3(\Gamma, T^*X\setminus 0)$. It is then easy to see that all the mappings $\underline{A_{\rho,\mu}}$ are real . Therefore the Lagrangean planes $\Lambda_\rho = A_{\rho,\mu}(\Lambda_\mu)$ are all positive semidefinite with $\Lambda_{\rho\mathbb{R}} = T_\rho(\mathbb{R}_+\Gamma)$. If $\mathcal{H}_p$, $\mathcal{H}_{\bar{p}}$, $\nu_3, \dots, \nu_{n-1}, \frac{\partial}{\partial\lambda}$ is a basis for Λ_μ , we can extend ν_j to sections in $T_\Gamma(\widetilde{T^*X\setminus 0})$, using the mappings $A_{\rho,\mu}$, and we get sections $u_3, \dots, u_{n-1} \in C^\infty(\Gamma ; \widetilde{\mathcal{N}_\Gamma^*})$ satisfying (II) exactly as in the proof of the necessety above .

Proof of Proposition 3. We now consider the situation when we have a bicharacteristic strip, so we assume that (0.1)-(0.6) are valid as well as the other conditions of Proposition 3. Let $A_{\gamma(t),\gamma(s)} : T_{\gamma(s)}(\widetilde{T^*X\setminus 0}) \longrightarrow T_{\gamma(t)}(\widetilde{T^*X\setminus 0})$ be the maps induced by the almost analytic Hamilton field $\mathcal{H}_p$. Put $\nu(t) = A_{\gamma(t),\gamma(t_0)}(\nu)$, where ν is some vector in $T_{\gamma(t_0)}(\widetilde{T^*X\setminus 0})$. Fix $t \in \mathbb{R}$ and let (x,ξ) be local real symplectic coordinates near $\gamma(t)$. Then for small $s \in \mathbb{R}$ we have in the chosen coordinates

$$(3.9) \qquad \nu(t+s) = \left(\frac{\partial \Phi_s(x,\xi)}{\partial (x,\xi)}\right)_{\gamma(t)} (\nu(t)) \quad ,$$

where $\Phi_s : \widetilde{T^*X\setminus 0} \longrightarrow \widetilde{T^*X\setminus 0}$ are the locally defined diffeomorphisms given by

$$(3.10) \qquad \frac{\partial \Phi_s(x,\xi)}{\partial s} = \mathcal{H}_p(\Phi_s(x,\xi)) \quad , \qquad \Phi_0(x,\xi) = (x,\xi) \quad ,$$

$$-\varepsilon < s < \varepsilon \quad , \qquad (x,\xi) \in \mathbb{C}^{2n} \quad .$$

Differentiating (3.10) with respect to (x,ξ) for $s = 0$, we get

$$(3.11) \qquad \left(\frac{\partial}{\partial s}\frac{\partial \Phi_s}{\partial (x,\xi)}\right)_{s=0} = \frac{\partial \mathcal{H}_p(x,\xi)}{\partial (x,\xi)} = \begin{pmatrix} \frac{\partial^2 p}{\partial x \partial \xi} & \frac{\partial^2 p}{\partial \xi^2} \\ -\frac{\partial^2 p}{\partial x^2} & -\frac{\partial^2 p}{\partial \xi \partial x} \end{pmatrix} .$$

Combining (3.9) and (3.11) we get

$$\frac{\partial}{\partial t}\, i^{-1} \sigma(\nu(t), \overline{\nu(t)}) = \left(\frac{\partial}{\partial s}\right)_{s=0} i^{-1} \sigma(\nu(t+s), \overline{\nu(t+s)}) =$$

$$i^{-1}\left(\sigma\left(\left(\frac{\partial \nu(t+s)}{\partial s}\right)_{s=0}, \overline{\nu(t)}\right) + \sigma\left(\nu(t), \overline{\left(\frac{\partial \nu(t+s)}{\partial s}\right)_{s=0}}\right)\right) =$$

$$= 2 \operatorname{Im} \sigma\left(\frac{\partial \mathcal{H}_p}{\partial (x,\xi)}(\nu(t)), \overline{\nu(t)}\right) = -2 \operatorname{Im} \operatorname{Hess} p\,(\nu(t), \overline{\nu(t)}) \quad .$$

By the assumption in Propostion 3 , we therefore have

$$(3.12)\qquad \frac{\partial}{\partial t}\ i^{-1}\sigma\,(\gamma(t),\overline{\gamma(t)})\quad \begin{cases} \geq 0 & \text{for}\quad t \geq t_0 \\ \leq 0 & \text{for}\quad t \leq t_0 \end{cases} .$$

Now it is easy to find a positive semidefinite Lagrangean plane $\Lambda_{\gamma(t_0)} \subset T_{\gamma(t_0)}(p^{-1}(0))$ with $\Lambda_{\gamma(t_0)\,\mathbb{R}}$ spanned by $\frac{\partial}{\partial\lambda}$, $\mathcal{H}_p$. It follows then from (3.12) that $\Lambda_{\gamma(t)} = A_{\gamma(t),\gamma(t_0)}(\Lambda_{\gamma(t_0)})$ are all positive semidefinite Lagrangean planes with $\Lambda_{\gamma(t)\,\mathbb{R}} = T_{\gamma(t)}(\mathbb{R}_+\gamma)$. By the arguments above we then see that (I) is satisfied .

4. Proof of Theorems 2 and 5 .

We only prove Theorem 5 since the proof of Theorem 2 is almost the same . Assume that (0.11)-(0.15) are satisfied and that $\Lambda \subset p^{-1}(0)$ is a strictly positive conic Lagrangean manifold with $\Lambda_{\mathbb{R}} = \mathbb{R}_+\Gamma$. Let $\mathcal{L}$ be the " Maslov linebundle " on Λ , defined in section 6 in [4] and let $\Gamma^m(\Lambda;\mathcal{L})$ be the space of sections of $\mathcal{L}$ which are positively homogeneous of degree m . Let $u \in I_c^k(X,\Lambda)$ have the principal symbol $a \in \Gamma^{k+n/4}(\Lambda;\mathcal{L})$. Then by Proposition 8.3 in [4] , we know that $Pu \in I_c^k(X,\Lambda)$ and has the principal symbol $i^{-1}\mathcal{L}_{\mathcal{H}_p} a + c\,a$ where $\mathcal{L}_{\mathcal{H}_p} a$ is the Lie derivative of a along $\mathcal{H}_p$, defined in [4 , section 8] and c is the subprincipal symbol of P , positively homogeneous of degree 0 .

<u>Lemma 4.1.</u> There exists $a_0 \in \Gamma^{n/4}(\Lambda;\mathcal{L})$, vanishing only on a discrete set of rays in $\mathbb{R}_+\Gamma$, such that $i^{-1}\mathcal{L}_{\mathcal{H}_p} a_0 + c\,a_0 = 0$. Moreover for every $k \in \mathbb{R}$ the operator $i^{-1}\mathcal{L}_{\mathcal{H}_p} + c \;:\; \Gamma^k(\Lambda;\mathcal{L}) \longrightarrow \Gamma^k(\Lambda;\mathcal{L})$ is surjective .

Proof of the Lemma. As in the proof of Lemma 3.3 we take a submanifold $\widetilde{S^*X} \subset T^*X\setminus 0$ of codimension 1, transversal to $\frac{\partial}{\partial\lambda}$ and containing Γ. Let $\Lambda' = \Lambda \cap \widetilde{S^*X}$. For every $k \in \mathbb{R}$, the operator $i^{-1}\mathscr{L}_{\mathcal{H}_p} + c \; : \; \Gamma^k(\Lambda\,;\mathscr{L}) \longrightarrow \Gamma^k(\Lambda\,;\mathscr{L})$ is by restriction to Λ' equivalent to an operator $\mathscr{L}_k : C^{aa}(\Lambda'\,;\mathscr{L}|_{\Lambda'}) \rightarrow C^{aa}(\Lambda'\,;\mathscr{L}|_{\Lambda'})$ which satisfies the assumptions of Theorem 2.4. The first part of the lemma then follows from Theorem 2.4 (b) and the fact that $\mathscr{L}_{n/4}$ is locally the Cauchy - Riemann operator on Γ. The second part follows from Theorem 2.4 (a).

Now take $u_0 \in I_c^0(X, \Lambda)$ with principal symbol $a_0 \in \Gamma^{n/4}(\Lambda\,;\mathscr{L})$ as in the lemma. Then $P u_0 \in I_c^{-1}(X, \Lambda)$ and we let $b_{-1} \in \Gamma^{n/4-1}$ be the principal symbol. Let $a_{-1} \in \Gamma^{n/4-1}$ be a solution of $(i^{-1}\mathscr{L}_{\mathcal{H}_p} + c)\, a_{-1} = - b_{-1}$. Then if $u_{-1} \in I_c^{-1}(X, \Lambda)$ has a_{-1} as principal symbol, it is clear that $P(u_0 + u_{-1}) \in I_c^{-2}(X, \Lambda)$. Repeating this construction, we find $u \in I_c^0(X, \Lambda)$ as an asymptotic sum $u \sim u_0 + u_{-1} + \dots$, $u_j \in I_c^{-j}(X, \Lambda)$, such that $P u \in C^\infty(X, \Omega_{1/2})$. (This is just the usual geometrical optics construction.) The proof is complete.

References.

[1] Duistermaat, J. J, Hörmander, L. Fourier integral operators II. Acta math., 128(1972), 183-269.

[2] Hörmander, L. On the existence and the regularity of solutions of linear pseudo-differential equations. L'Ens. Math. 17, (1971), 99-163.

[3] Malgrange, B. Existence et approximation des solutions des équations aux derivées partielles et des équations de convolution. Annales de L'inst. de Fourier 6(1955-56), 271-356.

[4] Melin, A., Sjöstrand, J. Fourier integral operators with complex valued phasefunctions. Proceedings of this conference.

SECOND-ORDER FUCHSIAN ELLIPTIC EQUATIONS AND EIGENVALUE ASYMPTOTICS (*)

François Trèves

CONTENTS

(*) Research supported in part under NSF Grant 27671

Introduction

The purpose of this work is to call the attention of students of degenerate elliptic equations to the role of eigenvalue asymptotics (in the sense of Maslov [12]) in determining the solvability and hypoellipticity properties of such equations. We look at second-order operators which, in local coordinates near a point of the boundary of a C^∞ manifold with boundary, Ω, can be put in the form

$$(*) \qquad P = tP_2(x,t,D_x,D_t) + P_1(x,t,D_x,D_t) \quad .$$

Here $x = (x^1,\dots,x^n)$ denote the "tangential"variables, t the "normal" variable (to the boundary); P_2 is a second-order differential operator, uniformly elliptic in $\overline{\Omega}$ (actually, strongly elliptic: for every (x,t,ξ), $|\xi|$ large, the polynomial in τ , $P_2(x,t,\xi,\tau)$, has one root in the upper half-plane and one in the lower half-plane); P_1 is a first-order operator. The difficulty is not increased by allowing the coefficients of ∂_t , in both P_1 and P_2 , to be pseudodifferential operators with respect to x, of the proper degree, depending smoothly on t (up to and including the boundary $\partial\Omega$, which will correspond to t = 0). Pseudodifferential operators sneak into the theory anyway, even when dealing with differential operators, in the guise of the eigenvalue asymptotics.

We consider a very limited question: the possibility of solving the equation $Pu = f$ in the space $C^\infty_{(t)}(\mathcal{D}'_{(x)})$ of smooth functions of $t \geq 0$ valued in the space of distributions with respect to x. We also look at the "parallel" hypoellipticity question: knowing that f is smooth in all (i. e. normal and tangential) variables, can we conclude that this is also true of u (a priori smooth in the normal variable)? No difficult questions, like those arising from jumps in the degeneracy degree (studied in [11]), are approached.

To the operator P can be associated a sequence of pseudodifferential operators of order one on the boundary, $\gamma^j(x,D_x)$ ($j = 0, 1,\dots$) - its eigenvalue asymptotics. At each point of the cotangent bundle over the boundary (minus the zero section) everyone of them, except possibly one, is elliptic. Whether P is solvable (resp., tangentially hypoelliptic) depends on whether this is true of the "exceptional" eigenvalue at each point of $T^*(\partial\Omega)$. We use the method of concatenations (introduced in [15]; see also [6]) and show how to construct the eigenvalues $\gamma^j(x,D_x)$. In order to show that their solvability and/or hypoellipticity are equivalent to those of P we apply the main theorem (for second-order operators) of Bolley and Camus [4], [5].

Needless to say, the phenomena uncovered in the present work are not radically new! Similar "discrete" phenomena have been described for wide classes of pseudo-differential equations with double characteritics - beginning with the work of Grushin and Vishik (for references see [9]). Some of the results in this area are somewhat limited in so far as they presume a certain kind of homogeneity - or else the ellipticity (of order one) of the eigenvalue asymptotics (the two hypotheses are closely related). The concatenation method rids us of those strictures and makes it possible to compute the total symbols of the eigenvalues.

The extension of the theory of eigenvalue asymptotics to Fuchsian partial differential equations is hardly surprising. The theory of eigenvalues of Fuchsian ordinary differential equations is one of the classics of Analysis. Recently Baouendi and Goulaouic (see e. g. [3]) have circumscribed the Fuchsian PDE (degenerating on a hypersurface) as a good class to study from a variety of viewpoints. On the other hand, the translation of certain problems about regular equations into related ones for Fuchsian equations is quite natural. It has been systematically exploited by Gilioli and Treves in [8]. The prototype of the "translatable" operators is

$$L = \partial_t^2 + t^{2k}\partial_x^2 + i\lambda t^{k-1}\partial_x \quad (k \in \mathbb{Z}_+,\ \lambda \in \mathbb{C}), \tag{@}$$

which can be transformed into $(k+1)t^{k-1}P$, with

$$P = s(\partial_s^2 + \partial_x^2) + \frac{k}{k+1}\partial_s + i\,\frac{\lambda}{k+1}\partial_x \ , \tag{@@}$$

by the change of variables $s = t^{k+1}/(k+1)$. Actually, all the pseudodifferential operators with double characteristics for which discrete spectral properties have been established can be (microlocally) transformed into systems of Fuchs type, and their solvability and hypoellipticity properties reformulated in suitable analoguous properties of these systems. The Fuchsian systems thus obtained are rather special among all Fuchsian systems - their indicial equations (see Def. I.2.1) are special: for instance the indicial equation of @@ has the roots 0, $1/(k+1)$. This leads one to believe that Fuchsian systems are the proper context for the study of discrete eigenvalue asymptotics. On the other hand, the discrete character appears at the level of the principal symbols and the latters can be derived from the Maslov quantization relations, as we show (following Rockland [13]) in the last paragraphs of this text. There has been considerable interest, lately, in the meaning of Maslov quantization relations (by Leray, Weinstein, Guillemin, etc.) and one can hope that the situation will be satisfactorily clarified in the near future.

The two first chapters of the paper are devoted to the abstract evolution equations of Fuchs type. Most of what is presented in Ch. I is well known (and in fact, on some points, better results are known); its aim is simply to lay the groundwork for the statements and proofs of necessary and sufficient conditions of solvability and hypoellipticity in Ch. II. In Ch. III we study "true" pseudodifferential-differential operators of the kind (*). The notation is more or less standard; we write most of the time ∂_t for $\partial/\partial t$ and $D_x = -\sqrt{-1}\,\partial_x$.

One last word about the scope of the methods and the results we present. The author is the first to concede that they are purely introductory and that they shall, and should, be superseded by better ones: for instance the concatenation method seems to work well for second-order operators and poorly for higher-order ones; our methods, although nicely microlocalizable, do not enable us to construct parametrices. For this one should develop a calculus based on generalized Laplace transform - akin to the generalized Fourier transform of Maslov [12]. When good homogeneity properties hold, it would reduce to the ordinary Laplace (or Fourier) transform as shown in [4], [5]. Otherwise it must be modified, a little bit in the direction of the integral operators used in Section III.2 (under the name of Laplace integral operators). In fact it is not very difficult to compute the total symbols of the eigenvalues by this method. It should also enable one to relinquish the requirement of regularity in the normal direction and describe kernels and cokernels in all the relevant spaces of distributions, as Bolley and Camus have done under the assumption that their condition (C) is violated - that is to say, under the assumption that all the eigenvalues are elliptic of order one.

CHAPTER I

GENERALITIES ABOUT FUCHSIAN EVOLUTION EQUATIONS

I.1 NOTATION

We describe the basic concepts entering in the evolution equations which we shall study. As in [15] A will denote, throughout, a densely defined, unbounded, positive -definite operator in a (complex) Hilbert space H, with bounded inverse A^{-1}. By H_A^s we denote the s-th Sobolev space built by means of A : s is an arbitrary real number; H_A^s is the completion of the domain $\mathcal{D}(A^s)$ of A^s in H for the (Hilbert) norm $\|h\|_s = \|A^s h\|$ ($\| \ \|$ always stands for the norm in $H = H^0$; if $s \geq 0$, $\mathcal{D}(A^s)$ is automatically complete). Whatever $s \in \mathbb{R}$, H_A^s and H_A^{-s} can be regarded as the antidual of one another, via the bracket $(u,v) = (A^s u, A^{-s} v)_H$. Given any $m \in \mathbb{R}$, A^m is a linear isometry of H_A^s onto H_A^{s-m} ; if $m = 2s$, it is the canonical isometry of H_A^s onto its antidual. If $s' < s$, H_A^s is continuously embedded and dense in $H_A^{s'}$.

We shall also use the spaces H_A^{∞} and $H_A^{-\infty}$, respectively intersection and union of the H_A^s ($s \in \mathbb{R}$). They may be equipped with their natural locally convex topologies: the projective and inductive limits, respectively, of those of the H_A^s . Thus H_A^{∞} becomes a Fréchet space, $H_A^{-\infty}$ a nonstrict inductive limit of Fréchet spaces, and these two spaces can be regarded as the dual of one another.

In practice H will be a space of functions (or of distributions) with respect to space variables $x = (x^1,\dots,x^n)$ varying in an open subset Ω of $\mathbb{R}^n$, or in some C^{∞} (or analytic) manifold. For instance H will be $L^2(\Omega)$ and A some elliptic differential or pseudodifferential operator, such as $-\Delta_x$ or some fractional power of $-\Delta_x$, more correctly a self-adjoint extension of such an operator. Then H_A^{∞} can be interpreted as the space of C^{∞} functions of x, up to the boundary if there is a boundary, and $H_A^{-\infty}$ as the space of distributions in Ω (regarded as a manifold with boundary if it has one).

We also introduce the analogue of analytic functions and functionals with respect to the space variables: this is the scale of Hilbert spaces E_A^s ($s \in \mathbb{R}$) built like the H_A^s but on the operator e^A instead of A (observe that e^A enjoys essentially the same properties as A: the only difference is that its spectrum is contained in a half-line $[e^{\lambda_0}, +\infty[$ with $\lambda_0 > 0$). The union $E_A^{-\infty}$ of the E_A^s can be thought of as the space of analytic functionals in x, whereas E_A^{∞} should be thought of as the space of entire analytic functions of x (cf., [14]).

We are going to handle functions and distributions defined in intervals of the real line and valued in the spaces H, H_A^s, E_A^s ($-\infty \leq s \leq +\infty$); the variable in the interval will be denoted by t and regarded as the time. Most often the interval of definition will be closed and bounded, of the form $[0,T]$ with $T > 0$. We shall denote by $\mathcal{E}$ the space of C^{∞} functions in $[0,T]$ (i. e., restrictions of C^{∞} functions in $\mathbb{R}^1$) valued in H_A^{∞}; and by $\mathcal{F}$ the space of C^{∞} functions in $[0,T]$ valued in $H_A^{-\infty}$: $f \in \mathcal{F}$ if there is a sequence of real numbers $s_0 \geq s_1 \geq \dots \geq s_k \geq \dots$ such that, for each integer k = 0, 1,..., $f^{(k)} = \partial_t^k f$ is a continuous function $[0,T] \to H_A^{s_k}$. The dual of $\mathcal{E}$ (equipped with its natural topology) will be denoted by $\mathcal{D}'$: it is the space of <u>distributions</u> in the closed interval $[0,T]$ (regarded as a manifold with boundary) valued in $H_A^{-\infty}$. Customarily we shall refer to them, simply, as distributions.

We are also going to need the spaces of C^{∞} functions of t, $0 \leq t \leq T$, which are "flat" , i. e., , which vanish of infinite order at t = 0, valued either in H_A^{∞} or in $H_A^{-\infty}$ and denoted respectively by $\mathcal{E}_{flat}$ or $\mathcal{F}_{flat}$: by definition, $f \in \mathcal{F}$ belongs to $\mathcal{F}_{flat}$ if it has the following property:

(I.1.1) <u>to every integer</u> $N \geq 0$ <u>there is a number</u> $s \in \mathbb{R}$ <u>such that</u> f <u>is a continuous mapping</u> $[0,T] \to H_A^s$ <u>and such that</u> $\sup_{0<t\leq T} t^{-N} \|f(t)\|_s < +\infty$.

On the other hand, $f \in \mathcal{E}$ belongs to $\mathcal{E}_{flat}$ <u>if</u> $\sup_{0<t\leq T} t^{-N} \|f(t)\|_s < +\infty$ <u>whatever</u> $N \in \mathbb{Z}_+$, $s \in \mathbb{R}$.

Next we introduce the <u>pseudodifferential operators</u> built by means of A : they are the formal power series of the kind:

$$(I.1.2) \qquad a(t,A) = \sum_{j\in\mathbb{Z}} a_j(t)A^{-j} ,$$

where the coefficients $a_j(t) \in C^\infty([0,T];\mathbb{C})$ and vanish identically for $j < -d$. For us d will always be an integer; it is the <u>degree</u> of $a(t,A)$ if $a_d(t)$ does not vanish identically. We denote by $\Psi_A^d([0,T])$ (or Ψ^d, in short) the space of the series (I.1.2) whose degree does not exceed d. With each element $a(t,A) \in \Psi^d$ we may associate a class of "actual" operators $H_A^{-\infty} \to H_A^{-\infty}$, depending smoothly on t, in the following manner: select, which is feasible, a sequence of continuous functions χ_j in $\mathbb{R}_+$, $0 \le \chi_j \le 1$ and $\chi_j(\rho) = 1$ for $\rho > \rho_j$, where the ρ_j form an increasing sequence of positive numbers (usually converging to $+\infty$), such that

$$(I.1.3) \qquad a_\chi(t,A) = \sum a_j(t)\,\chi_j(A)\,A^{-j}$$

actually converges in $C^\infty([0,T];L(H^s;H^{s-d}))$ (if it converges for some real number s it does for every one of them). It is checked at once that if $\chi' = \{\chi'_j\}$ is another sequence of cut-off functions such as χ, we have

$$(I.1.4) \qquad a_\chi - a_{\chi'} \in C^\infty([0,T];L(H_A^{-\infty};H_A^{\infty})) ,$$

i. e., is <u>regularizing</u> (and smooth with respect to t). As it is now standard, we identify the formal series (I.1.3) with the equivalence class (modulo regularizing operators) $\{a_\chi\}$, and refer to it as a pseudodifferential operator. Most of the time we shall deal with pseudodifferential operators of degree <u>zero</u>, i. e., elements of Ψ^0.

As a matter of fact, in the forthcoming, functions and distributions will often be <u>vector-valued</u>: they will be N-vectors whose entries belong to $\mathcal{E}, \mathcal{F}, \mathcal{D}'$, etc. Accordingly the operators will be matrices with entries in Ψ^d. But in order to alleviate the notation we shall disregard this fact and continue to refer to the functions and distributions as if they were scalar, elements of $\mathcal{E}$, $\mathcal{F}$, $\mathcal{D}'$, etc., and to the operators as if they were members of Ψ^d, hoping that this will not engender confusion.

I.2 FUCHSIAN EVOLUTION SYSTEMS. AN EXTENSION OF THE CAUCHY-KOVALEVSKA THEOREM

We deal here with first-order Fuchsian systems of evolution equations,

$$(I.2.1) \qquad L = t(I\partial_t - a(t,A)A) - \alpha(A) ,$$

where I stands for the $N \times N$ identity matrix (N, an integer ≥ 1), $a(t,A)$ and $\alpha(A)$ are $N \times N$ matrices with entries in Ψ^0 (see Section I.1). Each entry in $\alpha(A)$ is a formal power series, in the powers of A^{-1}; no generality would have been gained by making them depend on t, since $t^{-1}\{\alpha(t,A) - \alpha(A)\}$ could have been incorporated in $a(t,A)A$.

Definition I.2.1.- The characteristic equation of $\alpha(A)$,

$$(I.2.2) \qquad \det(zI - \alpha(A)) = 0,$$

is called the indicial equation of the system (I.2.1).

Eq. (I.2.2) is polynomial of degree N with respect to z with coefficients in $\mathbb{C}[[A^{-1}]]$. Puiseux' theorem states that it has N roots in the field $\mathbb{C}\{A^{-1}\}$ of formal series in the fractional powers of A^{-1}. We shall be very much concerned with the case where it has roots in $\mathbb{Z}_+$; the constants can of course be regarded as elements of $\mathbb{C}\{A^{-1}\}$.

We now state and prove a generalization of the Cauchy-Kovalevska theorem (see [17], Section 17; for a more general version than the one presented here, see [3])

We consider the following diagram :

$$
\text{(I.2.3)} \qquad
\begin{array}{ccc}
C^{\infty}([0,T];E_A^{-\infty}) & \xrightarrow{L} & C^{\infty}([0,T];E_A^{-\infty}) \\
\mathcal{T} \downarrow & & \downarrow \mathcal{T} \\
E_A^{-\infty}[[t]] & \xrightarrow[L^{\#}]{} & E_A^{-\infty}[[t]]
\end{array} \quad .^{(*)}
$$

The space $E_A^{-\infty}$ has been defined in Section I.1 ; $E_A^{-\infty}[[t]]$ denotes the complex vector space of formal power series in t with coefficients in $E_A^{-\infty}$. The vertical arrows $\mathcal{T}$ denote the mapping "Taylor expansion at $t = 0$ " . The lower horizontal arrow $L^{\#}$ stands for the operator L acting, in the obvious manner, on the formal power series which belong to $E_A^{-\infty}[[t]]$.

<u>Theorem I.2.1.- The Taylor expansion map $\mathcal{T}$ establishes an isomorphism between the kernel (resp., the cokernel) of L and that of $L^{\#}$</u> (in Diagram (I.2.3)).

<u>In particular, L defines an automorphism of $C^{\infty}([0,T];E_A^{-\infty})$ if and only if</u>

(I.2.4) <u>the indicial equation</u> (I.2.2) <u>has no root belonging to</u> $\mathbb{Z}_+$.

<u>Proof</u>: It is well-known, and easy to check, that (I.2.4) is necessary and sufficient in order that $L^{\#}$ define an automorphism of $E_A^{-\infty}[[t]]$.

<u>Case # 1</u> : <u>the following condition is satisfied</u> :

(I.2.5) <u>the real part of each eigenvalue of a_0 , the leading coefficient in $a(A)$ is strictly negative.</u>

Since (I.2.5) $\Longrightarrow$ (I.2.4) trivially, we know already that $L^{\#}$ is bijective.

(*) All the function spaces in this diagram, and in the statements below, should be tensored with $\mathbb{C}^N$ (cf., remark at the end of Section I. 1).

If (I.2.5) holds and if we call c_0 the infimum of $-\operatorname{Re}\chi$, $\chi \in \operatorname{Spec}\alpha_0$, we have, for some $C_N > 0$, depending only on N,

$$(I.2.6) \qquad |\exp(-\alpha_0 \log y)| \leq C_N(1 + |\log y||\alpha_0|)^N \exp(-c_0|\log y|) , \; 0 \leq y \leq 1,$$

where $|\ |$ denotes both the matrix norm and the absolute value of numbers. From (I.2.6) it follows that we may find representatives of $\alpha(A)$ (defined in the fashion of (I.1.3)) such that

$$y^{-1-\alpha(A)} = y^{-1} e^{-\alpha(A)\log y}$$

is an integrable function, and $y^{-\alpha(A)}$ a continuous function of y in the unit interval $0 \leq y \leq 1$, valued in the space of bounded linear operators on anyone of the Hilbert spaces H_A^s, E_A^s ($s \in \mathbb{R}$). (Incidentally note also that

$$(I.2.7) \qquad y^{-1-\alpha(A)} = \sum_{k=0}^{+\infty} \frac{1}{k!} \beta(A)^k A^{-k}(\log y)^k y^{-1-\alpha_0} ,$$

where $\beta(A) = \{\alpha_0 - \alpha(A)\}A^{-1}$, defines a bona fide pseudodifferential operator of degree zero depending on y, $0 \leq y \leq 1$, in an "integrable" manner.)

We set now to solve

$$(I.2.8) \qquad Lu = f \in C^\infty([0,T];E_A^{-\infty}).$$

We take $u = u_0 + u_1 + \dots$, requiring

$$(I.2.9)_0 \qquad (t\partial_t - \alpha(A))u_0 = f ,$$

$$(I.2.9)_{j>0} \qquad (t\partial_t - \alpha(A))u_j = ta(t,A)Au_{j-1} .$$

Using the remark at the beginning of the proof, we take:

$$(I.2.10)_0 \qquad u_0(t) = \int_0^1 y^{-1-\alpha(A)} f(yt)\, dy ,$$

$$(I.2.10)_{j>0} \qquad u_j(t) = t\int_0^1 y^{-\alpha(A)} a(yt,A)Au_{j-1}(yt)\, dy .$$

Obviously we can find $s \in \mathbb{R}$ such that $A^m f \in C^0([0,T];E^s)$ for every $m \in \mathbb{Z}_+$. It implies that $A^m u_0 \in C^0([0,T];E^s)$ for all m, and, by virtue of $(I.2.10)_0$,

$$(I.2.11) \qquad \| A^m u_0(t) \|_{E^s} \leq C_0 \sup_{0 \leq \tau \leq t} \| A^m f(\tau) \|_{E^s} .$$

By induction on both m and j we derive from $(I.2.10)_j$, whatever $s' < s$,

$$(I.2.12) \qquad \| A^m u_{j+1}(t) \|_{E^{s'}} \leq C_0 \frac{(Ct)^j}{j!} \sup_{0 \leq \tau \leq t} \| A^{m+j} u_0(\tau) \|_{E^{s'}} .$$

Let us set (for $s > s'$) :

$$(I.2.13) \qquad T(s,s') = \inf(T, C^{-1}e(s - s')) .$$

If $0 \leq t < \theta T(s,s')$, where $0 < \theta < 1$,

$$(I.2.14) \qquad \| A^m u_{j+1}(t) \|_{E^{s'}} \leq C_0 \, \theta^j \sup_{0 \leq \tau \leq t} \| A^m u_0(\tau) \|_{E^s} ,$$

which implies that the series $u_0(t) + u_1(t) + \dots$ converges uniformly in the space $C^0([0,T(s,s')];E_A^{s'})$ (and in fact the same is true of the series $A^m u_0(t) + A^m u_1(t) + \dots$ whatever $m \in \mathbb{Z}_+$). Its sum is the solution $u(t)$ of (I.2.8). By using additional information about the derivatives of f we derive corresponding information about those of u (existence, continuity and estimates of these derivatives). The uniqueness of the solution u also follows: when $f \equiv 0$, we may take $u_j = u_{j-1} = u$ in $(I.2.9)_{j>0}$ and $(I.2.10)_{j>0}$. But then, taking $j \to +\infty$, we derive from (I.2.14) that $u(t) = 0$ if $t < T(s,s')$. If s' is taken close enough to $-\infty$, we have $T(s,s') = T$.

Case # 2 : Condition (I.2.4) is satisfied

As we have already pointed out, it is well-known and obvious that, in this case, $L^\#$ is bijective. We want to prove that Eq. (I.2.8) has a unique solution.

Its existence is obtained by using the following simple "concatenation (cf. [15]):

$$(I.2.15) \qquad t(L + kI) = (L + (k-1)I)t \qquad , k = 1, 2, \ldots.$$

By virtue of (I.2.4) $p(A) = (k-1)I - \alpha(A)$ is invertible. Set then $v = p(A)^{-1}f$, $w = -(\partial_t - a(t,A)A)v$. By the first part of the proof (Case # 1) we know that, if k is large enough, we can find a unique $w_1 \in C^\infty([0,T];E_A^{-\infty})$ such that $(L + kw_1) = w$, whence:

$$(L + (k-1)I)(v + tw_1) = (L + (k-1)I)v + tw = p(A)v = f .$$

Let us now use the concatenation

$$(I.2.16) \qquad X\, p(A)^{-1}(L + (k-1)I) = (L + kI)p(A)^{-1}X .$$

where

$$(I.2.17) \qquad X = \partial_t - a(t,A)A .$$

Let then $h \in C^\infty([0,T];E_A^{-\infty})$ satisfy

$$(I.2.18) \qquad (L + (k-1)I)h = 0 .$$

By (I.2.16) we must have $Xh = 0$ and if we combine this with (I.2.18) (recalling that $L = tX - \alpha(A)$) we obtain $p(A)h = 0$, hence $h = 0$.

We have thus shown that $L + (k-1)I$, acting on $C^\infty([0,T];E_A^{-\infty})$, is bijective. By descending induction on k we conclude that this is also true of L .

<u>Case # 3</u> : <u>Condition (I.2.4) is not satisfied</u>

We shall denote by k_o the largest nonnegative integer which is a root of the indicial equation (I.2.2). We use once more the concatenation (I.2.15), in the form:

$$\text{(I.2.19)} \qquad L\, t^k = t^k\,(L + kI)\ .$$

For $k > k_0$, $L + kI$ is an automorphism of $C^\infty([0,T];E_A^{-\infty})$ and defines an automorphism of $E_A^{-\infty}[[t]]$, $L^\# + kI$. Consequently, in view of (I.2.19), L induces an automorphism of $t^k C^\infty([0,T];E_A^{-\infty})$, $L^\#$ one of $t^k E_A^{-\infty}[[t]]$. Observe then that the Taylor expansion map $\mathcal{T}$ induces an isomorphism of

$$\mathcal{G}_k = C^\infty([0,T];E_A^{-\infty}) / \, t^k C^\infty([0,T];E_A^{-\infty})$$

onto $E_A^{-\infty}[[t]] / \, t^k E^{-\infty}[[t]] \;\cong\; \mathcal{P}_k \otimes E_A^{-\infty}$, where $\mathcal{P}_k$ denotes the space of polynomials of degree $\leq k - 1$ (here with coefficients in $\mathbb{C}^N$). In view of these remarks the conclusion in Th. I.2.1 is equivalent to the commutativity of the diagram:

$$\text{(I.2.20)} \qquad \begin{array}{ccc} \mathcal{G}_k & \xrightarrow{\dot{L}} & \mathcal{G}_k \\ \mathcal{T}\big\downarrow & & \big\downarrow\mathcal{T} \\ \mathcal{P}_k \otimes E_A^{-\infty} & \xrightarrow[\dot{L}^\#]{} & \mathcal{P}_k \otimes E_A^{-\infty} \end{array} ,$$

where the upper dots indicate that the maps are obtained by going to the quotient spaces. As the commutativity of (I.2.20) is evident, Th. I.2.1 is proved.

Remark I.2.1.- It follows at once from the last part (Case # 3) of the proof that the kernel and cokernel of L, or equivalently, those of $L^\#$, are "finite dimensional", in the sense that they are of the form $E_A^{-\infty} \otimes M$, where M is a finite dimensional space of functions (or of formal power series). Actually, their dimensions are equal, that is, the dimensions of the spaces M corresponding to the kernel and to the cokernel respectively, are equal. In other words the index of L, and that of $L^\#$, are equal to zero. This is because our system L has Fuchsian order equal to zero (cf. [2]).

Remark I.2.2.- Inspection of the first part of the proof (Case # 1) shows that we have obtained a more precise result that the one stated, namely the following one: Let s be any real number and let the right-hand side f in Eq. (I.2.8) belong to $C^\infty([0,T];E_A^s)$. There is a solution u of (I.2.8) with the following property:

(I.2.21) Given any real number $s' < s$, $u \in C^\infty([0,T(s,s')];E_A^{s'})$ (where $T(s,s')$ is the number defined in (I.2.13)).

It is of course not true that every solution u of (I.2.8) has property (I.2.21), except in the case where the solution u is unique, obviously, which is equivalent with the validity of (I.2.4).

Corollary I.2.1.- The operator L defines an automorphism of $C^\infty_{flat}([0,T];E_A^{-\infty})$ (subspace of $C^\infty([0,T];E_A^{-\infty})$ consisting of the functions which vanish of infinite order at $t = 0$; cf. Section I.1).

Corollary I.2.2.- Let X be the regular operator (I.2.17); $u \mapsto Xu$ is an isomorphism of $tC^\infty([0,T];E_A^{-\infty})$ onto $C^\infty([0,T];E_A^{-\infty})$.

Proof: It suffices to apply Th. I.2.1 to $L = Xt$, observing that (I.2.4) is then satisfied.

Cor. I.2.2 is a particular case of the abstract Cauchy-Kovalevska theorem.

I.3 SOLVABILITY OF FUCHSIAN SYSTEMS

Th. I.2.1 makes it clear that the solvability in $\mathcal{F} = C^\infty([0,T];H_A^{-\infty})$ of the equation $Lu = f$ cannot be taken to mean that, to every $f \in \mathcal{F}$, there is a solution $u \in \mathcal{F}$ (this would exclude, for instance, that the operator $t\partial_t$ be solvable!). We must consider the diagram:

$$\begin{array}{ccc} \mathcal{F} & \xrightarrow{L} & \mathcal{F} \\ \mathcal{C}\downarrow & & \downarrow\mathcal{C} \\ H_A^{-\infty}[[t]] & \xrightarrow[L^{\#}]{} & H_A^{-\infty}[[t]] \end{array} \qquad (I.3.1)$$

where $H_A^{-\infty}[[t]]$ stands for the space of formal power series in t with coefficients in $H_A^{-\infty}$; otherwise the arrows have the obvious meaning (cf. Diagram (I.2.3)).

<u>Definition I.3.1</u>.- <u>We say that</u> L <u>is</u> <u>solvable</u> <u>if</u>, <u>in Diagram</u> (I.3.1), <u>we have</u>:

$$\text{Im } L = \overset{-1}{\mathcal{C}}(\text{Im } L^{\#}) \quad . \qquad (I.3.2)$$

<u>We say that</u> L <u>is</u> <u>locally solvable at</u> $t = 0$ <u>if there is a number</u> T', $0 < T' \leq T$, <u>such that</u> L <u>is solvable after we replace</u> T <u>by</u> T'.

Observe that, if L is solvable, we must have (cf. p. I-1-02):

$$L\mathcal{F} \supset \text{Ker } \mathcal{C} = \mathcal{F}_{\text{flat}} \ . \qquad (I.3.3)$$

<u>Lemma I.3.1</u>.- <u>The Taylor expansion map at</u> $t = 0$, $\mathcal{C}$, <u>induces an isomorphism of</u> $\mathcal{F}/\mathcal{F}_{\text{flat}}$ <u>onto</u> $H_A^{-\infty}[[t]]$.

<u>Proof</u>: The injectivity is trivial. In order to prove the surjectivity, we select a C^{∞} function $g(\rho)$ on the real line, $g(\rho) = 1$ for $\rho < 1$, $g(\rho) = 0$ for $\rho > 2$. Let then $\dot{f} = \sum_{j=0}^{+\infty} f_j t^j$ be a formal power series with coefficients $f_j \in H_A^{s_j}$, $j = 0, 1, \ldots$. We may find two sequences of strictly positive numbers m_j , r_j ($j = 0, 1, \ldots$) such that the series $\sum_j t^j g(r_j t A^{m_j}) f_j$ converges in $\mathcal{F}$, necessarily to a function whose Taylor expansion at $t = 0$ is equal to $\dot{f}$.

Theorem I.3.1.- *The Taylor expansion map* $\mathcal{T}$ *establishes an isomorphism between the kernel (resp., the range) of* $\dot{L}$ *and that of* $L^{\#}$ *in the diagram*

$$
(I.3.4) \qquad
\begin{array}{ccc}
\mathcal{F}/\mathcal{F}_{\text{flat}} & \xrightarrow{\dot{L}} & \mathcal{F}/\mathcal{F}_{\text{flat}} \\
\mathcal{T}\downarrow & & \downarrow\mathcal{T} \\
H_A^{-\infty}[[t]] & \xrightarrow[L^{\#}]{} & H_A^{-\infty}[[t]]
\end{array}
$$

(where $\dot{L}$ is obtained by going to the quotient spaces in the upper line of (I.3.1)).

In particular, if Condition (I.2.4) *is satisfied,* $\dot{L}$ *is an automorphism of* $\mathcal{F}/\mathcal{F}_{\text{flat}}$.

Th. I.3.1 is an immediate consequence of Lemma I.3.1 and of the well-known fact that, when (I.2.4) holds, $L^{\#}$ is an automorphism of $H_A^{-\infty}[[t]]$.

Corollary I.3.1.- *If Condition* (I.2.4) *is satisfied, the following properties are equivalent*: i) L *is solvable* (Def. I.3.1); ii) L *is an automorphism of* $\mathcal{F}$; iii) L *maps* $\mathcal{F}_{\text{flat}}$ *onto itself* ; iv) L *is an automorphism of* $\mathcal{F}_{\text{flat}}$.

Indeed, by Cor. I.2.1, L is injective on $\mathcal{F}_{\text{flat}}$. By Th. I.2.1, L is injective on $\mathcal{F}$ when (I.2.4) holds.

Corollary I.3.2.- *Whether* (I.2.4) *holds or not, the following conditions are equivalent*: i) L *is an automorphism of* $\mathcal{F}_{\text{flat}}$; ii) $L + rI$ *is solvable for some (or for every) real number* r *such that* $\alpha(A)$ *has no eigenvalue belonging to* $r + \mathbb{Z}_+$.

Proof: We have $t^r(L + rI) = L\, t^r$ and $u \mapsto t^r u$ is an automorphism of $\mathcal{F}_{\text{flat}}$. It follows that Property i) is equivalent with the same one for $L + rI$ in the place of L. The equivalence with ii) follows then from Cor. I.3.2.

Remark I.3.1.- It is not true that L solvable implies $L\mathcal{F}_{flat} = \mathcal{F}_{flat}$: the scalar operator $L = t(\partial_t - A)$, for instance, is solvable (cf. Part ii) of Th. I.3.3) but $L + 1$ is not, according to Th. I.3.2, hence $L\mathcal{F}_{flat} \neq \mathcal{F}_{flat}$ by Cor. I.3.2.

Theorem I.3.2.- In order that the equivalent conditions in Cor. I.3.2 hold, it is necessary (but not sufficient!) that the following condition be satisfied:

(I.3.5) the real part of each eigenvalue of $a_0(t)$, leading coefficient in $a(t,A)$, is ≤ 0, whatever $t \in [0,T]$.

Proof: It is a simplified version of an argument of Alinhac in [1]. Let t_0 be any number in $]0,T]$ and denote by $\mathcal{F}_{flat,t_0}$ the subspace of $\mathcal{F}$ consisting of the functions which vanish identically for $t < t_0$. Since $a_0(t)$ is the same for L and for $L + rI$ with r as in Property ii) of Cor. I.3.2, we may as well assume that L enjoys Property (I.2.4) and apply Cor. I.3.1. Given any $f \in \mathcal{F}_{flat,t_0}$ there is a unique $u \in \mathcal{F}$, which in fact belongs to $\mathcal{F}_{flat}$, solution of $Lu = f$. By Cor. I.2.1 we derive that $u \in \mathcal{F}_{flat,t_0}$. Thus $L\mathcal{F}_{flat,t_0} = \mathcal{F}_{flat,t_0}$. But this in turn implies that $M = L(t-t_0)$ defines an automorphism of $\mathcal{F}_{flat,t_0}$. Its indicial equation (as a Fuchsian operator on $[t_0,T]$) has no integral root. Consequently, if we apply once more Cor. I.3.1, we see that

$$L\left\{(t-t_0)C^{\infty}([t_0,T];H_A^{-\infty})\right\} = C^{\infty}([t_0,T];H_A^{-\infty}) ,$$

which means that the Cauchy problem for L in $[t_0,T]$ is well-posed. A standard result tells us then that the eigenvalues of $a_0(t)$, for $t_0 \leq t \leq T$, must all lie in the closed lower half-plane. Since t_0 is arbitrarily close to zero, we get (I.3.5).

Remark I.3.2.- All the preceding statements remain true if we substitute everywhere $\mathcal{E}$ for $\mathcal{F}$.

Remark I.3.3.- Suppose that (I.2.4) holds but not (I.3.5). Then $L\mathcal{E}$ is dense but not closed in $\mathcal{E}$, nor is $L\mathcal{F}$ closed in $\mathcal{F}$. Indeed, by Th. I.2.1, we know that $C^{\infty}([0,T];E_A^{\infty})$ is contained in $L\mathcal{E}$ (cf. Remark I.2.2); it is of course dense in $\mathcal{E}$ and $\mathcal{F}$.

Remark I.3.4.- If (I.2.4) holds but not (I.3.5), it is possible to show that $L\mathcal{F}$ does not contain $\mathcal{E}$ (a stronger property than not being closed in $\mathcal{F}$).

Next we discuss the solvability property of $L = tX - \alpha(A)$ in the scalar case: here $N = 1$, i. e., $X = \partial_t - a(t,A)A$, and $a(t,A)$ and $\alpha(A)$ are scalar pseudodifferential operators of order zero. We shall only consider the following two cases: either

(I.3.6) $$\operatorname{Re} a_0(t) > 0 , \quad 0 \leq t \leq T ,$$

or else

(I.3.7) $$\operatorname{Re} a_0(t) \leq 0 , \quad 0 \leq t \leq T .$$

In all cases we set

(I.3.8) $$U(t,t') = \exp(A \int_{t'}^{t} a(\tau,A)d\tau) , \qquad 0 \leq t,\ t' \leq T .$$

Theorem I.3.3.- *Let* $L = t(\partial_t - a(t,A)A) - \alpha(A)$ *be a scalar first-order operator of Fuchs type.*

i) *Suppose that* $\alpha(A) \notin \mathbb{Z}_+$. *Then* $L : \mathcal{F} \to \mathcal{F}$ *is injective. If* (I.3.6) *holds, its range is dense but not closed in* $\mathcal{F}$. *If* (I.3.7) *holds,* $L\mathcal{F} = \mathcal{F}$.

ii) *Suppose that* $\alpha(A) = p \in \mathbb{Z}_+$. *Then the kernel of* L *in* $\mathcal{F}$ *consists of the functions*

(I.3.9) $$h(t) = t^p U(t,T)h_0 , \quad h_0 \in H_A^{-\infty} , \text{ in Case (I.3.6)} ,$$

(I.3.10) $$h(t) = t^p U(0,t)h_0 , \quad \text{in Case (I.3.7)}.$$

The range of L *consists in all cases of the functions* $f \in \mathcal{F}$ *satisfying*

(I.3.11) $$X^p f \big|_{t=0} = 0 .$$

Proof: i) $\alpha(A) \notin \mathbb{Z}_+$. The injectivity of L follows from Th. I.2.1. Th. I.3.2 (cf. also Remark I.3.3) implies that $L\mathcal{F}$ is not closed in $\mathcal{F}$ when (I.3.6) holds. Suppose now that (I.3.7) holds. Suppose first that

$$\text{(I.3.12)} \qquad \operatorname{Re} \alpha_0 < 0 \qquad \text{(cf. (I.2.5))}.$$

Given any $f \in \mathcal{F}$, the solution of $Lu = f$ is given by

$$\text{(I.3.13)} \qquad u(t) = \int_0^1 y^{-1-\alpha(A)}\, U(t,yt) f(yt)\, dy \quad .$$

When (I.3.12) does holdt (while (I.3.7) does) one reasons exactly as in Case # 2 in the proof of Th. I.2.1.

ii) $\alpha(A) = p \in \mathbb{Z}_+$. The assertion concerning the kernel of L follows at once from Th. I.2.1 and the description of Ker $L^{\#}$. We prove the assertion about the range of L.

Suppose that $f = Lu$, $u \in \mathcal{F}$. Then $X^p f = tX^{p+1}u$ (X is defined in (I.2.17)) whence the necessity of (I.3.11). Suppose now that (I.3.11) holds and set $X^p f = tg$. Let us define an operator G_X on $w \in \mathcal{F}$ in the following manner:

$$\text{(I.3.14)} \qquad G_X w(t) = -\int_t^T U(t,t')w(t')dt' \qquad \underline{\text{in Case}}\ \text{(I.3.6)},$$

$$\text{(I.3.15)} \qquad G_X w(t) = \int_0^t U(t,t')w(t')dt' \qquad \underline{\text{in Case}}\ \text{(I.3.7)},$$

where $U(t,t')$ is defined in (I.3.8). In all cases it is checked at once that G_X defines an endomorphism of $\mathcal{F}$ and that $X\, G_X$ = Identity of $\mathcal{F}$. We take then:

$$v = G_X^{p+1} g \quad ,$$

whence $tX^{p+1}v = X^p f$, i. e., $X^p(Lv - f) = 0$, which implies

$$v - f = U(t,T)h \quad \text{in Case (I.3.6)}, \qquad v - f = U(0,t)h \quad \text{in Case (I.3.7)},$$

with $h(t)$ a polynomial of degree $\leq p - 1$ with coefficients in $H_A^{-\infty}$. Let us determine $w(t)$ by the equation:

$$t\,\partial_t w - pw = t^{p+1}\partial_t(t^{-p}w) = h = \sum_{j=0}^{p-1} h_j t^j \quad ,$$

that is,

$$w(t) = - \sum_{j=0}^{p-1} (p-j)^{-1} h_j t^j \quad .$$

We take $u = v - U(t,T)w$ in Case (I.3.6), $u = v - U(0,t)w$ in Case (I.3.7).

Q. E. D.

So far we have regarded the systems under study as true operators. But in Ch. II we shall often think of its coefficients as asymptotic series of the kind (I.1.2), in other words, as equivalence classes modulo regularizing (in the sense of A) operators (on the scale of Hilbert spaces H_A^s). What is then one to make of assertions concerning the solvability of such a class of operators ? Note that there is no problem with hypoellipticity. Different approaches may be followed, to circumvent this ambiguity. First of all we may look at solvability in $\mathcal{K}/\mathcal{E}$ instead of $\mathcal{K}$. This lifts the ambiguity but weakens somewhat the results. It is the viewpoint of pseudodifferential operators theory, which we shall adopt in Ch. III. The quotient $\mathcal{K}/\mathcal{E}$ can be regarded as the space of "tangential" singularities (in the microlocalized version of Ch. III, it will rather be the space of "tangential" microfunctions).

Actually the preceding approach may well lead to precise solvability results - in spaces of functions and distributions, not only of singularities. This is due to the fact that, at some stage, the solvability derives from _a priori_ estimates involving the adjoint operator, and under favorable circumstances, the latter will be stable under perturbations due to the use of different cut-off functions along the spectrum of A. The article [6] contains detailed description of how this can come about.

Another possible approach, particularly adapted to evolution equations such as those considered here or as those in [15], is to restrict the meaning of the equivalence relation between coefficients: one requires that the difference, say R, between two equivalent coefficients, not only be regularizing (_i. e._, map $\mathcal{K}$ into $\mathcal{E}$), but be in a sense analytic-regularizing, that is, $RC^{\infty}([0,T];E_A^s) \subset C^{\infty}([0,T];E_A^{s+d})$ for every $s \in \mathbb{R}$, with d a number > 0, independent of s. This of course restricts substantially the choice of the cut-off functions, along the spectrum of A, appearing as multipliers of the powers of A^{-1} in the expansion of the differences R. If L, M are two equivalent systems (in this restricted sense) and if Mu = f, we will have Lu = f + (L - M)u. But by applying the version of the Cauchy-Kovalevska theorem stated here as Th. I.2.1 we may solve Lv = (L - M)u with $v \in \mathcal{E}$ (possibly after some decreasing of the length T of the interval $[0,T]$); then L(u - v) = f. An advantage of this approach is that it is compatible with the concatenation method we are going to apply in Ch. II. For the details see [14].

CHAPTER II

SECOND-ORDER ELLIPTIC FUCHSIAN EVOLUTION EQUATIONS

II.1 REDUCTION OF SECOND-ORDER FUCHSIAN EQUATIONS WITH DISTINCT CHARACTERISTICS. THE OPERATORS UNDER STUDY AND THEIR CONCATENATIONS

We shall be concerned with second-order Fuchsian scalar evolution operators whose principal part is of the form tXtY, where

(II.1.1) $$X = \partial_t - a(t,A)A, \qquad Y = \partial_t - b(t,A)A,$$

with $a(t,A)$, $b(t,A) \in \Psi^0$. This notation will be systematically used, and also the following one:

(II.1.2) $$\vartheta(t,A) = a(t,A) - b(t,A).$$

The simple characteristics hypothesis, alluded to in the section title, can be stated

(II.1.3) the leading coefficient $\vartheta_0(t)$ of $\vartheta(t,A)$ does not vanish at any point of $(0,T)$.

It follows from this that $\vartheta(t,A)$ is invertible in Ψ^0 and that consequently the most general second-order Fuchsian operator with principal part tXtY can be written in the form:

(II.1.4) $$Q = (tX - \alpha(t,A))(tY - \beta(t,A)) + \gamma(t,A), \quad \alpha, \beta, \gamma \in \Psi^0.$$

Lemma II.1.1.- Let Q be the operator (II.1.4) and suppose that (II.1.3) holds. There is an operator $w(t,A) \in \Psi^{-1}$ and an operator $\gamma^{\#}(A) \in \Psi^0$, independent of t, such that

$$(II.1.5) \qquad Q = (tX - \alpha(t,A) - w(t,A))(tY - \beta(t,A) + w(t,A)) + \gamma^{\#}(A) .$$

Proof: The formulas (II.1.4) and (II.1.5) imply

$$t(X - Y)w - tw_t - (\alpha - \beta)w + w^2 + \gamma^{\#} = \gamma ,$$

that is to say:

$$(II.1.6) \qquad t(\vartheta Aw + w_t) = w^2 - (\alpha - \beta)w - (\gamma - \gamma^{\#}) .$$

Suppose that we have determined $w_0 = 0, w_1, \ldots, w_j, \gamma^{\#}_{-1} = 0, \gamma^{\#}_0, \ldots, \gamma^{\#}_{j-1}$. We equate then the coefficient of A^{-j} on both sides of (II.1.6) and select $^{\#}_{j}$ so that the coefficient in the right-hand side vanishes at t = 0 ; thus $\gamma^{\#}_0 = \gamma_0(0)$. And this obviously enables us to determine w_{j+1}.

Remark II.1.1.- w and $\gamma^{\#}$ in Lemma II.1.1 are obviously unique.

Setting $X^{\#} = X - \frac{1}{t}\left[(\alpha + w)(t,A) - (\alpha + w)(0,A)\right]$, $Y^{\#} = Y - \frac{1}{t}\left[(\beta - w)(t,A) - (\beta - w)(0,A)\right]$, $\alpha^{\#}(A) = (\alpha + w)(0,A)$, $\beta^{\#}(A) = (\beta - w)(0,A)$, we may write $Q = (tX - \alpha^{\#})(tY - \beta^{\#}) + \gamma^{\#}$. Finally, dropping the superscripts #, we see that

$$(II.1.7) \qquad Q = (tX - \alpha(A))(tY - \beta(A)) + \gamma(A) \qquad (\alpha, \beta, \gamma \in \mathbb{C}[[A^{-1}]]).$$

Remark II.1.2.- An important feature of the reduction to (II.1.7) is that the leading part of X and Y have not been modified, nor the restriction to t = 0 of the leading coefficients of α, β, γ .

Remark II.1.3.- The operator Q is "equivalent to the first-order 2×2 system,

$$\text{(II.1.8)}\qquad L = t\left\{I\partial_t - \begin{pmatrix} a(t,A) & 0 \\ 0 & b(t,A)\end{pmatrix}A\right\} - \begin{pmatrix}\alpha(A) & -\gamma(A) \\ 1 & \beta(A)\end{pmatrix},$$

which is of the form studied in Ch. I. The indicial equation of L (and, by definition, that of Q) is then:

$$\text{(II.1.9)}\qquad (r - \alpha(A))(r - \beta(A)) + \gamma(A) = 0 .$$

We now make the assumption that Q is strongly elliptic, i. e., that the leading coefficients $a_0(t)$ and $b_0(t)$ have nonvanishing real parts with opposite signs, for every t, $0 \leq t \leq T$. Since we may exchange X and Y and redefine α, β, γ , we may as well assume that

$$\text{(II.1.10)}\qquad \operatorname{Re} a_0(t) > 0 , \qquad \operatorname{Re} b_0(t) < 0, \qquad \forall\, t \in [0,T].$$

The present chapter will be entirely devoted to finding necessary and sufficient conditions in order that, under Hypothesis (II.1.10) and possibly after decreasing the length T of the interval $[0,T]$, the operator Q be solvable (Def. I.3.1).

The first remark which can be made is that, under Hypothesis (II.1.10), if the indicial equation (II.1.9) has no root in $\mathbb{Z}_+$, then Q will not be solvable. Indeed it follows from Corollaries I.3.1, I.3.2 and Th. I.3.2 applied to the equivalent system (II.1.8).

From now on we make the hypothesis that

(II.1.11) the indicial equation (II.1.9) has at least one root which belongs to $\mathbb{Z}_+$.

We denote by r_0 the smallest nonnegative integer which is a root of (II.1.9).

Lemma II.1.2.- If $m \leq r_0$ ($m \in \mathbb{Z}_+$),

$$\text{(II.1.12)}\qquad \hat{Q}^{-1}(t^m \mathcal{K}) \cap (\mathcal{P}_m \otimes H_A^{-\infty}) = \{0\} .$$

We recall that $\mathcal{P}_m$ is the space of complex polynomials of degree $\leq m - 1$.

Proof: Let u be an element of the intersection at the left in (II.1.2) and write $u(t) = u_0 t^d + u_1 t^{d+1} + \ldots$, with $d \leq m - 1$. We must have $Q(u_0 t^d) \in t^{d+1}\mathcal{F}$, which in turn implies $\left[(d - \alpha(A))(d - \beta(A)) + \gamma(A)\right] u_0 t^d \in t^{d+1}\mathcal{F}$. Since d is not a root of (II.1.9), this is only possible if $u_0 = 0$.

Corollary II.1.1.- If $m \leq r_0$, Q induces a bijection of $\mathcal{P}_m \otimes H_A^{-\infty}$ onto a supplementary of $t^m \mathcal{F}$.

Set

$$(II.1.13) \qquad Q_m = (tX + m - \alpha(A))(tY + m - \beta(A)) + \gamma(A).$$

We have:

$$(II.1.14) \qquad Q(t^m u) = t^m Q_m u \quad ,$$

and if r is a root of the indicial equation of Q, r - m is one of that of Q_m. In particular, zero is a root of the indicial equation of Q_{r_0} and we may write

$$(II.1.15) \qquad Q_{r_0} = t\mathbb{P} \, , \qquad \mathbb{P} = XtY - (\alpha(A) - r_0)Y - (\beta(A) - r_0)X \, .$$

Theorem II.1.1.- If (II.1.11) holds, Q maps bijectively $\mathcal{P}_{r_0} \otimes H_A^{-\infty}$ onto a supplementary of $t^{r_0}\mathcal{F}$ in $\mathcal{F}$, and maps $t^{r_0}\mathcal{F}$ onto the subspace $t^{r_0+1}\mathbb{P}\mathcal{F}$.

Th. II.1.1 follows at once from Cor. II.1.1 and from (II.1.14)-(II.1.15). Of course we have:

$$(II.1.16) \qquad \mathcal{F} = (\mathcal{P}_{r_0} \otimes H_A^{-\infty}) \oplus (t^{r_0}\mathcal{F}) \quad .$$

Corollary II.1.2.- Q is solvable (Def. I.3.1) if and only if this is true of $\mathbb{P}$.

Remark II.1.4.- The solvability of an operator such as P (whose Fuchsian order is one) is not directly explained in Def. I.3.1. But its meaning is clear: it is equivalent to the solvability of tP (whose Fuchsian order is zero).

By virtue of Cor. II.1.2 we must study the solvability of P. If we use its expression (II.1.15) and the fact that $X = Y - \vartheta(t,A)A$, we see that

(II.1.17) $$P = XtY - \sigma(A)Y + \rho(t,A)A.$$

Lemma II.1.3.- Let P be the operator (II.1.17) and suppose that (II.1.3) holds. There is an operator $w(t,A) \in \Psi^0$ and a series $c(A) \in \mathbb{C}[[A^{-1}]]$ such that

(II.1.18) $$P = (X - w(t,A))t(Y + w(t,A)) - \sigma(A)(Y + w(t,A)) + c(A)A.$$

The series $w(t,A)$ and $c(A)$ are unique.

The proof duplicates that of Lemma II.1.1. By renaming X and Y we may assume that

(II.1.19) $$P = XtY - \sigma(A)Y + c(A)A .$$

Remark II.1.5.- The principal parts of X and Y have not been modified through the preceding reduction. Note also that, in the notation of (II.1.15), $\sigma(A) = (\alpha + \beta)(A) - 2r_0$, i. e., $\sigma(A)$ is the sum of the roots of the indicial equation of Q_{r_0} .

Next we construct the concatenation based on P ([15], Section II.4). We set $P^0 = P$ and

$(II.1.20)_j$ $$P^j = X^j t Y^j - \sigma^j(A)Y^j + c^j(A)A,$$

with

(II.1.21) $$X^j = \partial_t - a^j(t,A)A , \quad Y^j = \partial_t - b^j(t,A)A$$

(a^j , b^j , σ^j , $c^j \in \Psi^0$, the last two series being independent of t). We define P^{j+1} by the identity

(II.1.22) $$Y^j P^j = P^{j+1} Y^j \qquad (j = 0, 1,\ldots).$$

Setting $\vartheta^j = a^j - b^j$, we see that

$$P^{j+1} = Y^j X^j t - \sigma^j Y^j + c^j A \tag{II.1.23}$$

$$= X^j t Y^j + X^j - \sigma^j Y^j + c^j A - [X^j, Y^j] t$$

$$= X^j t Y^j - (\sigma^j - 1) Y^j + \{c^j - (\vartheta^j + t\vartheta^j_t)\} A ,$$

which shows that P^{j+1} is of the same general form (II.1.17) as P, and can therefore be brought into the reduced form (II.1.19), i. e., $(\text{II.1.20})_{j+1}$.

For future reference the following properties of the operators P^j ought to be emphasized:

(II.1.24) the principal parts of X^j and Y^j are equal to $\partial_t - a_0(t)A$ and to $\partial_t - b_0(t)A$ respectively (they are thus the same as those of X and Y);

(II.1.25) $\sigma^j(A) = \sigma(A) - j$;

(II.1.26) $c_0^j = c_0 - j\vartheta_0(0)$.

These assertions follow from (II.1.23) and from the transformation for a general operator of the form (II.1.17) into the form (II.1.19) (cf. proof of Lemma II.1.1). In (II.1.26) c_0^j denotes the leading coefficient in $c^j(A)$.

II.2 THE SOLVABILITY PROPERTIES OF THE EVOLUTION OPERATOR P

The introduction of the formal power series $c^j(A) \in \mathbb{C}[[A^{-1}]]$ enables us to describe completely the solvability properties of the operator P of (II.1.19), as we are now going to show. Thus we use the concatenation (II.1.22).

Lemma II.2.1.- The operator P^j induces an endomorphism of Ker Y^j . It induces an automorphism of Ker Y^j if and only if $c^j(A) \neq 0$. (*)

Proof: The first part follows from (II.1.22) and the second part from the fact that the restriction of P^j to Ker Y^j is equal to $c^j(A)A$ (by $(\text{II.1.20})_j$).

Lemma II.2.2.- The operator Y^j induces an isomorphism of Ker P^j onto Ker P^{j+1} if and only if $c^j(A) \neq 0$.

Proof: First suppose that $c^j(A) \neq 0$. If $P^j h = 0$, $h \neq 0$, we have $P^{j+1}(Y^j h) = 0$ but $Y^j h \neq 0$ by Lemma II.2.1. If $P^{j+1} h_1 = 0$, $h_1 \neq 0$, we apply Th. I.3.3 (part i) and (II.1.24): there is $w \in \mathcal{H}$ such that $Y^j w = h_1$. We have therefore $P^j w \in$ Ker Y^j and, again by Lemma II.2.1, there is $w_1 \in$ Ker Y^j such that $P^j w_1 = P^j w$. We have $w - w_1$ Ker P^j and $Y^j(w - w_1) = h_1$.

Suppose now that $c^j(A) = 0$. There is $w \neq 0$ such that $Y^j w = 0$ by Th. I.3.3 (part ii) with $p = 0$). Q. E. D.

Let us introduce the following notation:

$$(II.2.1) \qquad S^0 = I , \qquad S^j = Y^{j-1} \dots Y^0 \quad (j = 1, 2, \dots).$$

We have, by (II.1.22),

$$(II.2.2) \qquad S^j P = P^j S^j \quad (j = 0, 1, \dots).$$

Lemma II.2.3.- If $c^k(A) \neq 0$ for every $k < j$, P induces an automorphism of Ker S^j.

Proof: Induction on the length of S^j , i.e., on j. The assertion has already been proved when $j = 1$. We set $S^{1,j-1} = Y^{j-1} \dots Y^1$, assuming $j > 0$. We know then that P^1 induces an automorphism of Ker $S^{1,j-1}$. Observe that

$$(II.2.3) \qquad h \in \text{Ker } S^j \implies Y^0 h \in \text{Ker } S^{1,j-1} .$$

Suppose that $Ph = 0$: then $P^1(Y^0 h) = Y^0 Ph = 0$, hence, because of (II.2.3), $Y^0 h = 0$, hence $h = 0$ by virtue of Lemma II.1.1.

(*) In the present sections, kernels and ranges will always be relative to the space $\mathcal{H}$, unless other wise specified.

On the other hand, whatever $h \in \text{Ker } S^j$, there is $v \in \text{Ker } S^{1,j-1}$ such that $P^1 v = Y^0 h$. By Part i) of Th. I.3.3 there is $w \in \mathcal{F}$ such that $Y^0 w = v$, hence $w \in \text{Ker } S^j$ by (II.2.3). We have $Y^0 Pw = P^1 Y^0 w = Y^0 h$, hence $Pw - h \in \text{Ker } Y^0$. By Lemma III.1.1 there is $w_1 \in \text{Ker } Y^0$ such that $h = P(w - w_1)$.

Q. E. D.

Lemma II.2.4.- *If $c^k(A) \neq 0$ for every $k < j$, S^j induces an isomorphism of* Ker P *onto* Ker P^j.

Proof: Straightforward induction from the case $j = 1$ which is Lemma II.1.2.

Theorem II.2.1.- *Suppose that $c^j(A) = 0$ for some $j \in \mathbb{Z}_+$* (then necessarily unique in view of (II.1.26)).

#1) *If $\sigma(A)$ is not equal to an integer $> j$, $P\mathcal{F}$ is not closed in $\mathcal{F}$. Furthermore, S^j induces an isomorphism of* Ker P *onto* Ker Y^j.

2) *If $\sigma(A) - j - 1 = p \in \mathbb{Z}_+$, the range of P consists exactly of the functions $f \in \mathcal{F}$ satisfying:*

(II.2.4)
$$(X^j)^p S^j f \Big|_{t=0} = 0 .$$

Furthermore S^j induces an isomorphism of Ker P *onto the preimage of* $\text{Ker}(tX^j - p)$ *under* Y^j.

Proof: By virtue of $(\text{II.1.20})_j$ and (II.1.25) the hypothesis in Th. II.2.1 implies

(II.2.5)
$$P^j = (X^j t - \sigma(A) + j)Y^j .$$

Therefore in view of (II.2.2), we have:

(II.2.6)
$$S^j P = (tX^j - \sigma(A) + j + 1)S^{j+1} .$$

By Part i) of Th. I.3.3 we know that $S^{j+1}\mathcal{F} = \mathcal{F}$. Consequently, if we apply Lemma II.2.3, we see that the range of P consists of the functions f such that $S^j f$ belongs to the range of $(tX^j - \sigma(A) + j)$. The assertions in Th. II.2.1 about the range of P follow then from Th. I.3.3. The latter also implies the assertions about Ker P if we apply Lemma II.2.4.

Remark II.2.1.- In #1), not only $P\mathcal{F}$ is not closed in $\mathcal{F}$, but it does not even contain $\mathcal{E}$; it is, however, dense in $\mathcal{F}$ (cf. Remarks I.3.3, I.3.4).

Example II.2.1.- Take $P = tXY$, $X = \partial_t - A$, $Y = \partial_t + A$. Then (in Th. II.2.1) $j = 0$, $p = 0$. The range of P is equal to $\{ f \in \mathcal{F} \; ; \; f(0) = 0 \} = t\mathcal{F}$ (thus P is solvable!). The kernel of P consists of the functions $h \in \mathcal{F}$ which can be written:

$$\text{(II.2.7)} \qquad h(t) = e^{-tA} h_o - \tfrac{1}{2}\left[e^{-(T-t)A} - e^{-(T+t)A}\right] h_1 \qquad (h_o \, , \, h_1 \in H_A^{-\infty}) \, .$$

If we express h_1 in terms of $h(0) = h_o$ and of $h(T)$, we get a formula analoguous to that for harmonic functions in a slab:

$$\text{(II.2.8)} \qquad h(t) = \left[\sinh(TA)\right]^{-1} \left\{ \sinh\left[(T-t)A\right] \, h(0) + \sinh(tA)\, h(T) \right\} .$$

Remark II.2.2.- All the preceding statements remain true if we replace systematically $\mathcal{F}$ by $\mathcal{E}$ and $H_A^{-\infty}$ by H_A^{∞} .

Theorem II.2.2.- Suppose that $c^j(A) \neq 0$ for every $j \in \mathbb{Z}_+$.

Then, if $T > 0$ is small enough, the operator P maps $\mathcal{F} = C^{\infty}([0,T]; H_A^{-\infty})$ onto itself.

Proof: We write $\mathcal{F} \cong (\mathrm{Ker}\, S^j) \oplus (\mathcal{F}/\mathrm{Ker}\, S^j)$ (cf. (II.2.2)); note also that S^j defines an isomorphism $\dot{S}^j$ of $\mathcal{F}/\mathrm{Ker}\, S^j$ onto $\mathcal{F}$ (by Th. I.3.3). According to our hypothesis and to Lemma II.2.3, P induces an automorphism of $\mathrm{Ker}\, S^j$ (whatever $j \in \mathbb{Z}_+$). If $P^j\mathcal{F} = \mathcal{F}$ we see that $\dot{P} = (\dot{S}^j)^{-1} P \, \dot{S}^j$ defines an epimorphism of $\mathcal{F}/\mathrm{Ker}\, S^j$ onto itself. In other words, in order to prove Th. II.2.2, it suffices to prove that $P^j\mathcal{F} = \mathcal{F}$ for some integer $j \geq 0$. This will follow easily from results of

Bolley & Camus ([4], [5]). If we denote by $W^m(\mathbb{R}_+)$ the m-th Sobolev space on the positive half-line, both with respect to t and to A, that is to say, the space of functions u(t) valued in H, such that

$$A^p \partial_t^q u \in L^2(\mathbb{R}_+;H)\ ,\quad \forall\, p,q \in \mathbb{Z}_+\ ,\ p+q \leq m,$$

they introduce the following function spaces: (*)

$$W_1^m(\mathbb{R}_+) = \left\{ u \in W^{m-1}(\mathbb{R}_+)\ ;\quad tu \in W^m(\mathbb{R}_+) \right\} .$$

It is immediately seen that operators such as P (or P^j) define continuous linear mappings $W_1^{m+2}(\mathbb{R}_+) \longrightarrow W^m(\mathbb{R}_+)$. Consider then the following operator:

$$P_0^j = (\partial_t - a_0(0)A)t(\partial_t - b_0(0)A) - \sigma^j(A)(\partial_t - b_0(0)A) + c_0^j A .$$

We are regarding here the P^j as "true" operators, not as classes of operators. Thus $\sigma^j(A)$ denotes a convergent representative of the formal power series so denoted. Let λ_o be the infimum of the numbers λ belonging to the spectrum of A, and set

$$\text{(II.2.9)} \qquad m^j = \sup_{\lambda \geq \lambda_0} \left\{ \operatorname{Re} \sigma^j(\lambda) \right\} - 1/2 .$$

<u>Lemma II.2.5.- If</u> j <u>is large enough and if</u> $m > m^j$, P_0^j <u>induces an isomorphism of</u> $W_1^{m+2}(\mathbb{R}_+)$ <u>onto</u> $W^m(\mathbb{R}_+)$.

<u>Proof of Lemma II.2.5</u> : If we use the spectral decomposition of A, P_o^j gets transformed into the same operator but where A is now replaced by the real variable λ ranging over the spectrum of A and can then be put into the form $\lambda \wp_o^j$ by the change of variables $s = \lambda t$, setting

$$\wp_o^j = (\partial_s - a_0(0))s(\partial_s - b_0(0)) - \sigma^j(\lambda)(\partial_s - b_0(0)) + c_0^j .$$

(*) the special case, in one variable, of the spaces $W_k^m(R_+^{n+1})$ introduced in the same context by Baouendi and Goulaouic [2].

At this point we apply Th. 2.1 of [4]. The condition (C) in Th. 2.1, [4], requires that $c_0^j/\vartheta_0(0) \in \mathbb{Z}_+$ and, according to (II.1.26), is violated as soon as

$$j > \mathrm{Re}\left[c_0/\vartheta_0(0)\right] . \tag{II.2.10}$$

Let us then introduce the analogues in one variable (that is, without the presence of the operator A) of the spaces $W_1^m(\mathbb{R}_+)$: the space $\mathcal{W}_1^m(\mathbb{R}_+)$ of complex-valued functions $u \in H^{m-1}(\mathbb{R}_+)$ (which means that $\partial_s^p u \in L^2(\mathbb{R}_+)$ for all $p \leq m - 1$) such that $su \in H^m(\mathbb{R}_+)$. Th. 2.1, [4], states that if $\lambda \geq \lambda_0$ and $m > m^j$, $\mathcal{P}_0^j$ defines an isomorphism of $\mathcal{W}_1^{m+2}(\mathbb{R}_+)$ onto $H^m(\mathbb{R}_+)$. Since ∂_s maps continuously $\mathcal{W}_1^{m+2}(\mathbb{R}_+)$ into $H^m(\mathbb{R}_+)$, this isomorphism depends continuously on $\lambda \geq \lambda_0$. Reverting from λ to A yields the sought result.

Let now $g \in C^\infty(\mathbb{R})$, $g(\rho) = 1$ for $\rho < 1$, $g(\rho) = 0$ for $\rho > 2$. If we exploit the fact that $P^j - P_0^j$ is a polynomial of degree ≤ 2, with coefficients in Ψ^0, in tA and $t\partial_t$, we see that, given any $m = 0, 1, \ldots$, there is a constant $C_m > 0$ such that the norm of the bounded linear operator

$$g(t/\varepsilon)(P^j - P_0^j) : W_1^{m+2}(\mathbb{R}_+) \longrightarrow W^m(\mathbb{R}_+)$$

does not exceed $C_m\varepsilon$. It follows from this that there is a number $\varepsilon_m > 0$ such that, if $\varepsilon < \varepsilon_m$, the operator

$$P_0^j + g(t/\varepsilon)(P^j - P_0^j) \tag{II.2.11}$$

defines an isomorphism of $W_1^{m+2}(\mathbb{R}_+)$ onto $W^m(\mathbb{R}_+)$ $(m > m^j$; cf. Lemma II.2.5). Since (II.2.11) coincides with P^j for $t < \varepsilon$, we reach the following conclusion. Take $T < \varepsilon_m$ and $f \in \mathcal{F}$. Select $s \in \mathbb{R}$ such that $A^s f$ can be extended to $\mathbb{R}_+$ as an element of $W^m(\mathbb{R}_+)$ (this is always possible). Then there is $u \in \mathcal{D}'$ such that $A^s u \in W_1^{m+2}(\mathbb{R}_+)$ and that

$$P^j u = f \quad , \qquad 0 < t < T. \tag{II.2.12}$$

(We are supposing here that (II.2.10) holds and that $m > m^j$.)

The last step in the proof is a <u>hypoellipticity</u> result: we show that the solution u in (II.2.12) belongs to $\mathcal{F}$. Take $\varepsilon < T$, $s' < s$ and consider the equation

$$\text{(II.2.13)} \qquad \left[P_0^j + g(t/\varepsilon)(P^j - P_0^j)\right] v = P^j\left[g(2t/\varepsilon)A^{s'}u\right] .$$

Because of the ellipticity of P^j in the open half-line R_+ we know that $u \in C^\infty(]0,T];H_A^{-\infty})$. It follows that given any $m' \geq m$, we may choose s' so as to achieve that the right-hand side in (II.2.13) belongs to $W^{m'}(R_+)$. Take then $\mathcal{E} < \mathcal{E}_{m'}$ (which we may assume $\leq \varepsilon_m$). According to what was said earlier, there is a unique element v of $W_1^{m'+2}(R_+)$ satisfying (II.2.13). But since P^j is equal to (II.2.11) on the support of $g(2t/\mathcal{E})$, and in view of the uniqueness in $W_1^{m+2}(R_+)$, we must have $g(2t/\mathcal{E})A^{s'}u = v$ and therefore $\partial_t^p u \in L^2(0,T;H_A^{-s'})$, $\forall\ p \leq m'$. Letting m' go to $+\infty$ proves our contention.

<u>Remark II.2.3.</u>- It is obvious on inspection of the proof of Th. II.2.2 that the statement remains valid if we substitute $\mathcal{E}$ for $\mathcal{F}$. Actually we have the following hypoellipticity result, which does not require decreasing of T :

<u>Theorem II.2.3.</u>- <u>Suppose that</u> $c^j(A) \neq 0$ <u>for every</u> $j \in \mathbb{Z}_+$.

<u>Given any function</u> $u \in \mathcal{F}$, <u>if</u> $Pu \in C^\infty([0,T[;H_A^\infty)$, <u>the same is true of</u> u.

<u>Proof:</u> We use again the concatenation (II.1.22). Suppose that P^{j+1} has the property stated in Th. II.2.3 and let $u \in \mathcal{F}$ be such that $P^j u \in C^\infty([0,T[;H_A^\infty)$. Then by (II.1.22) we must have $Y^j u \in C^\infty([0,T[;H_A^\infty)$, but then this space also contains $c^j(A)Au = P^j u - (X^j t - \sigma^j(A))Y^j u$, hence u as $c^j(A)$ is invertible.

Thus it suffices to prove the statement with P^j substituted for P and for j large. This is done by duplicating the end of the proof of Th. II.2.2 (but taking now s and s' arbitrarily close to $+\infty$).

<u>Corollary II.2.1.</u>- <u>Same hypothesis as in Th. II.2.3. Let</u> $u \in \mathcal{F}$ <u>be such that</u> $u(T) \in H_A^\infty$. <u>Then, if</u> Pu <u>belongs to</u> $\mathcal{E}$, <u>so does</u> u.

Proof: Indeed, since P is regular elliptic in $]0,T]$, if $u(T) \in H_A^{\infty}$ and $Pu \in \mathcal{E}$, we have $u \in C^{\infty}(]0,T]; H_A^{\infty})$. Combining this with the conclusion in Th. II.2.3 yields the corollary.

Corollary II.2.2.- Same hypothesis as in Th. II.2.3. Every $u \in \mathrm{Ker}_{\mathcal{F}} P$ such that $u(T) \in H_A^{\infty}$ belongs to $\mathcal{E}$.

Remark II.2.4.- The hypoellipticity results in Th. II.2.3 and Corollaries II.2.1, II.2.2 could be made more precise, by relating the "tangential" regularity (i. e., the regularity in the sense of the operator A) of Pu and that of u. Indeed, the following result holds:

Theorem II.2.4.- Suppose that $c_0/\vartheta_0(0)$ is not a nonnegative integer.

Given any function $u \in \mathcal{F}$, if $Pu \in C^{\infty}([0,T[; H_A^s)$ for some $s \in \mathbb{R}$, then $u \in C^{\infty}([0,T[; H_A^{s+1})$.

The property in the conclusion of Th. II.2.4 could be called "tangential hypoellipticity with loss of one derivative" (since P is of order two; on related results valid for regular operator see [6]). Note also that the hypothesis that $\vartheta_0(0)^{-1}c_0 \notin \mathbb{Z}_+$ is equivalent with saying that every $c^j(A)$ is elliptic, i. e., invertible in $\mathbb{C}[[A^{-1}]]$. Anologuous results, but where the loss of tangential regularity exceeds one derivative, can be established on the hypothesis that for some $j \in \mathbb{Z}_+$, $c^j(A) = A^{-m}\gamma(A)$, where $\gamma(A)$ is invertible (of order zero).

II.3 THE SERIES $c^j(A)A$ AS EIGENVALUE ASYMPTOTICS

In general the power series $c^j(A) = \sum_{k=0}^{+\infty} c_k^j A^{-k}$ are not convergent. This is true even when the operator P (given by (II.1.17)) is a very simple evolution operator, for instance:

Example II.3.1 : $P = (\partial_t^2 - A^2)t + \rho(t)$,

where $\rho(t)$ is a real C^∞ function satisfying:

(II.3.1) $\rho^{(j)}(0) \geq 0$, $j = 1, 2, \ldots$.

The nonconvergence of the series $c^j(A)$ can occur at the very first step, that is when $j = 0$: the radius of convergence of the series $c(A)$ in (II.1.19) can already be zero. Indeed, inspection of the reduction to the form (II.1.19) shows that

(II.3.2) $c_0 = c_2 = 0$, $c_1 = \rho(0)$,

and for $j \geq 3$,

(II.3.3) $$c_j \geq \sum_{i=1}^{j-2} \frac{1}{(j-i-1)!} \rho^{(i)}(0)\, \rho^{(j-i-1)}(0) .$$

It is then easy to choose $\rho(t)$ satisfying (II.3.1) such that the radius of convergence of the series $c^0(A)$ be zero. One can even take ρ to be analytic in a neighborhood of the origin.

The reader will note that Example II.3.1 includes a nontrivial class of degenerate second-order elliptic partial differential equations: it suffices to take for A^2 a suitable self-adjoint extension of $-\Delta_x$ (x denotes the space variable). (On this, cf. [15], Example II.4.2.)

Following Maslov ([11], Part 2 ; see also [13]) we shall say that the formal power series $\lambda(A)A$, where $\lambda(z) \in \mathbb{C}[[z^{-1}]]$, is an eigenvalue asymptotic of the operator P if there is a function $u \in \mathcal{F}$ such that the following is true:

(II.3.4) $u(T) \in H_A^\infty$, $u \notin \mathcal{E}$;

(II.3.5) $Pu - \lambda(A)Au \in \mathcal{E}$.

In Eq. (II.3.5), P and $\lambda(A)$ may be interpreted as anyone of their convergent representatives: it does not matter which ones, since any change of representatives modifies the left-hand side by terms which belong to $\mathcal{E}$. Note also that the ellipticity of P for $0 < t \leq T$ implies that, if $u(T) \in H_A^\infty$, we have $u \in C^\infty(]0,T]; H_A^\infty)$.

Since $u \notin \mathcal{E}$, (II.3.4) means that u is singular in the sense of the operator A, "tangentially" as it were, at $t = 0$, whereas (II.3.5) states that $Pu - \lambda(A)Au$ is regular (both in the normal direction, i. e., with respect to t, and in the tangential one, i. e., as mesured in the spaces H_A^s) everywhere.

Theorem II.3.1.- The series $c^j(A)A$ $(j = 0, 1, \ldots)$ make up the set of all the eigenvalue asymptotics of P (given by (II.1.17)).

Proof: Let us first prove that $c^j(A)A$ is an eigenvalue asymptotic of P. Take h_0 in $H_A^{-\infty}$ but $h_0 \notin H_A^{\infty}$. Then, by virtue of (I.3.8), (II.1.10) and Remark II.1.5,

$$u(t) = \exp\left(\int_0^t b^j(s,A)A\, ds\right) h_0 \tag{II.3.6}$$

is a solution of $Y^j u = 0$ and satisfies (II.3.4). We have $P^j u = c^j(A)Au$, and thus we see that $c^j(A)A$ is an eigenvalue asymptotic of P^j. Let us use now the operator S^j defined in (II.2.1). By applying Th. I.3.3 we can find $v \in \mathcal{F}$ such that

$$S^j v = u, \qquad v(T) \in H_A^{\infty} \tag{II.3.7}$$

(it suffices to take $v = G_{Y^{j-1}} \ldots G_{Y^0} u$, using the notation (I.3.15)). By (II.2.2) we have

$$S^j \left[P - c^j(A)A\right] v = 0 . \tag{II.3.8}$$

We apply now Lemma II.2.3 not to P itself but rather to $P - c^j(A)A = P_{\#}$; the concatenation built on $P_{\#}$ is seen at once to be $P_{\#}^k = P^k - c^j(A)A$ and therefore, by virtue of (II.1.26), the hypothesis in Lemma II.2.3 is satisfied (S^j is not modified when we go from P to $P_{\#}$). If follows from this that there is $w \in \operatorname{Ker} S^j$ such that

$$\left[P - c^j(A)A\right](v - w) = 0. \tag{II.3.9}$$

It is seen at once that any function $f \in \operatorname{Ker} S^j$ satisfies $f(T) \in H_A^{\infty}$; since $S^j(v - w) = u$, we cannot have $v - w \in \mathcal{E}$. We have thus proved what we wanted.

Let us now show that every eigenvalue asymptotic of P is of the form $c^j(A)A$. Let $\lambda(A) \in \mathbb{C}[[A^{-1}]]$ be distinct from all the $c^j(A)$ and set $P_\lambda = P - \lambda(A)A$. The concatenation built on P_λ is $P^j_\lambda = P^j - \lambda(A)A$. It satisfies the hypothesis in Cor. II.2.1: if $Pu - \lambda(A)Au \in \mathcal{E}$ and $u(T) \in H^\infty_A$ we must have $u \in \mathcal{E}$, and thus $\lambda(A)A$ is not an eigenvalue asymptotic of P.

CHAPTER III

THE NONCOMMUTATIVE CASE

III.1 SECOND-ORDER ELLIPTIC FUCHSIAN OPERATORS WHOSE COEFFICIENTS ARE PSEUDODIFFERENTIAL OPERATORS IN SPACE VARIABLES

In the present chapter we take a quick look at operators like (II.1.17) but where the coefficients are "true" pseudodifferential operators, defined in a smooth manifold Γ. The variable in Γ will be denoted by x, the variable along the fibres in $T^*\Gamma$ by ξ ; in local coordinates, $x = (x^1,\dots,x^n)$, $\xi = (\xi_1,\dots,\xi_n)$. The operators under scrutiny will be of the kind:

$$\text{(III.1.1)} \qquad P = (\partial_t - a(x,t,D_x))t(\partial_t - b(x,t,D_x)) - \\ - \sigma(x,D_x)(\partial_t - b(x,t,D_x)) + c(x,t,D_x) .$$

We assume here that $a(x,t,D_x)$, $b(x,t,D_x)$, $c(x,t,D_x)$ are pseudodifferential operators on Γ, of order ≤ 1, depending smoothly on $t \geq 0$; $\sigma(x,D_x)$ is a ψdo on Γ of order zero. Other assumptions are possible, e. g., that a, b, c and σ have degrees different from one and zero respectively, but this introduces complications not especially illuminating. All ψdo's will be sums of homogeneous terms with integral degrees (≥ 0 or < 0); here again greater generality could be achieved, but we shall not try to do it.

We denote by $a_0(x,t,\xi)$, $b_0(x,t,\xi)$ the principal symbols of $a(x,t,D_x)$ and $b(x,t,D_x)$ respectively. We make the strong ellipticity hypothesis analoguous to (II.1.10):

$$\text{(III.1.2)} \qquad \mathrm{Re}\, a_0(x,t,\xi) > 0, \quad \mathrm{Re}\, b_0(x,t,\xi) < 0, \quad \forall (x,\xi) \in T^*\Gamma, \ \xi \neq 0, \forall t \geq 0$$

Example III.1.1: We may have $\Gamma = \mathbb{R}^n$, identified with the hyperplane $t = 0$ in $\mathbb{R}^{n+1}$, and $P = P(x,t,D_x,D_t)$ a second-order linear partial differential operator with smooth coefficients in $\overline{\mathbb{R}}^{n+1}_+ = \{(x,t) \in \mathbb{R}^{n+1} ; \ t \geq 0\}$. If $P_0(x,t,\xi,\tau)$ denotes the principal symbol of (III.1.2) (which is a polynomial in (ξ,τ) homogeneous of degree two), our hypotheses mean that $P_0(x,0,\xi,\tau) \equiv 0$ and that $\frac{1}{t} P_0(x,t,\xi,\tau)$ is strongly elliptic. Every partial differential operator with these properties can be put in the form (III.1.1); note that in general the coefficients a, b, c and σ will not be <u>differential</u> operators on $\mathbb{R}^n$, merely ψdo's.

Example III.1.2: This is the local version of Example III.1.1. Suppose that one studies a second-order degenerate (strongly) elliptic operator $L(y,D_y)$ in an open bounded subset Ω of $\mathbb{R}^{n+1}$. We may then take Γ to be an open submanifold of the boundary $\partial\Omega$ of Ω, on which the principal symbol $L_0(y,\eta)$ of $L(y,D_y)$ vanishes of order one, while $d(y,\eta)^{-1}L_0(y,\eta)$ is strongly elliptic in some neighborhood of Γ (in $\overline{\Omega}$). Then, in a neighborhood of each point of Γ and up to a nonvanishing factor, $L(y,D_y)$ will have the form (III.1.1) in which t will denote the normal variable, and x stand for a choice of the tangential variables.

We are concerned here with problems of solvability and hypoellipticity of the operator P, (III.1.1), in which the solution and the data are assumed to be smooth with respect to the time (or "transversal") variable. The "tangential" distribution spaces, playing the role of the spaces H^s_A of Chapters I & II, will be the Sobolev spaces $H^s_{loc}(\Gamma)$. This lead us to the spaces:

$$\text{(III.1.3)} \qquad \mathcal{F} = C^\infty([0,T];\mathcal{D}'^F(\Gamma)) , \quad \mathcal{E} = C^\infty([0,T];C^\infty(\Gamma)) ,$$

where $\mathcal{D}'^F(\Gamma) = \bigcup_{s\in\mathbb{R}} H^s_{loc}(\Gamma)$ is the space of distributions in Γ whose order is finite. We may ask whether the equation

$$\text{(III.1.4)} \qquad Pu = f \quad \underline{\text{in}}\ [0,T] \times \Gamma ,$$

has a solution in $\mathcal{F}$ whatever $f \in \mathcal{F}$ or whether every solution belongs to $\mathcal{E}$ whenever $f \in \mathcal{E}$. Actually, from a viewpoint more closely related to the considerations in Section II.3, we could study the range and kernel of P in $\mathcal{F}/\mathcal{E}$, that is, modulo functions which are smooth, both transversaly and tangentially. This also fits well within the framework of pseudodifferential operators (in the variable x) and lends itself to microlocalization, that is, localization in the cotangent bundle $T^*\Gamma$. One would then seek microlocal estimates (cf. [6]) which, after patching together, yield also local solvability, in $\mathcal{F}$ or in $\mathcal{E}$.

In all this we assume that the length T of the time interval is finite (results which would be valid for $T = +\infty$, i. e. global in time, would require more precise information and methods: cf. [4], [5]). We shall however not specify the value of T (actually, it might have to be decreased at various stages); throughout the sequel we presume that the operators are acting on fuctions defined in the whole real line (valued in spaces of distributions with respect to x) but vanishing identically as soon as t exceeds some positive value (the value T !).

III.2 LAPLACE INTEGRAL OPERATORS

In the present section we describe a class of operators which includes parametrices for the initial value problem (assuming that (III.1.2) holds):

$$\text{(III.2.1)} \qquad u_t - b(x,t,D_x)u = 0 \;, \qquad u\big|_{t=0} = u_0 \,.$$

These operators are the analogues, in the present context, of the Hermite operators used in [6]. If we wish to keep up this kind of terminology, they should be called Laplace (integral) operators. The archetype of the Hermite operators is given by

$$\text{(III.2.2)} \qquad e^{-t^2|D_x|}\, u(x) = (2\pi)^{-n}\int e^{ix\cdot\xi - t^2|\xi|}\, \hat{u}(\xi)d\xi \;,$$

acting from $\mathcal{E}'(\mathbb{R}^n)$ to $\mathcal{D}'(\mathbb{R}^n)\,\hat{\otimes}\, C^\infty(\mathbb{R}^1)$. That of the Laplace integral operators is given by

$$\text{(III.2.3)} \qquad e^{-t|D_x|}\, u(x) = (2\pi)^{-n}\int e^{ix\cdot\xi - t|\xi|}\, \hat{u}(\xi)d\xi \;,$$

acting from $\mathcal{E}'(\mathbb{R}^n)$ to $\mathcal{D}'(\mathbb{R}^n)\hat{\otimes} C^\infty(\bar{\mathbb{R}}_+)$. Needless to say, such operators are not new. They have been used in the study of elliptic boundary problems in more or less disguised form, practically in the same manner in which we are going to use them (to construct parametrices of (III.2.1)). We have no intention of developping systematically their theory; we shall content ourselves with pointing out those among their properties which are of interest to us. We shall not even attempt to give the more general definitions.

The Laplace integral operators act from $\mathcal{D}'(\partial\Omega)$ to $\mathcal{D}'(\bar{\Omega})$ and are smooth in the direction transversal to the boundary $\partial\Omega$. In a local chart $(x^1,\dots,x^n,t)$ of the manifold with boundary $\bar{\Omega}$ (with t as the "normal" variable), such an operator will be given by a formula:

$$\text{(III.2.3)} \qquad Ku(x,t) = (2\pi)^{-n}\iint e^{i\Phi(x,y,t,\theta)}\, k(x,y,t,\theta)\, u(y)\, dy\, d\theta\ .$$

The "phase-function" Φ is complex-valued, positive-homogeneous of degree one with respect to θ and C^∞ with respect to $t \geq 0$. Its real part, $\phi = \operatorname{Re}\Phi$, is a phase-function in the sense of Fourier integral operators (see [10], [7]). Its imaginary part vanishes at $t = 0$. The basic assumption about Laplace integral operators is that $t^{-1}\operatorname{Im}\Phi$ is positive-elliptic, meaning that

(III.2.4) <u>to every compact subset</u> $\mathcal{K}$ <u>of</u> $\mathbb{R}^{2n}\times\bar{\mathbb{R}}_+$ <u>there is a constant</u> $c_{\mathcal{K}} > 0$ <u>such that, for every</u> (x,y,t) <u>in</u> $\mathcal{K}$ <u>and every</u> θ <u>in</u> $\mathbb{R}_n$,

$$\text{(III.2.5)} \qquad \operatorname{Im}\Phi(x,y,t,\theta) \geq c_{\mathcal{K}}\, t|\theta|\ .$$

The characterisation of the amplitudes is subtler. Let Σ^m denote the space of functions $k(x,y,t,\theta)\in C^\infty(\mathbb{R}^{2n}\times\bar{\mathbb{R}}_+\times\mathbb{R}_n)$ having the following property:

(III.2.6) <u>given any compact subset</u> $\mathcal{K}$ <u>of</u> $\mathbb{R}^{2n}\times\mathbb{R}_+$, <u>any triplet</u> $(p,q,r)\in(\mathbb{Z}^n_+)^3$ <u>there are two positive constants</u> C, N, <u>depending on</u> $\mathcal{K}$, p, q, r, <u>such that, for every</u> (x,y,t) <u>in</u> $\mathcal{K}$ <u>and every</u> θ <u>in</u> $\mathbb{R}_n$,

$$\text{(III.2.7)} \qquad \left|D_x^p D_y^q D_\theta^r k(x,y,t,\theta)\right| \leq C(1+|t\theta|)^N(1+|\theta|)^{m-|r|}\ .$$

Actually we shall not limit ourselves to amplitudes belonging to Σ^m for some $m \in \mathbb{R}$ but shall consider asymptotic sums

(III.2.8) $$k = \sum_{j=0}^{\infty} k_j ,$$

with $k_j \in \Sigma^{m_j}$ for each j, where $\{m_j\}$ is a strictly decreasing sequence of real numbers, converging to $-\infty$ (note that the numbers N entering in (III.2.7) for each k_j - in the place of k - will in general depend on j and grow to $+\infty$).

It is permitted to deal with series (III.2.8) due to the following remark: if the function k in (III.2.3) belongs to Σ^m, K is a Fourier integral operator with phase $\phi = \mathrm{Re}\,\Phi$ and amplitude $k \exp(-\mathrm{Im}\,\Phi)$. Indeed, by virtue of (III.2.4) and (III.2.6), the latter is a "classical" symbol of degree m. In particular, if $\phi = (x - y)\cdot\theta$, K is a pseudodifferential operator of order m. In all cases, as m gets closer to $-\infty$, the operator K gets ever more regularizing. Consequently, by considering either partial sums of the whole infinite series in which each terms has been multiplied by a suitable cut-off function of θ and $t\theta$, the series (III.2.8) can be used to define an equivalence class of operators on $\mathcal{D}'(\mathbb{R}^n)$ modulo regularizing operators.

We shall assume that differentiation term by term of the series (III.2.8) with respect to t, of any order, yields another asymptotic series with similar properties (but possibly with different degrees m_j).

Let us say that $k = k(x,y,t,\theta)$ vanishes of infinite order at $t = 0$ if the following holds:

(III.2.9) <u>Given any j, $M \in \mathbb{Z}_+$ and any compact set $\mathcal{K} \subset \mathbb{R}^{2n} \times \bar{\mathbb{R}}_+$, any triple $(p,q,r) \in (\mathbb{Z}^n_+)^3$, there are constants C, N such that</u>

(III.2.10) $$|D^p_x D^q_y D^r_\theta k_j(x,y,t,\theta)| \leq C\, t^M (1 + |t\theta|)^N (1 + |\theta|)^{m-|r|} ,$$

$$\forall (x,y,t) \in \mathcal{K}, \quad \forall \theta \in \mathbb{R}_n .$$

Observe that (III.2.10) implies:

$$\text{(III.2.11)} \qquad |D_x^p D_y^q D_\theta^r k_j(x,y,t,\theta)| \leq C'(1+|t\theta|)^{M+N}(1+|\theta|)^{m-M-|r|},$$

and since M is arbitrary, we reach the following important conclusion:

Proposition III.2.1.- If the amplitude k vanishes of infinite order at $t = 0$, the operator K is regularizing, i. e., maps $\mathcal{E}'(\mathbb{R}^n)$ into $C^\infty(\mathbb{R}^n \times \bar{\mathbb{R}}_+)$.

Let now $S(x,D_x)$ be a classical pseudodifferential operator (of degree d) in $\mathbb{R}^n$. We note that the Taylor expansion,

$$\text{(III.2.12)} \qquad \sum_{\alpha \in \mathbb{Z}_+^n} \frac{1}{\alpha!}(\partial_\xi^\alpha S)(x,\phi_x)(i\operatorname{Im} \Phi_x)^\alpha ,$$

is an asymptotic series of the kind (III.2.8), since $|\operatorname{Im} \Phi_x| \leq \text{const.}t|\theta|$.
It is natural to denote it by

$$\text{(III.2.13)} \qquad S(x,\Phi_x) .$$

With this notation we may write:

$$\text{(III.2.14)} \qquad S(x,D_x)Ku(x,t) = (2\pi)^{-n}\iint e^{i\Phi(x,y,t,\theta)}\, k_S(x,y,t,\theta)u(y)dyd\theta ,$$

where:

$$\text{(III.2.15)} \qquad k_S(x,y,t,\theta) =$$

$$\sum_{\alpha \in \mathbb{Z}_+^n} \frac{1}{\alpha!}(\partial_\xi^\alpha S)(x,\Phi_x)\; D_{x'}^\alpha \left\{ e^{i\Phi_2(x',x,y,t,\theta)} k(x',y,t,\theta)\right\}\Big|_{x'=x} ,$$

$$\text{(III.2.16)} \qquad \Phi_2(x,x',y,t,\theta) = \Phi(x',y,t,\theta) - \Phi(x,y,t,\theta) - (x-x')\cdot\Phi_x(x,y,t,\theta) .$$

We have, exactly like in the case of real Φ,

$$\text{(III.2.17)} \qquad k_S = S(x,\Phi_x)k + (\partial_\xi S)(x,\Phi_x)\cdot D_x k + \tfrac{1}{2}\{(\partial_\xi^2 S)(x,\Phi_x)\cdot\Phi_{xx}\}k + \dots .$$

We use this formulas right away to determine a parametrix of Problem (III.2.1). We seek a Laplace integral operator K such that

(III.2.18) $$(\partial_t - b(x,t,D_x))K \sim 0 \quad , \qquad K\big|_{t=0} - I \sim 0 ,$$

the equivalence meaning that the left-hand sides are regularizing, the first one in (x,t), the second one in x alone. We know that $b(x,t,\xi) = \sum_{j\in \mathbb{Z}_+} b_j(x,t,\xi)$ where, for each j, b_j is positive-homogeneous with respect to ξ of degree $1 - j$. We shall take the amplitude k of K to be a series (III.2.8). First of all we require:

(III.2.19) $$\Phi_t + ib_0(x,t,\Phi_x) = 0 ,$$

(III.2.20) $$\Phi(x,y,0,\theta) = (x - y)\cdot\theta.$$

Since ib_0 is not real, by virtue of (III.1.2), in general Eq. (III.2.19) cannot be solved exactly (the initial value problem (III.2.19)-(III.2.20) can be solved exactly when $b_0(x,t,\xi)$ is an analytic function of (x,ξ)). But it always has a solution modulo C^∞ functions (in $\mathbb{R}^{2n} \times \bar{\mathbb{R}}_+ \times (\mathbb{R}_n \setminus \{0\})$) which vanish of infinite order at $t = 0$. It is convenient to reason when $|\theta| = 1$ and then extend the solution by homogeneity (of degree one) in θ. Clearly we may achieve that

(III.2.21) $$\left| \Phi(x,y,t,\theta) - (x - y)\cdot\theta + itb_0(x,0,\theta)\right| \leq \tfrac{1}{2}\left|\mathrm{Re}\ b_0(x,0,\theta)\right| t,$$

which insures that Condition (III.2.4) is satisfied.

Next we require

(III.2.22) $$\partial_t k_0 - (\partial_\xi b_0)(x,t,\Phi_x)\cdot D_x k_0 - \tilde{b}_1(x,t,\Phi_x,\Phi_{xx})k_0 = 0 ,$$

(III.2.23) $$k_0(x,y,0,\theta) = 1,$$

where

(III.2.24) $$\tilde{b}_1(x,t,\Phi_x,\Phi_{xx}) = b_1(x,t,\Phi_x) + \tfrac{1}{2}(\partial_\xi^2 b_0)(x,t,\Phi_x)\cdot\Phi_{xx} \quad .$$

For $j = 1, 2, \ldots$, we require:

(III.2.25) $$\partial_t k_j - (\partial_\xi b_0)(x,t,\Phi_x)\cdot D_x k_j - \widetilde{b_1}(x,t,\Phi_x,\Phi_{xx})k_j = F_j(x,t,k_0,\ldots,k_{j-1}),$$

(III.2.26) $$k_j(x,y,0,\theta) = 0,$$

where F_j is the expression one derives from Formula (III.2.15) (needless to underline, Eq. (III.2.19) is formally the eikonal equation, while Eqq. (III.2.22) and (III.2.25) are the transport equations, standard in geometrical optics - when all the functions are real valued). Here again we reason when $|\theta| = 1$; we solve the problems (III.2.22)-(III.2.23), (III.2.25)-(III.2.26), modulo functions which vanish of infinite order at $t = 0$, and then we extend the solution by homogeneity (of the proper degree) with respect to θ. Thus we see that, for each j, the term k_j in (III.2.8) can be taken to be positive-homogeneous of degree j in θ. In this manner one can choose Φ and k so as to satisfy

(III.2.27) $$(\partial_t - b(x,t,D_x))(e^{i\Phi}\, k) \sim 0\,, \qquad e^{i\Phi}\, k \Big|_{t=0} = e^{i(x-y)\cdot\theta}\,,$$

where, by Prop. III.2.1, the equivalence in the first equation is to be understood modulo symbols of degree $-\infty$ (depending smoothly on $t \geq 0$).

There are two notions of adjoint (or of transpose) associated with the operator K, according to whether we regard it as an operator on $\mathcal{E}'(\mathbb{R}^n)$, depending on the parameter $t \geq 0$, or else as an operator $\mathcal{E}'(\mathbb{R}^n) \to \mathcal{D}'(\mathbb{R}^n)\hat{\otimes} C^\infty(\bar{\mathbb{R}}_+)$. We denote by $\widetilde{K}(t)$ the adjoint of the former, by K^* that of the latter. We extend both as operators acting on distributions (the strict definition of adjoint makes them act only on smooth functions). Observe that $u \mapsto \widetilde{K}(t)u$ can also be regarded as mapping $\mathcal{E}'(\mathbb{R}^n)$ into $\mathcal{D}'(\mathbb{R}^n)\hat{\otimes} C^\infty(\bar{\mathbb{R}}_+)$. It is then a Laplace integral operator, namely:

(III.2.28) $$\widetilde{K}u(x,t) = (2\pi)^{-n}\iint e^{-i\overline{\Phi(y,x,t,\theta)}}\, \overline{k(y,x,t,\theta)}\, u(y)\, dy d\theta\ .$$

It is verified at once that

(III.2.29) $$(K^* v)(x) = \int_0^{+\infty} \widetilde{K}(t)v(x,t)\, dt \qquad (v \in \mathcal{E}'(\mathbb{R}^n)\hat{\otimes} C_c^\infty(\bar{\mathbb{R}}_+))\ .$$

If then $S(x,D_x)$ is a classical pseudodifferential operator on $\mathbb{R}^n$, it is easy to check that $KS(x,D_x)$ is also a Laplace integral operator with phase Φ (the computation of its amplitude is left to the reader). Indeed, it suffices to write:

$$K\,S(x,D_x) = (S(x,D_x)^* \tilde{K})^{\sim} , \tag{III.2.30}$$

and we have already seen that $S(x,D_x)^* \tilde{K}$ is a Laplace integral operator.

Actually we shall also need a modification of the operator K. We note that

$$K^*Ku(x) = (2\pi)^{-2n} \int_0^{+\infty} \iiiint e^{-i\overline{\Phi(y,x,t,\theta)}+i\Phi(y,z,t,\theta')} \times \overline{k(y,x,t,\theta)}k(y,z,t,\theta)\; u(z)\; dydz\; d\theta\; d\theta'\; dt. \tag{III.2.31}$$

A straightforward computation shows that K^*K <u>is a positive-elliptic pseudodifferential operator of order one on</u> R^n, which can be written $\rho(x,D_x)^*\rho(x,D_x)$, where $\rho(x,D_x)$ is positive-elliptic of order 1/2. We define:

$$K_Y = K\,\rho(x,D_x)^{-1/2} . \tag{III.2.32}$$

We have:

$$K_Y^*K_Y = I , \tag{III.2.33}$$

and

$$K_Y K_Y^* = \pi_Y \tag{III.2.34}$$

is essentially an orthogonal projector (in $L^2(\mathbb{R}^n \times \bar{\mathbb{R}}_+)$) on Ker Y, if we write

$$Y = \partial_t - b(x,t,D_x) . \tag{III.2.35}$$

Clearly we may solve the initial value problem (III.2.18) with $b(x,t+t',D_x)$ substituted for $b(x,t,D_x)$. The solution, which we denote by $K(t,t')$, satisfies

$$YK(t,t') \sim 0 \quad \underline{\text{for}}\ t \geq t' , \qquad K(t',t') \sim I. \tag{III.2.37}$$

This enables us to construct a <u>right inverse</u> to Y :

$$G_Y f(t) = \int_0^t K(t,t')f(t')dt' . \tag{III.2.38}$$

Thus, if

$$\text{(III.2.39)} \qquad Yu = f \quad \underline{\text{for}}\ t \geq 0\ , \qquad u\big|_{t=0} = u_0\ ,$$

we have

$$\text{(III.2.40)} \qquad u = K_Y u_0 + G_Y f\ .$$

Proposition III.2.2.- Let $m \in \mathbb{Z}_+$. Then $f \mapsto G_Y f$ is a continuous linear map of $H_c^m(\mathbb{R}^n \times \mathbb{R}_+)$ into $H_{loc}^{m+1}(\mathbb{R}^n \times \mathbb{R}_+)$.

Proof: It is a particular case of the main result (construction of a parametrix) in [16] and we shall content ourselves with indicating the basic reason for such a fact. Since $\partial_t G_Y f = b(x,t,D_x)G_Y f + f$, it suffices to show that $f \mapsto |D_x| G_Y f$ maps continuously $H_c^m(\mathbb{R}^n \times \mathbb{R}_+)$ into $H_{loc}^m(\mathbb{R}^n \times \mathbb{R}_+)$. We have

$$\text{(III.2.41)} \qquad \int_0^{+\infty} \| \, |D_x| G_Y f \|^2_{H^m(\mathbb{R}^n)} \, dt \leq C^2 \int_0^{+\infty} \| f \|^2_{H^m(\mathbb{R}^n)} \, dt\ ,$$

where:

$$\text{(III.2.42)} \qquad C = \sup_{0 \leq t \leq +\infty} \int_0^t ||| \, |D_x| K(t,t') \, ||| \, dt'\ ,$$

where $||| \ \ |||$ stands for the operator norm on $H^m(\mathbb{R}^n)$. It is not difficult to prove that C is finite, and the proposition follows then from (III.2.41).

III.3 THE EIGENVALUE ASYMPTOTICS . MASLOV'S QUANTIZATION RELATIONS

We return to the operator (III.1.1) under Hypothesis (III.1.2). We shall make use of an extension of Lemma II.1.3. Let us denote by Ψ^m ($m \in \mathbb{R}$) the space of pseudodifferential operators of order $\leq m$ in the x variables, which depend in C^∞ fashion on $t \geq 0$ (if needed, the variation of t may be restricted to a bounded closed interval $[0,T]$, $T > 0$). By pseudodifferential operators we always mean "classical" ones and in fact, asymptotic sums of homogeneous terms whose degrees differ by an integer. We set $X = \partial_t - a(x,t,D_x)$ (cf. (III.2.35)).

Lemma III.3.1.- There are two operators $w \in \Psi^0$, $\gamma \in \Psi^1$ such that

(III.3.1) $$P = \left[(X - w)t - \sigma\right](Y + w) + \gamma \;;$$

(III.3.2) $\left[Y + w, \gamma\right]$ vanishes of infinite order at $t = 0$.

Proof: In view of (III.1.1) we have $c = \gamma + X(tw) - twY - tw^2 - \sigma w$, i. e.,

(III.3.3) $$\gamma = c + tw\vartheta - \left[X, tw\right] + tw^2 + \sigma w.$$

We recall that $\vartheta = -(X - Y) = a - b$. We must solve

(III.3.4) $$\left[Y, c + tw\vartheta\right] = \left[c + tw\vartheta, w\right] + \left[Y + w, \left[X, tw\right] - tw^2 - \sigma w\right]$$

modulo elements of Ψ^1 flat at $t = 0$. The noteworthy feature of (III.3.4) is that the left-hand side has order one, whereas the right-hand side has order zero. If we translate Eq. (III.3.4) in terms of the homogeneous parts w_j of the symbol of w, setting $\psi_j = c_j + tw_j\vartheta_o$, we get a succession of equations

(III.3.5) $$\partial_t \psi_j + i\left\{b_o, \psi_j\right\} = F_j(w_0, \ldots, w_{j-1}) \, ,$$

where $\{\;,\;\}$ are the Poisson brackets and the functionals F_j can be computed out of (III.3.4) (in particular, $F_0 \equiv 0$). If we are to be able to extract the w_j out of the solution of (III.3.5), we must also impose the condition

(III.3.6) $\psi_j = c_j$ at $t = 0$.

In general the initial value problem (III.3.5)-(III.3.6) cannot be solved exactly (it can when all symbols are analytic with respect to (x, ξ)). It can always be solved modulo C^∞ functions of $(x, t, \xi) \in \mathbb{R}^n \times \bar{\mathbb{R}}_+ \times S_{n-1}$ flat at $t = 0$. We may then extend the solution by homogeneity (of the proper degree, $1 - j$) with respect to ξ. Q. E. D.

Remark III.3.1.- As already pointed out, when all intervening symbols are analytic with respect to (x, ξ) in $\mathbb{R}^n \times S_{n-1}$, instead of (III.3.2) we may obtain

(III.3.2') $\left[Y + w, \gamma\right] = 0$.

In this case, w and γ are unique (in the general C^∞ case there is no uniqueness).

Renaming X, Y, c in (III.1.1) we may now assume that

(III.3.7) $$P = (Xt - \sigma)Y + c,$$

(III.3.8) $[Y,c]$ <u>vanishes of infinite order at</u> $t = 0$.

It is easy now to define a concatenation starting from P ; indeed,

(III.3.9) $$YP = P^1Y + R$$

where $R = [Y, \gamma] \in \Psi^1$ is flat at $t = 0$ and

(III.3.10) $$P^1 = Y(Xt - \sigma) + c .$$

We put then P^1 under the form analoguous to (III.3.7) and repeat the operation. We define thus a sequence of operators

(III.3.11) $$P^j = (X^jt - \sigma^j)Y^j + c^j , \quad R^j = [Y^j, c^j] \ \underline{\text{flat at}}\ t = 0,$$

such that

(III.3.12) $$Y^jP^j = P^{j+1}Y^j + R^j \quad , \quad j = 0, 1, \ldots \ (P^0 = P,\ Y^0 = Y,\ R^0 = R).$$

We underline the fact that the remainder of the argument applies to microfunctions, i. e., to elements of $\mathcal{F}/\mathcal{E}$ (the microfunction nature of these elements relates to the tangential variables x : in the t direction they are smooth - up to the boundary $t = 0$). We decompose $\mathcal{F}/\mathcal{E}$ in the "orthogonal" sum

(III.3.13) $$\mathcal{F}/\mathcal{E} = (\mathrm{Ker}\ Y^j) \oplus (\mathrm{Im}\ Y^{j*}),$$

where Im Y^{j*} is identified with the range of the orthogonal projection $I - \pi_{Y^j}$ (see (III.2.34)). Note that Y^{j*}, the adjoint of Y^j, is hypoelliptic, in the sense that the preimage of $\mathcal{E}$ under Y^{j*} is contained in $\mathcal{E}$ (in this rather imprecise language we are focusing on the behaviour near $t = 0$ and disregarding the behavious for t large). By virtue of Prop. III.2.1, $R^j\ K_{Y^j}$, and therefore also

$R^j \pi_{Y^j}$, is regularizing. Consequently, (III.3.12) shows that (modulo $\mathcal{E}$!)

(III.3.14) $$P^j \operatorname{Ker} Y^j \subset \operatorname{Ker} Y^j .$$

Let us set $S^j = Y^{j-1} \dots Y^0$ ($j \geq 1$) and $S^0 = I$. We introduce the following subspaces of $\mathcal{F}/\mathcal{E}$:

(III.3.15) $$M^j = \operatorname{Im} S^{j*} \quad , \quad N^j = M^j \cap \operatorname{Ker} S^{j+1} \quad .$$

We see that

(III.3.16) $$M^j = N^j \oplus M^{j+1} ,$$

(III.3.17) S^j <u>is a bijection of</u> N^j <u>onto</u> $\operatorname{Ker} Y^j$ <u>and of</u> M^{j+1} <u>onto</u> $\operatorname{Im} Y^{j*}$ (hence of M^j onto $\mathcal{F}/\mathcal{E}$).

Whatever $j = 0, 1, \dots$,

(III.3.18) $$\mathcal{F}/\mathcal{E} = N^0 \oplus \dots \oplus N^j \oplus M^{j+1} \quad .$$

Note also that, for $j \geq 1$,

(III.3.19) $$S^j P = P^j S^j + R'^j ,$$

where:

(III.3.20) $$R'^j = \sum_{k=1}^{j} R_k^j \, \partial_t^{j-k} \quad , \quad R_k^j \in \Psi^k$$

We shall now prove that

(III.3.21) <u>the restriction of</u> R'^j <u>to</u> $\operatorname{Ker} S^k$ <u>is regularizing</u> ($j, k \geq 1$).

The proof is based on the following

<u>Lemma III.3.2.- Let</u> $w \in \Psi^0$, $Z = Y - w$. <u>Given any operator</u> $F \in \Psi^m$ ($m \in \mathbb{R}$) <u>there is another operator</u> $E \in \Psi^m$ <u>such that</u>

(III.3.22) $$YEK_Z = FK_Z \quad \text{(mod regularizing operators).}$$

Proof of Lemma III.3.2: We have $YEK_Z = (E_t - [b,E])K_Z + EYK_Z$. Since $YK_Z = wK_Z$ it suffices to solve

(III.3.23) $$E_t - [b,E] + Ew = F .$$

Again, in general this can only be done modulo elements of Ψ^m flat at $t = 0$. If R is such an element and E a solution of (III.3.23) with F + R instead of F, we have $YEK_Z = FK_Z + RK_Z$; but RK_Z is regularizing, according to Prop. III.2.1.

Lemma III.3.3.- Let $j, k \in \mathbb{Z}_+$. Given any operator $F \in \Psi^m$ there is another operator $E_{jk} \in \Psi^m$ such that

(III.3.24) $$S^j E_{jk} K_{Y^k} = FK_{Y^k} \quad \text{(mod regularizing operators).}$$

Proof of Lemma III.3.3: Induction on $j = 0, 1, \ldots$ (S^0 = identity). By Lemma III.3.2 we select G such that $Y^{j-1} GK_{Y^k} = FK_{Y^k}$; then we take E_{jk} satisfying $S^{j-1} E_{jk} K_{Y^k} = GK_{Y^k}$. Q. E. D.

Remark III.3.1.- We are free to choose the value at $t = 0$ of the operator E in Lemma III.3.2, for instance take E elliptic at $t = 0$ (and therefore also for $t \geqslant 0$ small). Similarly we may choose the values at $t = 0$ of the first $j-1$ t-derivatives of E_{jk} in Lemma III.3.3.

Proof of (III.3.21): Note that (III.3.21) can be extended trivially to $k = 0$. We reason by induction on $k \geq 0$. By Lemma III.3.3 any $u \in \text{Ker } S^{k+1}$, i. e., such that $S^k u \in \text{Ker } Y^k$, is of the form $u = E_k K_{Y^k} h + u_1$, with $E_k \in \Psi^0$, $u_1 \in \text{Ker } S^k$ and h a distribution in $\mathbb{R}^n$. The induction on k implies $R'^j u_1 \in \mathcal{E}$, and Prop. III.2.1 that $R'^j E_k K_{Y^k}$ is regularizing.

If we combine (III.3.20) and (III.3.21) we reach the conclusion that

(III.3.25) $$P \text{ Ker } S^j \subset \text{Ker } S^j \quad (j = 1, 2, \ldots).$$

This means that in the direct sum decomposition (III.3.18) of $\mathcal{F}/\mathcal{E}$, P is represented by a triangular matrix; in order that P be injective (resp. surjective) it is necessary and sufficient that every diagonal entry of this matrix be so. In other words (and speaking in the language of microfunctions), the hypoellipticity (resp., solvability) properties of P depend on those of its restrictions to N^j ($j = 0,\dots, J-1$) and to M^J - for J arbitrarily large.

If we use the fact that S^j is a bijection of N^j onto Ker Y^j ((III.3.17)) and the relation (III.3.19) (also taking (III.3.21) into account) we see that

$$S^jP(S^j)^{-1} = P^j \text{ \underline{on} Ker } Y^j \pmod{\mathcal{E}}. \tag{III.3.26}$$

We have

$$P^j K_{Y^j} = c^j(x,t,D_x)K_{Y^j}, \tag{III.3.27}$$

and we know that $R^j = \left[Y^j, c^j\right]$ is flat at $t = 0$.

<u>Lemma III.3.4.</u>- <u>If</u> $\left[Y,c\right]$ <u>vanishes of infinite order at</u> $t = 0$, <u>we have</u>:

$$c(x,t,D_x)K_Y = K_Y c(x,0,D_x) \quad \text{(mod regularizing operators).} \tag{III.3.28}$$

<u>Proof</u>: We have $Y(cK_Y) = [Y,c]K_Y$ since $YK_Y = 0$; by the hypothesis and by Prop. III.2.1, $[Y,c]\,K_Y$ is regularizing. Thus $Y(cK_Y) = 0$ and $cK_Y\big|_{t=0} = c(x,0,D_x)$ whence (III.3.28).

If we apply Lemma III.3.4 to (III.3.27) we reach the conclusion that

$$P^j K_{Y^j} = K_{Y^j}\, c^j(x,0,D_x), \tag{III.3.28}$$

which implies

$$K^*_{Y^j}\, P^j K_{Y^j} = c^j(x,0,D_x) \;. \tag{III.3.29}$$

This shows that the properties of P restricted to N^j are equivalent, via the isomorphism $K^*_{Y^j} S^j$, to those of $c^j(x,0,D_x)$ - as a pseudodifferential operator in $\mathbb{R}^n$

This is now completely analoguous to the role of the $c^j(A)A$ in Ch. II, and we may regard the operators $c^j(x,0,D_x)$ as the eigenvalue asymptotics of P. The very same argument which led us to (II.1.26) leads us here to the formula giving the principal symbols of these eigenvalues:

$$c_0^j(x,0,\xi) = c_0(x,0,\xi) - j\vartheta_0(x,0,\xi) \qquad (j = 0, 1,\dots). \tag{III.3.30}$$

We come now to the restriction of P to M^{j+1} (for j large). We use once more (III.3.17). Here, however, (III.3.19) does not lead to (III.3.26), but only to

$$S^jP(S^j)^{-1} = P^j + R'^j(S^j)^{-1}, \tag{III.3.31}$$

and $R'^j(S^j)^{-1}$ is not, outside of trivial cases, regularizing. But modulo regularizing operators, it is equivalent to $R'^jG_{Y^0}\dots G_{Y^{j-1}}$ (G_Y is defined in (III.2.38); see also (III.3.21)). Repeated application of Prop. III.2.2 shows that $R'^jG_{Y^0}\dots G_{Y^{j-1}}$ is a bounded operator in $H^m(\mathbb{R}^n \times [0,T])$ and since R'^j vanishes of infinite order at $t = 0$, it has arbitrarily small norm provided that $T > 0$ is small enough. At this point we use the main result of Bolley & Camus [5] (or rather its obvious extension to pseudodifferential operators) which states that, if j and m are large enough, P^j induces an isomorphism of $W_1^{m+2}(\mathbb{R}_+^{n+1})$ onto $H^m(\mathbb{R}_+^{n+1})$ (we assume that some coefficients of P^j have been suitably modified at infinity in $\mathbb{R}_+^{n+1}$). This is equivalent with a priori estimates bearing on P^j and on its adjoint. Microlocally such estimates will be stable by small perturbations such as $R'^j(S^j)^{-1}$ and lead to the bijectivity of the right-hand side of (III.3.31) - in the proper microlocal sense and context.

Thus, to a large extent, the (microlocal) solvability and/or hypoellipticity properties of P are equivalent to those of its eigenvalue asymptotics $c^j(x,0,D_x)$. It should be pointed out, however, that the situation is not as simple, here, as in the case of regular operators, such as those studied in [6] for instance. Indeed, as indicated in Th. II.2.1, even when $c^j(x,0,D_x)$ fails to be surjective for some $j \in \mathbb{Z}_+$, for instance when $c^j(x,0,D_x)$ vanishes identically, the operator

P might turn out to be locally (or microlocally) solvable - depending on the zero-order operator $\sigma(x,D_x)$. The phenomena are then much less simple (in the general noncommutative case) and we shall not attempt to analyze them here.

We conclude with a few remarks pointing to the link between Formula (III.3.30) and Maslov's quantization relations ([12]). We follow closely Rockland [13], while making the adaptations required by the degenerate nature of the operator P under study. As before we reason microlocally and identify the boundary Γ of Ω to R^n, its cotangent bundle minus the zero section, $\dot{T}^*\Gamma$, to $R^n \times (R_n \setminus \{0\})$. Maslov's quantization relations yield the principal symbols of the eigenvalue asymptotics and relate them to certain Lagrangian manifolds (curves in our case). In the present circumstances the relevant information lies in the dominant part of the operator near $t = 0$, namely

(III.3.32) $$(X_0 t - \sigma_0)Y_0 + c_0(x,0,D_x) ,$$

where:

(III.3.33) $$X_0 = \partial_t - a_0(x,0,D_x) , \quad Y_0 = \partial_t - b_0(x,0,D_x)$$

and $\sigma_0 = \sigma_0(x,D_x)$. Note that (III.3.32) is equal to P modulo $\Psi^0 + t\Psi^1 + t^2\Psi^2$. We shall make what is essentially a hypothesis of self-adjointness concerning the part $(X_0 t - \sigma_0)Y_0$: we assume that $b_0(x,t,\xi) = -\overline{a_0(x,0,\xi)}$ and that $\sigma_0 \equiv 0$. Then the operator under study, (III.3.32), is equal to

(III.3.34) $$- X_0 t X_0^* + c_0(x,0,D_x)$$

mod $\Psi^0 + t\Psi^1$. Its principal symbol is equal to

(III.3.35) $$p(x,t,\xi,\tau) = - t \left|\tau + i a_0(x,0,\xi)\right|^2 ,$$

while its subprincipal symbol is equal, at $t = 0$, to

(III.3.36) $$(\mathrm{Re}\, a_0 + c_0)(x,0,\xi) .$$

With each point (x,ξ) of $\dot{T}^*\Gamma$ we associate a one-parameter family of curves $\{\gamma_E\}$ $(E > 0)$; γ_E is defined by the equation:

(III.3.37) $$\rho(x,t,\xi,\tau) = -E\ , \quad t>0\ .$$

The curve γ_E is regarded as a Lagrangian submanifold of the symplectic plane $\Pi_{x,\xi}$, the (t,τ)-coordinate plane through (x,ξ) equipped with the symplectic form $dt\wedge d\tau$. Only certain values of E are admissible, those which satisfy Maslov's relation:

(III.3.38) $$\frac{1}{2\pi}\oint_{\gamma_E}\eta = \frac{1}{4}\,\mathrm{Ind}\,\gamma_E \quad \underline{\mathrm{mod}}\ \underline{\mathrm{one}}\ ,$$

where $\mathrm{Ind}\,\gamma_E$ is the Maslov index of γ_E (see [11], [7]) and η, a one-form such that $d\eta = dt\wedge d\tau$. As we shall see the admissible values of E form a discrete sequence; furthermore they depend smoothly on $(x,\xi) \in \dot{T}^*\Gamma$ and are positive-homogeneous of degree one with respect to ξ. If $E = E(x,\xi)$ satisfies (III.3.38) there will be an eigenvalue asymptotic of (III.3.34) and, in fact, of P itself, whose principal symbol is equal to

(III.3.39) $$-E(x,\xi) + (\mathrm{Re}\,a_0 + c_0)(x,0,\xi)\ .$$

It is convenient to choose $\eta = t\,d\tau$. We have

(III.3.40) $$\oint_{\gamma_E} t\,d\tau = \pi\,E/(\mathrm{Re}\,a_0(x,0,\xi))\ .$$

On the other hand, in computing the Maslov index of γ_E we encounter the difficulty that it is not a compact cycle. But the canonical transformation

(III.3.41) $$t = s^2\ , \quad \tau = \sigma/2s\ ,$$

transforms γ_E into the ellipse

(III.3.42) $$\left|\tfrac{1}{2}\sigma + i a_0(x,0,\xi)s\right|^2 = E$$

oriented counterclockwise. There are two points on this ellipse with vertical tangents, i. e., with tangent parallel to the σ-axis, those with coordinates

$$s = \pm\ E^{\frac{1}{2}}/\mathrm{Re}\,a_0(x,0,\xi)\ , \quad \sigma = 2s\ \mathrm{Im}\,a_0(x,0,\xi)\ .$$

At both points $ds/d\sigma$ changes sign from + to - as we run along the ellipse. Recalling that the fundamental symplectic form is $d\sigma \wedge ds$ (or $d\tau \wedge dt$), we see that the Maslov index of the ellipse is + 2. Accordingly we take Ind $\gamma_E = 2$ (which amounts to regard the vertex of γ_E, i. e., the point with coordinates

$$t = E/(\mathrm{Re}\, a_0(x,0,\xi))^2 \quad , \quad \tau = \mathrm{Im}\, a_0(x,0,\xi) \ ,$$

as a double singularity for the first coordinate map $(t,\tau) \to t$, with plus sign).

Finally, from (III.3.38) and (III.3.40) we conclude that

$$\text{(III.3.43)} \qquad E = (2j+1)\,\mathrm{Re}\, a_0(x,0,\xi) \quad , \quad j \in \mathbb{Z}_+ \, .$$

Replacing (III.3.43) into (III.3.39) shows that the latter is equal to

$$\text{(III.3.44)} \qquad (c_0 - 2j\,\mathrm{Re}\, a_0)(x,0,\xi) \ .$$

Since, under our present hypotheses, $2a_0 = \vartheta_0$, (III.3.44) is exactly equal to the right-hand side in (III.3.30).

Bibliographical references

[1] Alinhac, S.- Problème de Cauchy pour des opérateurs singuliers, to appear.

[2] Baouendi, S. M. & Goulaouic, Ch.- Regularité et théorie spectrale pour une classe d'opérateurs elliptiques et dégénérés, Arch. Rat. Mec. Anal. 34, n° 5 (1969), 361-379.

[3] Baouendi, S. M. & Goulaouic, Ch.- Cauchy problems with characteristic initial hypersurface, Comm. Pure Applied Math.,

[4] Bolley, P. & Camus, J.- Sur une classe d'opérateurs elliptiques et dégénérés à une variable, J. Math. pures et appl., 51 (1972), 429-463.

[5] Bolley, P. & Camus, J.- Sur une classe d'opérateurs elliptiques et dégénérés à plusieurs variables, Bull. Soc. math. France, Memoir 34 (1973) 55-140.

[6] Boutet de Monvel, L. & Treves, F.- On a class of pseudodifferential operators with double characteristics, Inventiones math. 24 (1974), 1-34.

[7] Duistermaat, J. J.- Fourier Integral Operators, lecture notes, Courant Institute Math. Sciences, N. Y. U. (1973)

[8] Gilioli, A. & Treves, F.- An example in the solvability theory of linear PDEs, to appear in Amer. J. of Math.

[9] Grushin, V. V.- On a class of hypoelliptic pseudodifferential operators degenerate on a submanifold, Mat. Sbornik 84 (126) (1971) 111-134 (Math. USSR Sbornik 13 (1971), 155-185).

[10] Hörmander, L.- Fourier Integral Operators I, Acta Math. 127 (1971), 79-183.

[11] Kohn, J. J. & Nirenberg, L.- Degenerate elliptic-parabolic equations of second order, Comm. Pure Applied Math. XX (1967), 797-872.

[12] Maslov, V. P.- Théorie des perturbations et méthodes asymptotiques (French transl.) Dunod & Gauthier-Villars, Paris (1972).

[13] Rockland, Ch.- Hypoellipticity and eigenvalue asymptotics, to appear.

[14] Tosques, M.- Analytic-hypoellipticity of certain second-order evolution with double characteristics, to appear in Trans. Amer. Math. Soc.

[15] Trèves, F.- Concatenations of Second-Order Evolution Equations applied to local solvability and hypoellipticity, Comm. Pure Applied Math. XXVI (1973), 201-250.

[16] Trèves, F.- A new method of proof of the subelliptic estimates, Comm. Pure Applied Math. XXIV (1971), 71-115.

[17] Trèves, F.- Basic linear partial differential equations, to appear Academic Press New York 1975.

ON MASLOV'S QUANTIZATION CONDITION

Alan Weinstein
University of California, Berkeley

0. INTRODUCTION.

Let X be a riemannian manifold, $K: T^*X \to \mathbb{R}$ the function assigning to each cotangent vector the square of its length. It has been known for some time [KE][MA] that the spectrum of the Laplace-Beltrami operator on X is related to the lagrangian submanifolds of T^*X on which K is constant. In a recent review article, Duistermaat [DU_2] has given a very clear exposition of part of Maslov's work, including a result of which the following theorem is a special case

0.1 <u>Eigenvalue Theorem.</u> Let $L \subseteq T^*X$ be a compact lagrangian submanifold and $E > 0$ a real number satisfying the following conditions:

(i) $K \equiv E$ on L;

(ii) the geodesic flow on T^*X, restricted to L, leaves invariant a non-zero density;

(iii) (Maslov's quantization condition) for every closed curve γ on L,

$$\frac{1}{2\pi} \int_\gamma \omega_X - \frac{1}{4} m_L(\gamma) \in \mathbb{Z} ,$$

where ω_X is the fundamental 1-form on T^*X, and $m_L \in H^1(L;\mathbb{Z})$ is the Keller-Maslov-Arnold-Hörmander characteristic class of L. ([HÖ],§3.3).

Let d be the smallest element of the set $\{1,2,4\}$ for which $dm_L(\gamma) \equiv 0 \pmod 4$ for all closed curves γ on L.

Then there is a sequence $\{\lambda_j\}_{j=0}^{\infty}$ of eigenvalues of the Laplace-Beltrami operator on X such that the sequence $\{|\lambda_j - E(dj+1)^2|\}_{j=0}^{\infty}$ is bounded.

Duistermaat's proof of this theorem involves the consideration of the asymptotic behavior of distributions with respect to a continuous parameter tending to infinity, although the statement of the theorem does not involve any such asymptotics. The purpose of this paper is to give a more "geometric" proof of the Eigenvalue Theorem, via an interpretation of Maslov's quantization condition which lies entirely within the theory of Fourier integral operators (henceforth to be called FIO's). To carry out this proof, it has been necessary to extend the calculus of FIO's to include products in which the transversality condition in [HÖ] is replaced by a weaker assumption of "clean intersection." I learned at this symposium that Guillemin had developed the same extension for another reason. I will refer, therefore, to his paper [GL] for the analytic details of the "clean product" construction, which should now be considered as an integral (pun intended) part of the calculus of FIO's.

It appears that the theory of FIO's with complex-valued phase functions, as described at this symposium by Melin and Sjostrand [M-S] should permit the extension of the eigenvalue theorem to certain cases in which L is allowed to be an isotropic submanifold of any dimension. In particular, it should be possible to associate a sequence of eigenvalues to a stable periodic orbit. (The stability enters when one looks for a positive lagrangian submanifold of which the periodic orbit is the real part.) This would provide a rigorous justification for work of Gutzwiller [GT] and Voros [VO]. Work on the details of this extension is currently in progress and will, hopefully, be reported upon in a future publication.

Besides containing a proof of the eigenvalue theorem and a discussion of clean products, this paper should be of some use, I hope, to people learning about FIO's who wish to see examples of calculations involving Maslov line bundles. I have tried not to skip over too many details in these calculations, with this aim in mind.

I would like to thank Professor Chazarain and the University of Nice for the invitation and financial assistance which made it possible for me to attend this symposium. Further research and travel support came from NSF grant GP-347-85X3 and the University of California committee on research. Finally, I would like to thank Hans Duistermaat and Victor Guillemin for numerous stimulating discussions.

1. CONSTRUCTION OF A CANONICAL RELATION

The principal geometric idea in this paper is that one may construct, starting with a lagrangian submanifold $L \subseteq T_0^*X$ satisfying Maslov's quantization condition, a conic lagrangian submanifold $\Lambda \subseteq T_0^*X \times T_0^*S^1$. (The subscript 0 means that the zero section has been deleted.) Fourier integral distributions associated with Λ may be interpreted in three different ways. As distributions on $X \times S^1$, they are candidates for periodic solutions of the wave equation on X. As operators from $\mathcal{D}'(X)$ to $\mathcal{D}'(S^1)$, they may be thought of as distributions on X with values in $\mathcal{D}'(S^1)$; this fits into Maslov's general setting of distributions with values in an infinite-dimensional vector space. Finally, as operators from

$\mathcal{D}'(S^1)$ to $\mathcal{D}'(X)$, they may be thought of as ways of identifying the "standard" space $\mathcal{D}'(S^1)$ with a space of distributions on X. This third interpretation will be the fundamental one for our purposes, with the wave equation viewpoint playing a brief role at the end of the argument. (Note: $\mathcal{D}'(X)$ should always be understood to mean distributional $\frac{1}{2}$-densities on X.)

We begin now with a differentiable manifold X of dimension $n \geq 1$ and a connected lagrangian submanifold $L \subseteq T_0^*X$. The class $m_L \in H^1(L;\mathbb{Z})$, reduced modulo 4, may be considered as a homomorphism from $\pi_1(L)$ to $\mathbb{Z}_4$. This determines a principal $\mathbb{Z}_4$ bundle $\tilde{L}$ over L having the given homomorphism as its holonomy map. Associated with $\tilde{L}$, via the homomorphism $\mathbb{Z}_4 \to U(1)$ which takes $g(\mathrm{mod}\ 4)$ to $e^{\frac{\pi}{2}ig}$, is a flat hermitian line bundle M_L called the Maslov bundle of L. (See [ST] for generalities on bundles and [HÖ] for particulars on Maslov classes and bundles.) The Maslov quantization condition says that

$$e^{i\int_\gamma \omega_X} = e^{\frac{\pi}{2}im_L(\gamma)}$$

for every closed curve γ on L, so one obtains the rotation angle for the holonomy of M_L around a closed curve γ on L by integrating ω_X around γ. We observe, in addition, that the pullback of the bundle $\tilde{L}$ to the space $\tilde{L}$ via the projection $\tilde{L} \to L$ is trivial (the pullback of any bundle to itself being trivial), so the pullback to $\tilde{L}$ of m_L is zero modulo 4.

For $d = 1,2,$ or 4, we identify the group $\mathbb{Z}_d \approx \mathbf{Z}/d\mathbf{Z}$ with the subgroup of $\mathbb{Z}_4$ generated by $\frac{4}{d}$ (mod 4).

1.1 <u>Lemma.</u> The following are equivalent, for $d = 1,2,$ or 4:

(i) the structure group of L can be reduced to $\mathbb{Z}_d$;

(ii) $dm_L \equiv 0 \pmod 4$

(iii) $\tilde{L}$ contains an open subset which is a d-fold covering of L.

Proof. (i) means that $\tilde{L}$ contains a principal $\mathbb{Z}_d$ bundle, which implies (iii). Next, assume (iii) and consider any closed curve γ on L. The holonomy of $\tilde{L}$ around γ is $m_L(\gamma) \pmod 4$. By (iii), this element of $\mathbb{Z}_4$, acting on $\mathbb{Z}_4$ by translations, leaves invariant a set having d elements, so it must contained in $\mathbb{Z}_d$. So $dm_L(\gamma) \equiv 0 \pmod 4$, implying (ii). Finally, it follows from (ii) that $m_L(\gamma) \pmod 4$ belongs to $\mathbb{Z}_d$ for all γ, which implies (i).

Q.E.D.

Let d be the least element of $\{1,2,4\}$ for which the equivalent conditions in Lemma 1.1 are true, and let $\bar{L} \subseteq \tilde{L}$ be a d-fold covering of L. (There are $\frac{4}{d}$ choices for $\bar{L}$, but they are all equivalent.) Since d is minimal, $\bar{L}$ must be connected. Denote the covering map from $\bar{L}$ to L by p. If we consider p as a map from $\bar{L}$ to T_0^*X, it is a lagrangian immersion, and $m_{\bar{L}} = p^*m_L$, which is zero modulo 4 since that is the case on $\tilde{L}$. The Maslov quantization condition implies that $\frac{1}{2\pi}\int_\gamma p^*\omega_X \in \mathbb{Z}$ for all closed curves on $\bar{L}$, so we may integrate $p^*\omega_X$ to obtain a mapping α from $\bar{L}$ to the circle $\mathbb{R}/2\pi\mathbb{Z} = S^1$ such that $d\alpha = p^*\omega_X$. The mapping α will be used shortly in the construction of Λ, but we pause to make a few observations.

Since we will be considering actions of the group $\mathbb{Z}_d$ on many spaces, it is convenient to use the notation of Souriau [SO] in which g_S denotes the transformation on the space S corresponding to the element g of $\mathbb{Z}_d$. In particular, $\mathbb{Z}_d$ acts on $\mathbb{C}$ in the usual way, and we will write $g_{\mathbb{C}}$ to denote the complex number

$e^{\frac{\pi}{2}ig}$. (Recall that, for any d, we are considering $\mathbb{Z}_d$ as a subgroup of $\mathbb{Z}/4\mathbb{Z}$.) If β is an element of a complex line bundle, $g_{\mathbb{C}}\cdot\beta$ will simply be the result of scalar multiplication of β by $g_{\mathbb{C}}$.

We also have an action of $\mathbb{Z}_d$ on S^1 by $g_{S^1}\cdot r \equiv r + \frac{\pi}{2}g$ (mod 2), and an action of $\mathbb{Z}_d$ on L as the group of covering transformations.

1.2 <u>Lemma.</u> $\alpha:\bar{L} \to S^1$ is $\mathbb{Z}_d$-equivariant; i.e., for any $\bar{\ell} \in \bar{L}$ and $g \in \mathbb{Z}_d$, $\alpha(g_{\bar{L}}\cdot\bar{\ell}) - \alpha(\bar{\ell}) \equiv \frac{\pi}{2}g$ (mod 2π).

<u>Proof.</u> Let $\bar{\gamma}$ be a path in $\bar{L}$ from $\bar{\ell}$ to $g_{\bar{L}}\cdot\bar{\ell}$. Then $\gamma = p\circ\bar{\gamma}$ is a closed curve in L for which $m_L(\gamma) \equiv g$ (mod 4) Now $\alpha(g_{\bar{L}}\cdot\bar{\ell}) - \alpha(g) = \int_{\bar{\gamma}} p^*\omega_X = \int_{\gamma}\omega_X$ which, by the Maslov quantization condition, is congruent modulo 2π to $\frac{\pi}{2}\, m_L(\gamma)$, which we have just seen to be congruent to $\frac{\pi}{2}g$. Q.E.D.

Since $m_{\bar{L}} = \rho^* m_L$, we may consider $M_{\bar{L}}$ as the pullback $\rho^* M_L$; the action of $\mathbb{Z}_d$ on $\bar{L}$ lifts in a natural way to $M_{\bar{L}}$ as the group of covering transformations of the d-fold covering $q:M_{\bar{L}} \to M_L$. Since $m_{\bar{L}} \equiv 0$ (mod 4), $M_{\bar{L}}$ admits sections which are constant in the local trivializations given by the $\mathbb{Z}_d$ structure. Such sections are called <u>parallel</u>; the next lemma shows how $\mathbb{Z}_d$ acts on them.

1.3 <u>Lemma.</u> Let σ be a parallel section of M_L, $\bar{\ell} \in \bar{L}$, and $g \in \mathbb{Z}_d$. Then $\sigma(g_{\bar{L}}\cdot\bar{\ell}) = g_{\mathbb{C}}\cdot g_{M_{\bar{L}}}\cdot\sigma(\bar{\ell})$.

<u>Proof.</u> Let $\bar{\gamma}:[0,1] \to \bar{L}$ be a path from $\bar{\ell}$ to $g_{\bar{L}}\cdot\bar{\ell}$. $\rho = q\circ\sigma\circ\bar{\gamma}$ is then a parallel section of M_L over $\gamma = \rho\circ\bar{\gamma}$, which is a closed curve in L based at $\ell = p(\bar{\ell})$. By the definition of

holonomy, $\rho(1) = e^{\frac{\pi}{2} i m_L(\gamma)} \cdot \rho(0)$. But $m_L(\gamma) \equiv g \pmod 4$, so

$e^{\frac{\pi}{2} i m_L(\gamma)} = g_{\mathbb{C}}$, and

$$q(\sigma(g_{\bar{L}} \cdot \bar{\ell})) = q(\sigma(\bar{\gamma}(1))) = \rho(1) = g_{\mathbb{C}} \cdot \rho(0) = g_{\mathbb{C}} \cdot q(\sigma(\bar{\gamma}(0)))$$

$$= g_{\mathbb{C}} \cdot q(\sigma(\bar{\ell})) = g_{\mathbb{C}} \cdot q(g_{M_{\bar{L}}} \cdot \sigma(\bar{\ell})) = q(g_{\mathbb{C}} \cdot g_{M_{\bar{L}}} \cdot \sigma(\bar{\ell})).$$

Since q is injective on the fibre of $M_{\bar{L}}$ over $g_{\bar{L}} \cdot \bar{\ell}$, we may delete q from both ends of the preceding chain of equalities to obtain the conclusion of the lemma. Q.E.D.

We finally arrive at the definition of the conic lagrangian manifold Λ. Let

$$j : \bar{L} \times \mathbb{R}^+ \to T_0^*X \times T_0^*S^1$$

be given by the formula

$$(1.4) \qquad j(\bar{\ell},\tau) = (\tau\ell;(\alpha(\bar{\ell}), -\tau)) ,$$

where $\ell = \rho(\bar{\ell})$, and we have identified T^*S^1 with $S^1 \times \mathbb{R}$. (The expression $\tau\ell$ is just a scalar multiplication in T^*X.) $\Lambda \subseteq T_0^*X \times T_0^*S^1$ is taken to be the image of j.

Duistermaat [DU$_2$] makes a similar construction in the case where ω_X is exact on L, and he asserts that the map corresponding to our j is a lagrangian immersion. We will prove here that j is a lagrangian immersion by direct investigation of its differential. Another proof, using phase functions, is given in the beginning of the next section.

Since j is homogeneous, it suffices to consider its differential at points where $\tau = 1$. Identify $T_{(\bar{\ell},1)}(\bar{L} \times \mathbb{R})$ with

$T_{\bar{\ell}}L \times \mathbb{R}$, $T_{j(\bar{\ell},1)}(T_0^*(X) \times T_0^*(S^1))$ with $T_\ell(T_0^*X) \times \mathbb{R} \times \mathbb{R}$, and the vertical subspace of $T_\ell(T_0^*X)$ with T_x^*X, where x is the "base-point" in X of ℓ. For $\bar{v} \in T_{\bar{\ell}}\bar{L}$, write v for $(T_{\bar{\ell}}p)(\bar{v}) \in T_\ell L$.

1.5 Lemma. (i) $(T_{(\bar{\ell},1)}j)(\bar{v},a) = (v + a\ell;(\omega_X(v), -a))$;

(ii) j is an immersion;

(iii) j is lagrangian.

Proof. (i) results from a straightforward calculation. If $(T_{(\bar{\ell},1)}j)(\bar{v},a)$ is zero, then a must be zero, and v must be zero as well. Since ρ is a local diffeomorphism, this implies $\bar{v} = 0$, and we have (ii). To prove (iii), we evaluate the fundamental 1-form $\omega_X \times S^1$ on $(T_{(\bar{\ell},1)}j)(\bar{v},a)$ to get

$$\omega_X(v) + a\omega_X(\ell) + \omega_{S^1}(\omega_X(v), -a).$$

Since we are at the point $(\alpha(\bar{\ell}), -1)$ in $T_0^*S^1$, the last term is $(-1)(\omega_X(v))$, which cancels the first one. The middle term is zero because ℓ is a vertical vector, which is annihilated by ω_X.

Q.E.D.

Remark. The canonical relation Λ' corresponding to Λ is the image of $j':\bar{L} \times \mathbb{R} \to T_0^*X \times T_0^*S^1$, where j' is defined by formula 1.4 with $-\tau$ replaced by τ.

2. THE SYMBOL BUNDLE OF Λ

To determine the Maslov bundle of Λ , we will give a description in terms of phase functions. This description will, incidentally, give another proof that Λ is a lagrangian submanifold.

Let $\bar{\ell} \in \bar{L}$, and suppose that a neighborhood of $\ell \in L$ is defined by a phase function ϕ on an open set $U \times V$ in $X \times \mathbb{R}^N$. In other words $(x,\theta) \longrightarrow \phi_\theta'(x,\theta)$ has zero as a regular value, and the manifold $\Sigma = \{(x,\theta) \in U \times V \mid \phi_\theta'(x,\theta) = 0\}$ is mapped onto a neighborhood of ℓ in L by the map $(x,\theta) \longmapsto (x,\phi_x'(x,\theta))$. We consider a lift $\lambda:\Sigma \to \bar{L}$ of this map and observe that $d(\phi|\Sigma) = \lambda^* p^* \omega_X = \lambda^*(d\alpha) = d(\alpha \circ \lambda)$. Adding a constant to ϕ , if necessary, we may assume that $\phi|\Sigma$ is (modulo 2π) actually equal to $\alpha\circ\lambda$.

Now consider the phase function ψ on $U \times S^1 \times V \times \mathbb{R}^+$ defined by $\psi(x,r,\theta,\tau) = \tau(\phi(x,\theta)-r)$. This is homogeneous if we consider the cone axis in $V \times \mathbb{R}^+$ to be given by the τ variable alone. The equations defining the critical set Ξ of ψ are

$$0 = \psi_\theta' = \tau\phi_\theta'(x,\theta)$$

and

$$0 = \psi_\tau' = \phi(x,\theta) - r \ .$$

Since $\tau \neq 0$, Ξ is $\{(x,r,\theta,\tau) \,\big|\, (x,\theta) \in \Sigma$ and $r = \phi(x,\theta)\}$, which we identify with $\Sigma \times \mathbb{R}^+$ by projecting onto the first, third, and fourth components. The mapping $\mu:\Xi \to T_0^*(X) \times T_0^*(S^1)$ is given by

$$(x,\phi(x,\theta),\theta,\tau) \longmapsto (x,\psi_x'(x,\phi(x,\theta),\theta,\tau);\phi(x,\theta),\psi_r'(x,\phi(x,\theta),\theta,\tau))$$

$$= (x,\tau\phi_x'(x,\theta);\phi(x,\theta),-\tau).$$

Comparing this with formula 1.4 and using the fact that $\phi = \alpha\circ\lambda$ on Σ, we find that $\mu(x,\phi(x,\theta),\theta,\tau) = j(\lambda(x,\theta),\tau)$. Using the identification of Ξ with $\Sigma \times \mathbb{R}^+$, we have $\mu = j\circ(\lambda \times \text{identity})$, so the phase function ψ defines the conic open subset $[j\circ(\lambda \times \text{identity})](\Sigma \times \mathbb{R}^+)$ in Λ. It is straightforward to check that ψ is non-degenerate and thereby to obtain another proof that Λ is a lagrangian submanifold.

To study the Maslov bundle of Λ, we may restrict our attention to those phase functions which arise in the manner just described from phase functions for $\bar{L}$. Assume, then, that we have another phase function $\tilde{\phi}$ on an open subset of $X \times \mathbb{R}^{\tilde{N}}$ defining a neighborhood of $\bar{\ell} \in \bar{L}$. $\tilde{\psi}(x,r,\tilde{\theta},\tau) = \tau(\tilde{\phi}(x,\tilde{\theta}) - r)$ is then another phase function for Λ, and we must compute the transition function which relates the trivializations of M_Λ arising from ψ and $\tilde{\psi}$. According to Chapter III of [HÖ] (see the last paragraph of that chapter), the transition function is the constant $e^{i\frac{\pi}{2}\sigma}$, where

$$\sigma = \operatorname{sgn}\begin{pmatrix} \psi''_{\theta\theta} & \psi''_{\theta\tau} \\ \psi''_{\tau\theta} & \psi''_{\tau\tau} \end{pmatrix} - \operatorname{sgn}\begin{pmatrix} \tilde{\psi}''_{\theta\theta} & \tilde{\psi}''_{\theta\tau} \\ \tilde{\psi}''_{\tau\theta} & \tilde{\psi}''_{\tau\tau} \end{pmatrix} .$$

The first matrix is equal to

$$\begin{pmatrix} \tau\phi''_{\theta\theta} & \phi'_\theta \\ \phi'_\theta & 0 \end{pmatrix} ,$$

which, since $\phi'_\theta = 0$ on Σ and $\tau > 0$, has the same signature as $\phi''_{\theta\theta}$. A similar argument holds for the second matrix, so $\sigma = \operatorname{sgn}\phi''_{\theta\theta} - \operatorname{sgn}\tilde{\phi}''_{\tilde{\theta}\tilde{\theta}}$, and the transition function for M_Λ is exactly that for $M_{\bar{L}}$, pulled back by the natural projection $\pi:\bar{L} \times \mathbb{R}^+ \to \bar{L}$. It follows that $M_\Lambda = \pi^* M_{\bar{L}}$. (Actually, we should write $(\pi \circ j^{-1})^* M_{\bar{L}}$.) Since, as we saw in Section 1, $M_{\bar{L}}$ is trivial, so is M_Λ .

Remark. The fact that M_Λ is the pullback of $M_{\bar{L}}$ can also be obtained from formula (i) of Lemma 1.5 and the definition of the Maslov class in terms of pairs of lagrangian subbundles.

We now consider the 1/2-density part of the symbol bundle. The following notation will be used. If V is any vector space, $|V|^s$ denotes the space of densities of order s on V. If E is a vector bundle, $|E|^s$ is the bundle of s-densities on the fibres of E. $\Gamma(E)$ denotes the space of C^∞ sections of E. If the base of E is a cone bundle, and the $\mathbb{R}^+$ action on the base lifts to E, then $\Gamma^k(E) \subseteq \Gamma(E)$ denotes the sections which are homogeneous of degree k.

Since $\Lambda = j(\bar{L} \times \mathbb{R}^+)$, there is a natural isomorphism of $|T\Lambda|^{1/2}$ with $(\pi \circ j^{-1})^* |T\bar{L}|^{1/2}$ obtained by multiplication with $|d\tau|^{1/2}$, τ being the coordinate on $\mathbb{R}^+$. Since $p^*(TL) \approx T\bar{L}$, we also have $|T\Lambda|^{1/2} \approx (p \circ \pi \circ j^{-1})^* |TL|^{1/2}$. As we have seen, similar pullback relations hold for the Maslov bundles, so $|T\Lambda|^{1/2} \otimes M_\Lambda \approx (\pi \circ j^{-1})^* (|T\bar{L}|^{1/2} \otimes M_{\bar{L}})$. We now fix the choice of a parallel section σ of $M_{\bar{L}}$, which determines a parallel section ρ of M_Λ . We choose σ (and hence ρ) to have unit norm with respect to the hermitian structure on the fibres of the Maslov bundle. Given any section of $|TL|^{1/2}$, we may pull it back to a section of $|T\Lambda|^{1/2}$ which, when multiplied by ρ, gives a section of $|T\Lambda|^{1/2} \otimes M_\Lambda$ which is homogeneous of degree 1/2 (since $|d\tau|^{1/2}$ is involved). If we then multiply by $\tau^{k-1/2}$, the section becomes homogeneous of degree k. By the steps just described, we have constructed mappings $\gamma^k : \Gamma(|TL|^{1/2}) \to \Gamma^k(|T\Lambda|^{1/2} \otimes M_\Lambda)$. The only arbitrariness in γ^k lies in the choice of σ : a different choice would alter γ^k by multiplication with a complex number of norm 1.

We conclude this section by examining some equivariance properties of Λ and the sections of its symbol bundle defined above. The action of $\mathbb{Z}_d$ on S^1 lifts to an action on $T_0^*S^1$ by homogeneous canonical transformations — $g_{T_0^*S^1}(r,t) = (g_{S^1}(r),t)$; we let $\mathbb{Z}_d$ act trivially on T_0^*X and $\mathbb{R}^+$ and obtain product actions on $\bar{L} \times \mathbb{R}$ and $T_0^*X \times T_0^*S^1$.

2.1 __Lemma.__ $j:\bar{L} \times \mathbb{R}^+ \to T_0^*X \times T_0^*S^1$ is $\mathbb{Z}_d$-equivariant, so $\Lambda \subseteq T_0^*X \times T_0^*S^1$ is $\mathbb{Z}_d$-invariant.

__Proof.__ $j(g_{\bar{L}}\cdot\bar{\ell},\tau) = (\tau\ell;(\alpha(g_{\bar{L}}\cdot\bar{\ell}), -\tau))$, since $p(g_{\bar{L}}\cdot\bar{\ell}) = p(\bar{\ell}) = \ell$. By Lemma 1.2, $\alpha(g_{\bar{L}}\cdot\ell) = g_{S^1}\cdot\alpha(\bar{\ell})$, so

$$j(g_{\bar{L} \times \mathbb{R}^+}\cdot(\bar{\ell},\tau)) = j(g_{\bar{L}}\cdot\bar{\ell},\tau) = (\tau\ell;(g_{S^1}\cdot\alpha(\bar{\ell}), -\tau))$$

$$= g_{T_0^*X \times T_0^*S^1}\cdot(\tau\ell;(\alpha(\bar{\ell}), -\tau))$$

$$= g_{T_0^*X \times T_0^*S^1}\cdot j(\bar{\ell},\tau).$$

Q.E.D.

Since the action of $\mathbb{Z}_d$ on $T_0^*X \times T_0^*S$ comes from an action on $X \times S^1$, the action of $\mathbb{Z}_d$ on Λ lifts to M_Λ, and hence to $|T\Lambda|^{1/2} \otimes M_\Lambda$.

2.2 __Lemma.__ Let $a \in \Gamma(|TL|^{1/2})$. Then, for $\lambda \in \Lambda$ and $g \in \mathbb{Z}_d$, we have $(\gamma^k a)(g_\Lambda\cdot\lambda) = g_{\mathbb{C}}\cdot g_{M_\Lambda}\cdot(\gamma^k a)(\lambda)$.

Proof. The pullback of a to $\Gamma(|T\Lambda|^{1/2})$ is invariant under $\mathbb{Z}_d$, as is the function $\tau^{k-1/2}$. It remains to examine the action of $\mathbb{Z}_d$ on the parallel section ρ. Lemma 1.3 and the fact that M_Λ is the pullback of $M_{\bar{L}}$ imply that $\rho(g_\Lambda\cdot\lambda) = g_{\mathbb{C}}\cdot g_{M_\Lambda}\cdot\rho(\lambda)$, from which the desired formula follows immediately. Q.E.D.

3. FOURIER INTEGRAL OPERATORS ASSOCIATED WITH Λ

The basic construction of [HÖ] gives a map from $\Gamma^k(|T\Lambda|^{1/2}\otimes M_\Lambda)$ to the space

$$I^{k-\frac{1}{4}(n+1)}(X\times S^1,\Lambda)/I^{k-1-\frac{1}{4}(n+1)}(X\times S^1,\Lambda)$$

of distributions on $X\times S^1$. Composing with γ^k, we obtain $F^k:\Gamma(|TL|^{1/2})\to I^{k-\frac{1}{4}(n+1)}(X\times S^1,\Lambda)/I^{k-1-\frac{1}{4}(n+1)}(X\times S^1,\Lambda)$. The purpose of this section is to show that we can find a representative in $I^{k-\frac{1}{4}(n+1)}(X\times S^1,\Lambda)$ of $F^k(a)$ which, considered as an operator from $\mathcal{D}'(S^1)$ to $\mathcal{D}'(X)$, has nice symmetry properties.

It seems worthwhile to point out an unusual but useful convention in the symbol calculus. If $C\subseteq T_0^*X\times T_0^*Y$ is a canonical relation, we think of C as a relation from T_0^*Y to T_0^*X. This is contrary to the way in which graphs of mappings are usually considered. If $D\subseteq T_0^*Y\times T_0^*Z$ is another canonical relation, then it is the composition $C\circ D$ which is defined (rather than $D\circ C$ as in the case of the usual convention). Similarly, the FIO's associated with C map from $\mathcal{D}'(Y)$ to $\mathcal{D}'(X)$, so that the calculus of products gives a covariant, rather than contravariant, functorial correspondence between canonical relations and operators. For example, if $f:X\to Y$ is a diffeomorphism, then its lift to cotangent bundles, which we call $T_0^*f \subseteq T_0^*X \times T_0^*Y$, is considered to map from T_0^*Y to T_0^*X, just as the pullback operator f^*, which is a FIO associated with T_0^*f ([DU_1], p. 55), maps $\mathcal{D}'(Y)$ to $\mathcal{D}'(X)$.

With these conventions in mind, we look at the actions of $\mathbb{Z}_d$ on S^1 and $T_0^*S^1$ discussed in Section 1. Recall that $g_{S^1}(r) \equiv r + \frac{\pi}{2}g \pmod{2\pi}$, while $g_{T_0^*S^1}(r,\tau) = (g_{S^1}(r),\tau)$. The canonical relation associated with $g_{T_0^*S^1}$ is, therefore, $T_0^*g_{S^1}^{-1}$, and the associated FIO is $g_{S^1}^{-1*}$. The principal symbol of $g_{S^1}^{-1*}$ is the function 1 with respect to the natural trivialization of the symbol bundle of a local canonical graph. (See p. 180 of [HÖ].) To keep track of the underlying canonical relation, we denote this symbol by g_σ.

We now interpret Lemmas 2.1 and 2.2 in terms of the symbol category (HÖ], Section 4.2).

3.1 Lemma. Let $a \in \Gamma(|TL|^{1/2})$, $g \in \mathbb{Z}_d$. Then $(\gamma^k a) \times g = g_{\mathbb{C}}^{-1}\cdot\gamma^k a$.

<u>Proof.</u> Lemma 2.1 implies that $\Lambda' \circ g_{T_0^*S^1} = \Lambda'$, so both sides of the equation to be proven are defined on the same lagrangian submanifold. Now, for $\lambda' = j'(\bar{\ell},\tau)$, $(\bar{\ell},\tau) \in \bar{L} \times \mathbb{R}^+$, to compute $(\gamma^k a \times g_\sigma)(\lambda')$, we must express λ' as an element of $\Lambda' \circ g_{T_0^*S^1}$. We have

(i) $\lambda' = (\tau\ell;(\alpha(\bar{\ell}),\tau))$

(ii) $((g_{S^1}^{-1}\cdot\alpha(\bar{\ell}),\tau)\,;(\alpha(\bar{\ell}),\tau)) \in g_{T_0^*S^1}$

(iii) $g_{\Lambda'}^{-1}\cdot\lambda' = (\tau\ell;(g_{S^1}^{-1}\cdot\alpha(\bar{\ell}),\tau))$, so $(\gamma^k a \times g_\sigma)(\lambda') = (\gamma^k a)(g_\Lambda^{-1}\cdot\lambda')\cdot g_\sigma((g_{S^1}^{-1}\cdot\alpha(\bar{\ell}),\tau);(\alpha(\bar{\ell}),\tau))$.

By Lemma 2.2, the first factor equals $g_{\mathbb{C}}^{-1}\cdot g_{M_{\Lambda'}}^{-1}\cdot(\gamma^k a)(\lambda')$, while the second factor is "1". This "1" should be interpreted as an isomorphism between the fibres of $M_{\Lambda'}$ over $g_{\Lambda'}^{-1}\cdot\lambda'$ and λ'. Since $g_{T_0^*S^1}$ is the lift of a transformation on S^1, this isomorphism is just $g_{M_{\Lambda'}}$, so we conclude that

$(\gamma^k a \times g_\sigma)(\lambda') = g_{\mathbb{C}}^{-1}\cdot(\gamma^k a)(\lambda')$. Q.E.D.

Lemma 3.1 means that, if $A \in I^{k-\frac{1}{4}(n+1)}(X \times S^1,\Lambda)$ is a representative of $F^k a$, then, modulo elements of $I^{k-1-\frac{1}{4}(n+1)}(X \times S^1,\Lambda)$, we have

$$A\circ g_{S^1}^{-1*} = g_{\mathbb{C}}^{-1}\cdot A\,, \quad \text{or}$$

(3.2)

$$A\circ g^*_{S^1} = g_{\mathbb{C}}\cdot A\,.$$

We will now show that a suitable choice of A makes equation 3.2 exactly true.

3.3 <u>Lemma.</u> There exists $A \in F^k a$ such that equation 3.2 is true.

<u>Proof.</u> Let A_0 be any element of $F^k a$, and define A to be

$$\frac{1}{d}\sum_{h\in\mathbb{Z}_d} h_{\mathbb{C}}^{-1}\cdot A_0\circ h^*_{S^1}\quad .$$

By Lemma 3.1, the symbol of $h_{\mathbb{C}}^{-1}\cdot A_0\circ h^*_{S^1}$ is $\gamma^k a$, so the symbol of A is also $\gamma^k a$. Furthermore,

$$\begin{aligned} A\circ g^*_{S^1} &= \frac{1}{d}\sum_{h\in\mathbb{Z}_d} h_{\mathbb{C}}^{-1}\cdot A_0\circ h^*_{S^1}\circ g^*_{S^1} \\ &= g_{\mathbb{C}}\cdot\frac{1}{d}\sum_{h\in\mathbb{Z}_d}(hg)_{\mathbb{C}}^{-1}\cdot A_0\circ(hg)^*_{S^1}\,. \end{aligned}$$

If we write k for hg (composition in $\mathbb{Z}_d$ being written multiplicatively), k runs over $\mathbb{Z}_d$ as h does, so the last expression becomes

$$g_{\mathbb{C}} \cdot \frac{1}{d} \sum_{k \in \mathbb{Z}_d} k_{\mathbb{C}}^{-1} \cdot A_0 \circ k^*_{S^1} = g_{\mathbb{C}} \cdot A \,.$$

Q.E.D.

The action of $\mathbb{Z}_d$ on S^1 splits $\mathcal{D}'(S^1)$ into d subspaces, each of them associated with a character of $\mathbb{Z}_d$. Namely, for $\lambda \in \mathbb{C}$, let

$$\mathcal{D}'_{d,\lambda}(S^1) = \{\psi \in \mathcal{D}'(S^1) \,|\, \left(\tfrac{4}{d}\right)^*_{S^1} \psi = \lambda\psi\} \,.$$

Then

$$\mathcal{D}'(S^1) = \bigoplus_{\{\lambda | \lambda^d = 1\}} \mathcal{D}'_{d,\lambda}(S^1).$$

In fact, let $\{\psi_k | k \in \mathbb{Z}\}$ be the standard Fourier basis of $\mathcal{D}'(S^1)$; i.e., $\psi_k = e^{ikr}|dr|^{1/2}$. For $\lambda = \frac{2\pi i m}{d}$, $\mathcal{D}'_{d,\lambda}(S^1)$ is just the subspace spanned by those ψ_k for which $k \equiv m \pmod d$.

If $\psi \in \mathcal{D}'_{d,\lambda}(S^1)$ and A satisfies equation 3.2, then $\lambda A\psi = A\lambda\psi = [A \circ \left(\tfrac{4}{d}\right)^*_{S^1}]\psi = \left(\tfrac{4}{d}\right)_{\mathbb{C}} \cdot A\psi$, so $A\psi = 0$ unless $\lambda = \left(\tfrac{4}{d}\right)_{\mathbb{C}}$, which is $e^{\frac{2\pi i}{d}}$. In other words, A annihilates all but the summand $\mathcal{D}'_{d,e^{\frac{2\pi i}{d}}}(S^1)$, which we write simply as $\mathcal{D}'_d(S^1)$. Note that $\mathcal{D}'_d(S^1)$ may also be described as

$\{\psi \in \mathcal{D}'(S^1) | g^*_{S^1}\psi = g_{\mathbb{C}} \cdot \psi$ for all $g \in \mathbb{Z}_d\}$.

Next, we will take into account the fact that the domain in $T^*_0S^1$ of Λ is contained in the "upper half cylinder" $T^*_+S^1 = \{(r,\tau) \in T^*S^1 | \tau > 0\}$. Let $P: \mathcal{D}'(S^1) \to \mathcal{D}'(S^1)$ be the projection operator which takes each distribution to its "holomorphic part", i.e., $P\psi_k = \begin{Bmatrix} \psi_k & k \geq 0 \\ 0 & k < 0 \end{Bmatrix}$. P is a pseudodifferential operator [SE_1] whose symbol is 1 on $T_+^*S^1$ and 0 on $T_-^*S^1$. If A is any element of $F^k a$, then AP has the same symbol as A, and $AP \in F^k a$ as well. If A satisfies equation 3.2, so does AP, because P commutes with the action of $\mathbb{Z}_d$. Note that $(AP)P = AP^2 = AP$. Writing A for AP, we have the following result.

3.4 __Lemma.__ There exists $A \in F^k a$ such that A satisfies equation 3.2 and $AP = A$.

Any A satisfying the conditions in Lemma 3.4 annihilates the kernel of P, so it must "live" on the subspace $\mathcal{D}'_{d,+}(S^1)$ of $\mathcal{D}'_d(S^1)$ spanned by $\{\psi_k | k > 0$ and $k \equiv 1 \pmod d\}$.

3.5 __Remark.__ The spectrum of the operator $-E\frac{\partial^2}{\partial r^2}$ on its invariant subspace $\mathcal{D}'_{d,+}(S^1)$ is just the sequence $\{E(dj+1)^2\}_{j=0}^{\infty}$ which appears in the statement of the eigenvalue theorem.

Our next goal will be to show that A may be chosen to be an isometric embedding of $L^2_{d,+}(S^1)$ (by which we mean $\mathcal{D}'_{d,+}(S^1) \cap L^2(S^1)$) into $L^2(X)$. An operator A satisfying the conditions of Lemma 3.4 induces such an isometry if any only if A^*A is the orthogonal projection Π of $\mathcal{D}'(S^1)$ onto $\mathcal{D}'_{d,+}(S^1)$. Our program for obtaining this will be first to show that, for suitable k and a, A^*A and Π are FIO's having the same symbol and then to modify A, without changing its symbol, to make A^*A equal to Π.

The calculus in [HÖ] cannot be applied to the product A^*A because the transversality condition is not satisfied; however, A^* and A have a weaker "clean intersection" property, and it turns out that the calculus of products extends in a natural way to this situation. The next section is a discussion of this extension; we return in Section 5 to the study of A^*A.

4. CLEAN INTERSECTIONS AND PRODUCTS OF FIO'S

Let P, Q, and R be symplectic manifolds, $C_1 \subseteq P \times Q$ and $C_2 \subseteq Q \times R$ canonical relations. To study the composition $C_1 \circ C_2 \subseteq P \times R$, we consider the intersection of the product $C_1 \times C_2$ with $P \times \Delta_Q \times R$ in $P \times Q \times Q \times R$. The image of $C_1 \times C_2 \cap P \times \Delta_Q \times R$ under the projection of $P \times Q \times Q \times R$ onto $P \times R$ is just $C_1 \circ C_2$.

To see when $C_1 \circ C_2$ is a canonical relation, we consider the passage from $C_1 \times C_2$ to $C_1 \circ C_2$ in a more general context. Let S be a symplectic manifold, $K \subseteq S$ a coisotropic submanifold; i.e., for each $k \in K$, $T_kK \supseteq T_kK^\perp$. The bundle $TK^\perp$ is thus a subbundle of TK, and it is easily seen to be integrable. (See [SO] §§5, 9 for this and the next fact. [DU_1] also contains a related discussion, in which the term "involutive" is used instead of coisotropic.) Suppose now that the integrable distribution TK on K is regular in the sense that the space S_K of integral manifolds is itself a manifold with respect to the quotient topology. S_K then inherits a symplectic structure from S.

Now let $L \subseteq S$ be a lagrangian submanifold which intersects K cleanly in the sense that $L \cap K$ is a manifold and, for each k in $L \cap K$, $T_KL \cap T_kK = T_k(L \cap K)$. Now $T_kL \cap T_kK^\perp = (T_kL^\perp + T_kK)^\perp = (T_kL + T_kK)^\perp$. Since the dimension of $T_kL \cap T_kK$ is independent of k, being equal to the dimension of $L \cap K$, so are $\dim(T_kL + T_kK)$ and $\dim(T_kL + T_kK)^\perp = \dim(T_kL \cap T_kK^\perp)$. It follows that the projection from $K \cap L$ to its image L_K in S_K has constant rank, and L_K is therefore an immersed submanifold of S_K. A simple calculation ([HÖ] p. 162 or [SL]) shows that L_K is lagrangian in S_K.

In the situation with which this section began, S is $P \times Q \times Q \times R$ with the symplectic structure $\Omega_P \times -\Omega_Q \times \Omega_Q \times -\Omega_R$, L is $C_1 \times C_2$, and K is $P \times \Delta_Q \times R$. It is evident that $T_{(p,q,q,r)}K^{\perp} = \{(0,v,v,0) | v \in T_q Q\}$, so that S_K may be identified with $P \times R$, the induced symplectic structure being $\Omega_P \times -\Omega_R$; L_K is $C_1 \circ C_2$. From the general considerations above we see that $C_1 \circ C_2$ is an immersed lagrangian submanifold, i.e., a canonical relation, whenever $C_1 \times C_2$ and $P \times \Delta_Q \times R$ intersect clearly. When this is the case, we say that C_1 and C_2 have a <u>clean product</u>. The codimension of $T(C_1 \times C_2) + T(P \times \Delta_Q \times R)$ in $T(P \times Q \times Q \times R)$ is denoted by e and called the <u>excess</u> of the clean product. $\underline{e}$ is also the dimension of the fibres of the projection from $C_1 \times C_2 \cap P \times \Delta_Q \times R$ onto $C_1 \circ C_2$ and is zero in the transversal case. We now have the following extension of Theorem 4.2.2 of [HÖ].

4.1 <u>Clean Product Theorem.</u> Let $C_1 \cap T_0^*X \times T_0^*Y$ and $C_2 \subseteq T_0^*Y \times T_0^*Z$ be canonical relations which have a clean product such that the projection from $C_1 \times C_2 \cap T^*X \times \Delta_{T^*Y} \times T^*Z$ to $C_1 \circ C_2$ is proper. If $A_1 \in I^{k_1}(X \times Y, C_1')$ and $A_2 \in I^{k_2}(Y \times Z, C_2')$ are properly supported, then $A_1 \circ A_2 \in I^{k_1+k_2+\frac{1}{2}e}(X \times Z, (C_1 \circ C_2)')$.

<u>Proof.</u> See [GL]. The basic idea is to do the computation locally, splitting off the excess variables as extra phase variables. (This accounts for the increase in order.) □

Next, we describe how the principal symbol of $A_1 \circ A_2$ is obtained from those of A_1 and A_2. We begin with the half-density part.

4.2 __Lemma.__ Let E and F be subspaces of a vector space. Then there is a natural isomorphism between $|E|^{1/2} \otimes |F|^{1/2}$ and $|E \cap F|^{1/2} \otimes |E + F|^{1/2}$.

__Proof.__ From the natural isomorphism $E/E \cap F \approx {}^{E+F}\!/_{F}$, we get $|E|^{1/2} \otimes |E \cap F|^{-1/2} \approx |E + F|^{1/2} \otimes |F|^{-1/2}$. Multiplying both sides by $|E \cap F|^{1/2} \otimes |F|^{1/2}$ gives the desired result. Q.E.D.

__Remark.__ The isomorphism arising from $F/E \cap F \approx {}^{E+F}\!/_{E}$ is the same as the one in the proof above. (To see this, choose a basis of $E + F$ which contains bases of $E \cap F$, E, and F. Is there a more conceptual proof?)

Applying Lemma 4.2 in our general setting of clean intersections, we obtain the formula

$$|T_kL|^{1/2} \otimes |T_kK|^{1/2} \approx |T_kL \cap T_kK|^{1/2} \otimes |T_kL + T_kK|^{1/2} \quad .$$

If $|T_kK|^{1/2}$ has a distinguished element, as it does when $K = P \times \Delta_Q \times R$, then we have

$$(4.3) \qquad |T_kL|^{1/2} \approx |T_kL \cap T_kK|^{1/2} \otimes |T_kL + T_kK|^{1/2} \; .$$

4.4 __Lemma.__ Let E be any subspace of a symplectic space V. Then $|E|^{1/2}$ is naturally isomorphic to $|E^{\perp}|^{1/2}$.

Proof. The isomorphism $V \to V^*$ given by the symplectic structure induces an isomorphism $V/E^\perp \to E^*$. Then $|V|^{1/2} \approx |E^\perp|^{1/2} \otimes |E^*|^{1/2}$. But $|V|^{1/2}$ has a distinguished element, and $|E^*|^{1/2} \approx |E|^{-1/2}$, so the lemma follows. Q.E.D.

Applying Lemma 4.4 to equation 4.3, we get

$$(4.5) \qquad |T_kL|^{1/2} \approx |T_kL \cap T_kK|^{1/2} \otimes |T_kL \cap T_kK^\perp|^{1/2} \quad .$$

Now $T_kL \cap T_kK/T_kL \cap T_kK^\perp \approx T_{\tilde{k}}L_K$, where $\tilde{k}$ is the image of k in L_K, so

$$(4.6) \qquad |T_kL \cap T_kK|^{1/2} \approx |T_kL \cap T_kK^\perp|^{1/2} \otimes |T_{\tilde{k}}L_K|^{1/2} \ .$$

Substituting 4.6 into 4.5, we get the natural isomorphism (depending on the distinguished element of $|T_kK|^{1/2}$)

$$(4.7) \qquad |T_kL|^{1/2} \approx |T_kL \cap T_kK^\perp|^1 \otimes |T_{\tilde{k}}L_K|^{1/2} \quad .$$

Now suppose we are given a section a of $|TL|^{1/2}$ and that the projection $p: L \cap K \to L_K$ is proper. If we fix $\tilde{k} \in L_K$ and apply 4.7 to all k in the compact manifold $p^{-1}(\tilde{k})$, using the fact $T_k(p^{-1}(\tilde{k})) = T_kL \cap T_kK^\perp$, we obtain from a a section of $|T(p^{-1}(k))|^1 \otimes |T_kL_k|^{1/2}$. Integrating this section over the fibre $p^{-1}(\tilde{k})$, we obtain an element of $|T_{\tilde{k}}L_K|^{1/2}$. Doing this for each $\tilde{k} \in L_K$, we obtain an operation from $\Gamma(|TL|)^{1/2}$ to $\Gamma(|TL_K|)^{1/2}$.

Applying this operation to the clean product setting and combining with the natural multiplication $\Gamma(|TC_1|^{1/2}) \otimes \Gamma(|TC_2|^{1/2}) \to \Gamma(|T(C_1 \times C_2)|^{1/2})$, we obtain product operation from $\Gamma(|TC_1|^{1/2} \otimes \Gamma(|TC_2|^{1/2})$ to $\Gamma(|T(C_1 \circ C_2)|^{1/2})$

which coincides with the product in [HÖ] in the transverse case. If $a_i \in \Gamma^{k_i}(|TC_i|^{1/2})$, then one may check that the product of a_1 and a_2 lies in $\Gamma^{k_1+k_2+\frac{1}{2}e-\frac{1}{2}\dim Y}(|T(C_1 \circ C_2)|^{1/2})$. (The term $\frac{1}{2}e$ enters in the isomorphism of $|T_kL + T_kK|^{1/2}$ with $|T_kL \cap T_kK^{\perp}|^{1/2}$.

As far as the Maslov bundles are concerned, a calculation with phase functions similar to the argument on p. 181 of [HÖ] shows that the pullback to $C_1 \times C_2 \cap T^*X \times \Delta_{T^*Y} \times T^*Z$ of $M_{C_1 \circ C_2}$ is isomorphic to the restriction of $M_{C_1 \times C_2}$, which is in turn isomorphic to the tensor product of M_{C_1} with M_{C_2}. Combining this with the multiplication of $\frac{1}{2}$-densities, we get a product operation

$$\times : \Gamma(|TC_1|^{1/2} \otimes M_{C_1}) \otimes \Gamma(|T_{C_2}|^{1/2} \otimes M_{C_2})$$

$$(|T(C_1 \circ C_2)|^{1/2} \otimes M_{C_1 \circ C_2}) ,$$

which maps $\Gamma^{k_1}(\cdot) \otimes \Gamma^{k_2}(\cdot)$ into $\Gamma^{k_1+k_2+\frac{1}{2}e-\frac{1}{2}\dim Y}(\cdot)$. Finally, we have the generalization of Theorem 4.2.3 of [HÖ] to clean products.

4.8 <u>Clean Product Symbol Theorem.</u> Let the hypotheses of the Clean Product Theorem be fulfilled. If a_1 and a_2 are principal symbols of A_1 and A_2, then $a_1 \times a_2$ is a principal symbol of the product A_1A_2.

<u>Proof.</u> See [GL]. The proof involves examining the local representaations of A_1, A_2, and A_1A_2 in terms of phase and amplitude functions. □

5. MAKING A ISOMETRIC

Note: In this section we shall write Λ for what we have previously called Λ' .

We begin by finding the canonical relation $\Lambda^{-1} \circ \Lambda \subseteq T_0^*S^1 \times T_0^*S^1$ associated with the operators A^*A for $A \in F^k a$, $a \in \Gamma(|TL|^{1/2})$.

Recall that $\Lambda = \{(\tau\ell;(\alpha(\bar{\ell}),\tau))|\bar{\ell} \in \bar{L},\ \tau \in \mathbb{R}^+\}$, so

$$\Lambda^{-1} \times \Lambda = \{(((\alpha(\bar{\ell}_1),\tau_1);\ \tau_1\ell_1);\ (\tau_2\ell_2;(\alpha(\bar{\ell}_2),\tau_2)))|\bar{\ell}_1 \in \bar{L}, \tau_i \in \mathbb{R}^+\}.$$

At this point, we assume that L is aconic in the sense that the projection of L into the cosphere bundle S^*X is an embedding. In other words, we require:

(5.1) no ray in T_0^*X intersects L in more than one point;

(5.2) no ray in T_0^*X is tangent to L.

Condition 5.1 means that $\tau_1\ell_1 = \tau_2\ell_2$ only if $\tau_1 = \tau_2$ and $\ell_1 = \ell_2$, so

$$\Lambda^{-1} \times \Lambda \cap T^*S^1 \times \Delta_{T^*X} \times T^*S^1 =$$

$$= \{(((\alpha(\bar{\ell}_1),\tau);\tau\ell);\ (\tau\ell;(\alpha(\bar{\ell}_2),\tau)))|\ \ell_1 = \ell_2 = \ell\}\ ,$$

and $\quad \Lambda^{-1} \circ \Lambda = \{((\alpha(\bar{\ell}_1),\tau);\ (\alpha(\bar{\ell}_2,\tau))\ \bar{\ell}_i \in \bar{L},\ \ell_1 = \ell_2,\ \tau > 0\}.$

Now $\ell_1 = \ell_2$ if and only if $\bar{\ell}_2 = g_L \cdot \bar{\ell}_1$ for some $g \in \mathbb{Z}_d$, in which case $\alpha(\bar{\ell}_2) = g_{S^1} \cdot \alpha(\bar{\ell}_1)$, by Lemma 1.2.

Condition 5.2 implies that $\alpha: \bar{L} \to S^1$ has no critical points. In fact, the generating vector ξ of the cone axes is associated by the symplectic structure of T^*_0X with ω_X. If ω_X vanished on $T_\ell L$, then $\xi(\ell)$ would have to lie in $T_\ell L^\perp = T_\ell L$, contradicting 5.2.

We now assume

(5.3) L is compact.

Then $\bar{L}$ is compact, and $\alpha(\bar{L})$ is closed in S^1. Since it is also open, $\alpha(\bar{L}) = S^1$, and we have

5.4 <u>Lemma.</u> $\Lambda^{-1}\circ\Lambda = \{((r,\tau), g_{S^1}(r,\tau)) \mid r \in S^1,\ g \in \mathbb{Z}_d,\ \tau > 0\}$.

In other words, $\Lambda^{-1}\circ\Lambda$ consists of d components, each of which is the graph of $g_{T^*_+S^1}$ for some $g \in \mathbb{Z}_d$.

5.5 <u>Lemma.</u> Λ^{-1} and Λ satisfy the hypotheses of the clean product theorem, with excess $n-1$.

<u>Proof.</u> At a point of $\Lambda^{-1} \times \Lambda \cap T^*S^1 \times \Delta_{T^*X} \times T^*S^1$, the tangent space to $\Lambda^{-1} \times \Lambda$ is

$$\{(((\omega_X(v_1),a_1);\tau_1v_1+a_1\ell_1);(\tau_2v_2+a_2\ell_2;(\omega_X(v_2),a_2))) \mid$$

$$v_1,v_2 \in T_\ell L,\ a_1,a_2 \in \mathbb{R}\}.$$

Recall that $\ell_1 = \ell_2 = \ell$ and $\tau_1 = \tau_2 = \tau$. The condition that a tangent vector of the form above belong to $T(\Lambda^{-1} \times \Lambda \cap T^*S^1 \times \Delta_{T^*X} \times T^*S^1)$ is

$$\tau v_1 + a_1\ell = \tau v_2 + a_2\ell ,$$

or $\tau(v_1-v_2) = (a_2-a_1)\ell$. Since $\tau \neq 0$, $v_1 - v_2 \in T_\ell L$ is a multiple of ℓ, considered as a vertical vector. By condition 5.2, we must have $v_1 - v_2 = 0$, $a_2 - a_1 = 0$, so that the tangent vector in question belongs to $T(\Lambda^{-1} \times \Lambda) \cap T(T^*S^1 \times \Delta_{T^*X} \times T^*S^1)$. The product of Λ^{-1} and Λ is therefore clean. That the projection into $\Lambda^{-1} \circ \Lambda$ is proper follows directly from condition 5.3. The inverse image of a point under this projection is diffeomorphic to a level surface of α, having dimension $n-1$. Q.E.D.

5.6 Remark. Conditions 5.1 and 5.2 follow directly from the hypothesis that a positive function on T_0^*X, homogeneous of non-zero degree, is constant on L. If X is compact, condition 5.3 also follows.

5.7 Remark. Condition 5.3 implies that we can take the operator $A \in F^k a$ to be properly supported. We will assume this to be the case from now on.

Given a density $a \in \Gamma(|TL|^{1/2})$, we now wish to compute the symbol $\overline{\gamma^k a} \times \gamma^k a$ of A^*A, for $A \in F^k a$. We will begin by doing the computation over the component $\Delta_{T_+^*S^1}$ of $\Lambda^{-1} \circ \Lambda$. All the other components are obtained from this one by the action of $\mathbb{Z}_d$, and we will use equivariance properties to compute the rest of the symbol.

Over $\Delta_{T_+^*S^1}$, there is a natural trivialization of the symbol bundle which enables us to consider $\overline{\gamma^k a} \times \gamma^k a$ as a complex-valued function ϕ. We will choose k so that, for $A \in F^k a$, A^*A is a FIO of order zero, so that the function ϕ is of degree zero. Recall that A has order $k - \frac{1}{4}(n+1)$. By the clean product theorem, A^*A has order $2(k - \frac{1}{4}(n+1)) + \frac{1}{2}(n-1) = 2k-1$, so we must choose $k = \frac{1}{2}$, and A has order $\frac{1}{4}(1-n)$.

Since ϕ is homogeneous of degree zero, we may consider it to be simply a function from S^1 to $\mathbb{R}$. Since A^*A is a non-negative self-adjoint operator, the function ϕ must be non-negative. In fact, given $r \in S^1$, $\phi(r)$ is given by the clean product symbol theorem as the integral over $\alpha^{-1}(r) \subseteq \bar{L}$ of a positive density obtained from a. (I suspect that is just $\frac{\bar{a}a}{|d\alpha|}$.)

We will call $a \in \Gamma(|TL|^{1/2})$ <u>uniform</u> if ϕ is a non-zero constant function and <u>unitary</u> if $\phi \equiv \frac{1}{d}$. The following facts are then clear.

5.8 <u>Lemma.</u> If a is nowhere zero (in fact, if ϕ is nowhere zero), one can multiply a by a positive function to make it uniform. If a is uniform, one can multiply it by a positive constant to make it unitary.

We will now compute the entire symbol $\overline{\gamma^{1/2}a} \times \gamma^{1/2}a$. We may write it as

$$\sum_{h \in \mathbb{Z}_d} \phi_h \times h_\sigma ,$$

where each ϕ_h is a function on $T_0^*S^1$, homogeneous of degree 0 and vanishing on $T_-^*S^1$, considered as the symbol of a pseudodifferential operator. In particular, ϕ_0 is the function ϕ discussed above. By Lemma 3.1, $\overline{\gamma^{1/2}a} \times \gamma^{1/2}a \times g_\sigma^{-1} = g_{\mathbb{C}} \cdot \overline{\gamma^{1/2}a} \times \gamma^{1/2}a$,

so
$$\sum_{h \in \mathbb{Z}_d} \phi_h \times h_\sigma \times g_\sigma^{-1} = \sum_{h \in \mathbb{Z}_d} g_{\mathbb{C}} \cdot \phi_h \times h_\sigma .$$

Equating the terms which are defined on $\Delta_{T_+^*S^1}$, we find $\phi_g = g_{\mathbb{C}} \cdot \phi_0$, so

$$\overline{\gamma^{1/2}a} \times \gamma^{1/2}a = d \cdot \phi_0 \cdot [\tfrac{1}{d} \sum_{g \in \mathbb{Z}_d} g_{\mathbb{C}} \cdot g_\sigma] .$$

The factor in brackets is the symbol of

$$\frac{1}{d} \sum_{g \in \mathbb{Z}_d} g_{\mathbb{C}} \cdot g_{S^1}^{-1*} \quad ,$$

which is the operator of projection on $\mathcal{D}'_d(S^1)$. It follows that A^*A has the same symbol as $\Psi\Pi$, where Ψ is some pseudodifferential operator and Π is the projection on $\mathcal{D}'_{d,+}(S^1)$. If a is unitary, we can take Ψ to be the identity, since $d \cdot \phi_0$ is then 1 on $T^*_+S^1$. We have now completed the first part of the program described at the end of Section 3. It remains to choose A, for a unitary, so that A^*A is precisely equal to Π, not just at the symbol level.

Assume now that a is unitary and $A \in F^k a$ satisfies the conditions in Lemma 3.4, and consider the operator $B = A^*A + (I-\Pi)$. It is a non-negative self-adjoint operator which leaves invariant $\mathcal{D}'_{d,+}(S^1)$ and is the identity on its orthogonal complement. B is of the form $I + F$, and $F = A^*A - \Pi$ belongs to $I^{-1}(S^1 \times S^1, \Lambda^{-1} \circ \Lambda)$. Since the powers of $\Lambda^{-1} \circ \Lambda$ are equal to $\Lambda^{-1} \circ \Lambda$, we can construct by the procedure of Theorem 2.5.1 of [HÖ] a two-sided parametrix for B which lies in $I^0(S^1 \times S^1, \Lambda^{-1} \circ \Lambda)$. It follows that the kernel of B, which is also that of A restricted to $\mathcal{D}'_{d,+}(S^1)$, is spanned by finitely many smooth functions. In addition, the eigenspaces of B corresponding to eigenvalues greater than 2 are all spanned by finitely many smooth functions. Since the range of A has infinite codimension (the canonical relation Λ is not surjective), we can alter A by a finite smoothing operator so that it retains all its former properties and the spectrum of B is now contained in the interval (0,2). F now has norm less than 1, and we can write $B^{-1/2} = (I + F)^{-1/2}$ as a convergent power series in F. Since the powers of $\Lambda^{-1} \circ \Lambda$ are all equal to

$\Lambda^{-1} \circ \Lambda$ and F has order -1, $B^{-1/2} \in I^0(S^1 \times S^1, \Lambda^{-1} \circ \Lambda)$, and its principal symbol is that of the identity. (Compare [SE_2].) Now we may replace A by $AB^{-1/2}$. $(AB^{-1/2})^*(AB^{-1/2}) = A^*AB^{-1} = \Pi$, and $AB^{-1/2}$ is an FIO having the same symbol and equivariance properties as A. We have proven, therefore, the following theorem.

5.9 Embedding Theorem. Let $L \subseteq T_0^*X$ be a lagrangian submanifold satisfying Maslov's quantization condition and 5.1 - 5.3. Let $a \in \Gamma(|TL^{1/2}|)$ be a unitary 1/2-density. (There always exists such an a.) Then there is an FIO $A \in F^{1/2}a$ from $\mathcal{D}'(S^1)$ to $\mathcal{D}'(X)$ such that A^*A is the projection onto $\mathcal{D}'_{d,+}(S^1)$. A induces an isometric embedding of $L^2_{d,+}(S^1)$ into $L^2(X)$. The elements of the range of A all have their wavefront sets contained in the cone over L.

We can give the following further interpretation of the embedding theorem. Let $\left\{\psi_{dj+1}\right\}_{j=0}^{\infty}$ be the standard basis of $\mathcal{D}'_{d,+}(S^1)$. Then the $A\psi_{dj+1}$ form an orthonormal sequence of smooth functions on X. Any combination $\sum_{j=0}^{\infty} c_j A\psi_{dj+1}$ in which the c_j's have polynomial growth exists as a distribution on X and has its wavefront set contained in the cone over L. The sequence $\{A\psi_{dj+1}\}$ is, therefore, rapidly decreasing outside this cone. (Compare this interpretation with [DU_2] where, instead of a sequence, one obtains a family $\psi(\tau)$ depending on a continuous parameter $\tau \to \infty$.)

6. PROOF OF THE EIGENVALUE THEOREM

Suppose that the manifold X now carries a riemannian metric. Let Δ_X be the Laplace-Beltrami operator on X, $K: T_0^*X \to \mathbb{R}$ its principal symbol. Suppose that $L \subsetneq T_0^*X$ is a lagrangian submanifold on which $K \equiv E$. We will consider S^1 to have the standard metric with circumference 2π, and $X \times S^1$ to have the indefinite product metric $dx^2 - E\,dr^2$. The Laplace-Beltrami operator $\Delta_{X \times S^1}$ is then the (hyperbolic) wave operator $\Delta_X \otimes 1 - E \otimes \Delta_{S^1}$, whose symbol $K_{X \times S^1}$ is $K_X - EK_{S^1}$.

Denote the hamiltonian vector field of K_Y by ξ_Y, for $Y = X, S^1$, or $X \times S^1$. It follows from the construction of Λ in Section 1 and the homogeneity of $K_{X \times S^1}$ that $K_{X \times S^1} \equiv 0$ on Λ, so $\xi_{X \times S^1}$ is tangent to Λ, and $(\xi_X, E\xi_{S^1})$ is tangent to Λ'. It follows that the canonical relation Λ' is equivariant with respect to the actions of $\mathbb{R}$ induced by ξ_X and $E \cdot \xi_{S^1}$. From this we obtain:

6.1 Lemma. If $a \in \Gamma(|TL|^{1/2})$ is invariant under the flow of ξ_X on $\bar{L}$, then $\gamma^{1/2}a$ is invariant under the flow of $\xi_{X \times S^1}$ on Λ, and $\gamma^{1/2}a$ is uniform as long as a is not zero.

Proof. If a is invariant under ξ_X, then the pullbacks of a to $\bar{L}$ and Λ must be invariant as well. By the equivariance of Λ', $\overline{\gamma^{1/2}a} \times \gamma^{1/2}a$ must be invariant under $E\xi_{S^1}$, so it is constant. Q.E.D.

The idea behind what follows is that, since Λ' is equivariant with respect to the "classical" flows generated by ξ_X and $E\xi_{S^1}$, then it seems reasonable that $A \in F^{1/2}(a)$ should be equivariant (i.e., a so-called "intertwining operator") with respect to the "quantized" flows on $L^2(X)$ and $L^2(S^1)$ generated by $i\Delta_X$ and $iE\Delta_{S^1}$. Since the group $\mathbb{R}$ involved is not compact, we cannot average to make A exactly equivariant, but it will be approximately so. (A similar argument appears in [WE].)

Suppose now that the hypotheses of the Eigenvalue Theorem 0.1 are satisfied. By Lemmas 5.1 and 5.8 and Remark 5.6, the hypotheses of the Embedding Theorem 5.9 are satisfied, so there is an isometric FIO $A \in F^{1/2}(a)$, where $a \in \Gamma(|TL|^{1/2})$ is invariant under ξ_X.

6.2 Lemma. $\Delta_X A - A(E\Delta_{S^1})$ is a bounded operator.

Proof. The Schwartz kernel of $R = \Delta_X A - A(E\Delta_{S^1})$ is the result of operating with $\Delta_{X \times S^1} = \Delta_X \otimes 1 - E \otimes \Delta_{S^1}$ on the Schwartz kernel of A. R is therefore an element of $I^{\frac{1}{4}(1-n)+2}(X \times S^1, \Lambda)$ whose principal symbol is zero since $K_{X \times S^1}$ vanishes on Λ. If we consider R as an element of $I^{\frac{1}{4}(1-n)+1}(X \times S^1, \Lambda)$, its principal symbol is given by Theorem 5.3.1 of [D-H]. Since the subprincipal symbol of $\Delta_{X \times S^1}$ is zero and the principal symbol $\gamma^{1/2}a$ of A is invariant under $\xi_{X \times S^1}$, by Lemma 6.1, the result of applying Theorem 5.3.1 of [D-H] is zero, so $R \in I^{\frac{1}{4}(1-n)}(X \times S^1, \Lambda)$. By the Clean Product Theorem applied to R^*R (or Theorem 4.3.2 of [HÖ]), R is bounded from $L^2(S^1)$ to $L^2(X)$. Q.E.D.

The conclusion of the Eigenvalue Theorem follows from the boundedness of R. Consider A as an isometric embedding of $L^2_{d,+}(S^1)$ into $L^2(X)$ and recall Remark 3.5 that the spectrum of $E\Delta_{S^1}$ on $L^2_{d,+}(S^1)$ is the sequence $\{E(dj+1)^2\}_{j=0}^{\infty}$, the unit eigenvectors being the ψ_{dj+1}. Write μ_j for $E(dj+1)^2$ and v_j for the unit vector $A\psi_{dj+1}$ in $L^2(X)$. Then we have

$$|(\Delta_X - \mu_j)v_j| = |\Delta_X A\psi_{dj+1} - \mu_j A\psi_{dj+1}|$$

$$= |\Delta_X A\psi_{dj+1} - A\mu_j\psi_{dj+1}|$$

$$= |\Delta_X A\psi_{dj+1} - A(E\Delta_{S^1})\psi_{dj+1}|$$

$$= |R\psi_{dj+1}| \leq |R| .$$

Since $\Delta_X - \mu_j$ is self-adjoint, it must have an eigenvalue in $[-|R|, |R|]$ (to prove this, diagonalize Δ_X), so Δ_X must have an eigenvalue in $[\mu_j - |R|, \mu_j + |R|]$. The proof of the eigenvalue theorem is complete.

Q.E.D.

REFERENCES

[DU_1] J. J. Duistermaat, Fourier Integral Operators, New York University, 1973.

[DU_2] ______________, Oscillatory integrals, Lagrange immersions, and unfoldings of singularities, Comm. Pure Appl. Math. (to appear).

[D-H] J. J. Duistermaat and L. Hörmander, Fourier integral operators. II, Acta. Math. 128 (1972) 183-269.

[GL] V. Guillemin, Clean intersection theory and Fourier integral operators, this volume.

[GT] M. Gutzwiller, Periodic orbits and classical quantization conditions, J. Math. Phys. 12 (1971), 343-358.

[HÖ] L. Hörmander, Fourier integral operators. I, Acta Math. 127 (1971), 79-183.

[KE] J. B. Keller, Corrected Bohr-Sommerfeld quantum conditions for nonseparable systems, Ann. of Phys. 4 (1958), 180-188.

[MA] V. P. Maslov, Theory of Perturbations and Asymptotic Methods (in Russian), Moscow State University, Moscow, 1965, translated as Théorie des Perturbutions et Méthodes Asymptotiques, Dunod, Gauthier-Villars, Paris, 1972.

[M-S] A. Melin and J. Sjöstrand, Fourier integral operators with complex-valued phase functions, this volume.

[SE_1] R. T. Seeley, Singular integrals and boundary value Problems, Amer. J. Math 88 (1967), 781-809.

[SE_2] __________, Complex powers of an elliptic operator, in Singular Integrals, Proc. Symp. Pure Math. vol. 10, Amer. Math. Soc., 1968, 288-307.

[SL] J. J. Slawionowski, Quantum relations remaining valid on the classical level, Reports Math. Phys. 2 (1971), 11-34.

[SO] J.-M. Souriau, Structure des Systèmes Dynamiques, Dunod, Paris, 1970.

[ST] N. Steenrod, The Topology of Fibre Bundles, Princeton University Press, Princeton, 1951.

[VO] A. Voros, The WKB-Maslov method for nonseparable systems, Colloque Internationale de Géométrie Symplectique et Physique Mathématique (June 1974), C.N.R.S., Paris (to appear).

[WE] A. Weinstein, Fourier integral operators, quantization, and the spectra of riemannian manifolds, Colloque Internationale de Géométrie Symplectique et Physique Mathématique (June 1974), C.N.R.S., Paris (to appear).

MIX
Papier aus verantwortungsvollen Quellen
Paper from responsible sources
FSC® C105338

If you have any concerns about our products,
you can contact us on
ProductSafety@springernature.com

In case Publisher is established outside the EU,
the EU authorized representative is:
Springer Nature Customer Service Center GmbH
Europaplatz 3, 69115 Heidelberg, Germany

Printed by Libri Plureos GmbH
in Hamburg, Germany